A HISTORY OF NEUROPHYSIOLOGY
IN THE 19TH CENTURY

Dr. Mary A. B. Brazier was born in England and educated at London University where she obtained a B.Sc. in Physiology, a Ph.D. in Biochemistry, and a D.Sc. in Neurophysiology. She has an Honorary M.D. degree from the University of Utrecht. Dr. Brazier first researched at the Maudsley Hospital in London until she transferred to the Massachusetts General Hospital in Boston, with appointments at the Harvard Medical School and the Massachusetts Institute of Technology. She is now at the University of California, Los Angeles.

Dr. Brazier is the author of many articles in scientific journals and of the following books: *The Electrical Activity of the Nervous System* (four editions, and translated into five languages), *A History of the Electric Activity of the Brain, The Neurophysiology of Anesthesia,* and *A History of Neurophysiology in the 17th and 18th Centuries: From Concept to Experiment.*

A History of Neurophysiology
in the 19th Century

Mary A. B. Brazier

Department of Anatomy and Physiology
UCLA School of Medicine
Los Angeles, California

Raven Press New York

Raven Press, 1185 Avenue of the Americas, New York, New York 10036

Library of Congress Cataloging-in-Publication Data

Brazier, Mary Agnes Burniston, 1904–
      A history of neurophysiology in the 19th
century.

            Includes bibliographies and indexes.
            1. Neurophysiology—History—19th century.
I. Title.   [DNLM: 1. History of Medicine, 19th Cent.
2. Neurophysiology—history.   WL 11.1 B827ha]
QP353.B74   1987   612'.8'09034   86–42600
ISBN 0-88167-300-5

9 8 7 6 5 4 3 2 1

Preface

In the 19th century, neurophysiology began to be recognized as a specific field (though it did not yet have its name). The roots of neurophysiology grew from knowledge garnered by her forerunners: well-established neuroanatomy, neurology, and neurosurgery. Neurochemistry was experiencing its most productive early years.

The text that follows reflects the approach of a laboratory scientist, herself a neurophysiologist, and it thus strives to emphasize the experimental findings of the major figures and, where possible, illustrate examples from the original works. The single examples from the works of these scientists are reproduced to provide a clue to where their further experimentation can be found. Only in the second half of the century do we begin to find photographic reproductions of the apparatus and recordings of those important investigators, who had laid the foundations for the great advances that were to enrich the 20th century.

MARY A. B. BRAZIER

v

Acknowledgments

Many scholars have been generous in sending material used in this book. Special acknowledgment and thanks should be made to the following: D. Albe-Fessard, Paris; J. A. V. Bates, London; N. P. Bechtereva, Leningrad; L. Bergamini, Pavia; D. Biesold, Leipzig; the late M. Bonelli, Florence; P. Buser, Paris; W. F. Bynum, London; the late Ann Caron, London; E. Clarke, Oxford; W. A. Cobb, London; H. van Duijn, Amsterdam; M. M. Khananashvili, Thilisi; the late J. Konorski, Warsaw; P. G. Kostuk, Kiev; the late K. S. Kostoyantz, Moscow; the late G. A. Lindeboom, Amsterdam; V. L. Merkulov, Leningrad; the late G. Moruzzi, Pisa; R. Naquet, Paris; A. Nicolai, Berlin-Karlshorst; the late C. D. O'Malley, Los Angeles; K. Pateisky, Vienna; H. Petsche, Vienna; K. E. Rothschuh, Münster; W. Storm van Leenwen; Bilthoveu E. Schubert, East Berlin; C. Tassinari, Bologna; J. Trabka, Cracow; R. Wolfe, Boston.

Of the many libraries and museums that have eased my search, the staff of the following have been among the most patient: Académie des Sciences, Paris; Bibliothèque Nationale, Paris; Biblioteca Nazionale Centrale, Florence; Biblioteca Universitaria, Pisa; Bodleian Library, Oxford; British Library (British Museum), London; Conservatoire des Arts et Métiers, Paris; Countway Library, Harvard Medical School, Boston; Deutsches Museum, Munich; Istituto di Fisiologia, Umana Pavia; Institut für Geschichte der Medizin der Universität, Vienna; Institute of Electrical Engineers, London; Musée Carnavalet, Paris; Museo di Storia della Scienza, Florence; Museum Boerhaave, Leyden; National Library of Medicine, Bethesda, Maryland; Neidersachsische Landesbibliothek, Hannover; Royal Society, London; Sachsische Akademie der Wissenschaften zu Leipzig; Science Museum, South Kensington, London; Scola Normale Superiore, Pisa; Teylers Museum, Haarlem; University Museum, Utrecht; Wellcome Historical Library, London. I am greatly indebted to the librarians, curators, and directors.

The research for this book has been greatly facilitated by the excellent collection of historical books and manuscripts in the Biomedical Library of the University of California, Los Angeles. I am also indebted to the National Library of Medicine, Bethesda, Maryland.

For aid with translations, special thanks goes to Philip Rees (German), Irene Agnew (Russian), Elizabeth Wolfe (Ukrainian), and W. Binek (Polish).

FORMAL ACKNOWLEDGMENT

Grateful thanks for continuing support go to the National Institutes of Health, Bethesda, Maryland (NS-20468).

MARY A. B. BRAZIER

Contents

List of Illustrations

A HISTORY OF NEUROPHYSIOLOGY
IN THE 19TH CENTURY

CHAPTER I

Prelude: The Lure of Medical Electricity

For students of the nervous system the 19th century opened with the shock of Galvani's claim for animal electricity.[1] This came to a world intensely interested in the power of electricity but unprepared, in spite of the work on the electric fish, for it to be the agent that flowed in our nerves and controlled our muscles. At this stage, only the peripheral nerves were implicated, but before the end of the century the brain too would be shown to be a source of electricity.

Galvani's findings came in a period when there was great interest in this powerful invisible agent. Unfortunately, when the striking influence of electricity on muscles of the animal and human body, the error was frequently made of equating muscle contraction with "life." This led to bizarre experiments on beheaded criminals and even on deceased patients.

Before the close of the century, attempts had been made to use electricity as a therapy. One of the many to espouse this new electrifying technique was the Abbé Bertholon,[2] who travelled widely in Europe bringing back reports of strange cures of diseases that others could not replicate. He was not alone in the variety of claims he made, for this form of "therapy" had spread throughout Europe. So diverse were the diseases for which cures were being claimed that academies in several countries offered prizes, including the Académie at Lyon, which offered in 1777 a prize for the answer to the following questions: "Quelles sont les maladies qui dépendent de la plus ou moins de grande quantité fluide électrique dans le corps humain, et quels sont les moyens de remédies à l'une à l'autre?"

These prizes were a spur to many, including Bertholon who, in his two volumes on the

---

[1] Luigi Galvani. (1737–1798) De viribus electricitatis in motu musculari. Commentarius. *De Bonoiensi Scientiarum et Artium Instituo atque Academie Commentarii*, 7:363–418 (1791).

[2] Nicole Bertholon. *De l'Électricité des Vertébrés*. Didot, Paris, 1783.

1

electricity of the human body in health and disease, claimed to examine his cases as to whether electrification was the only ameliorator of the patient's condition or an additive to other therapies. A great believer in a "latent electricity" within the body, he held that it was manipulation of this inherent electricity that formed the basis of the cures he claimed. This concept was a rewording of "animal spirits," and in no way did he foresee the intrinsic electricity of nerve and muscle found (but little understood) by Galvani.

The next step in the invention of sources of electricity concerned the controversy that developed with Volta[3] over the explanation of Galvani's results; the voltaic pile, which soon replaced the Leyden jar in the hands of those espousing electrotherapy, was the ancestor of the batteries of today. By the turn of the century, books were beginning to appear on the history of medical electricity—for example, those by Vivenzio[4] in 1784 and by Sue[5] in 1802. Many early experimenters, Fontana[6] and Galdani,[7] for example, had noted the convulsions of their frogs when electricity was applied to their brains, though Galvani attempted this without success ("Si enim conductores non dissectae spinale medullae, aut nervis, ut consueviums, sed vel celebro—contractiones vel nullae, vel admodum extiguae sunt").

Giovanni Aldini (1762–1834)

In the first decade of the 19th century, Galvani's nephew, Aldini,[8] experimented with electroshock in man. Impressed by the muscular contractions he obtained on stimulating animals and cadavers, he stood close by the guillotine to receive heads of criminals so that they were in as fresh condition as possible for his experiments. He found that passing the current, either through the ear and mouth or through the exposed brain and mouth, evoked facial grimace. He then proceeded to apply electrical stimulation from a voltaic pile to the living. His concept was that the contractions were excited by "le développement d'un fluide dans la machine animale," and this he held to be conveyed by the nerves to the muscles. This is also the explanation popularized by Bertholon.

One set of these early experiments on man reaches into the 20th century, for Aldini applied galvanism to the mentally ill (Fig. 1). Having experimented on himself with electrodes in both ears or in one ear and his mouth or on his forehead and nose, he experienced a strong reaction ("une forte action"), followed by prolonged insomnia lasting several days. He found the experience very disagreeable, but he thought the changes it produced in the brain might be salutary in the psychoses ("la folie"). Passing the current between the ears produced violent convulsions and pain, but he claimed good results in patients suffering from melancholia. Aldini had no instrument to tell him the amount of current passed; he recorded only the number of copper and zinc disks in the voltaic pile.

[3]Alessandro Giuseppe Antonio Volta. (1745–1827). Letter to Sir Joseph Banks, March 20, 1800, on electricity excited by the mere contact of conducting substances of different kinds. *Phil. Trans. R. Soc.*, 90:403 (1800)

[4]G. Vivenzio. *Istoria dell'Elettricitá Medica*. Naples, 1784.

[5]Pierre Sue (1739–1816). *Historie du Galvanisme*. 4 vols. Bernard, Paris, 1802–1805.

[6]Felice Gaspar Ferdinand Fontana (1730–1805). Letter to Urbain Tosetti. In: *Mémoires sur les Parties Sensibles et Irritables du Corps Animal*, edited by A. Haller, 3:159 (1760).

[7]Leopoldo Galdani (1725–1813). *Institutiones Physiologicae et Pathologicae*. Luchtmans, Leyden, 1784.

[8]Giovanni Aldini. *Essai Théorique et Expérimental sur le Galvanisme*. 2 vols. Fournier Fils, Paris, 1804.

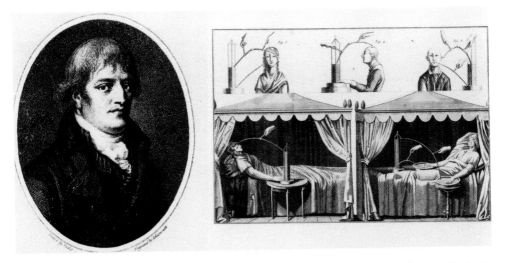

FIG. 1. Left: Giovanni Aldini (1762–1834), nephew of Galvani. **Right:** Aldini's experiments with electric shock "therapy" in man. **Above:** Mental patients with electrodes in various positions connecting to voltaic piles for stimulation. **Below:** Two recently dead patients connected directly or by saline baths to voltaic piles (*bottom*). (From: G. Aldini, *Essai Théorique et Expérimental sur le Galvanisme.* 2 vols. Fournier, Paris, 1804.)

Jean Paul Marat (1743–1793)

There were many other less spectacular attempts to bring electricity into medicine. Toward the end of the 18th century, even many now remembered for entirely different careers had experimented and searched for the properties of electricity. For example, in France, Swiss-born Marat, schooled in medicine in Bordeaux, Paris, and London, wrote an *Essay of Man* in 1773, and in 1775 he was granted a Scottish degree in medicine. While in England he published several works, including his studies on electricity,[9] which frequently refer to Benjamin Franklin. On returning to France, he was appointed Médecin des Gardes du Comte d'Artois (who was later to become Charles X, King of France), but he found time to write more than once on electricity,[10] including one on medical electricity. He also translated into French Newton's *Optiks*. Marat, as a scientist, visited America in 1776, for him a prophetic year, for within a decade in his own country he was himself deeply involved in revolution. His name is remembered for his journal *L'Ami du Peuple,*[11] which reflected his drive for emancipation of the French people from their oppression by those in power, a position that led to his assassination. Immortalized by the painter Jacques-Louis David for his death at the hands of Charlotte Corday in the turmoil of mixed goals in the French Revolution (Fig. 2) his remains were interned as those of a hero in the Pantheon in Paris. But on restoration of the monarch, they were expelled from there to find a resting place in the cemetery of Saint-Étienne-du-Mont (in the distinguished company of Blaise

[9]Jean Paul Marat. *Recherches Physiques sur l'Électricité.* Gloucier, Paris, 1783.

[10]Jean Paul Marat. *Mémoire sur l'Électricité médicale couronné 6 Aout 1783,* par *l'Académic Royales des Sciences, Belles-Lettres aux Arts de Rouen.* Paris, 1784.

[11]Jean Paul Marat. *L'Ami du Peuple* (1789–1793).

FIG. 2.   Left: Jean Paul Marat (1743–1793), physician, writer, and translator of Newton. His murder in the French Revolution was immortalized by the painter David. **Right:** Marat's treatise on electricity, the first edition being published in 1782.

Pascal and Jean Baptiste Racine). A fine medallion of Marat by David d'Angers (a fellow Jacobin) can be seen at the Musée des Beaux Arts in Angers.

In England another unexpected contributor was writer and naturalist Oliver Goldsmith[12] who had studied medicine at Leyden. In 1776 he wrote of the importance of an understanding of magnetism and electricity by the general reader. It is, however, not for this that he is remembered, but for the poetry and comedies he was writing at the same time (*She Stoops to Conquer* was published in this same period).

A perhaps stranger enthusiast for the role of electricity was John Wesley, the founder of the Methodist religion. His text reveals a somewhat confused understanding, but he comes out boldly to declare that electricity is "the noblest medicine yet known in the world."[13] He died one year before what Galvani was to reveal, but he was much impressed by cures affected by electrical treatment. His treatise ran into many editions (Fig. 3).

These examples from the fringe world of medicine are quoted to show the general interest in electricity, an interest that made those concerned with the nervous system ready to applaud or attack the proposal that Galvani had found the agent of nervous transmission. On a strictly scientific level, none of these reached the contribution of Joseph Priestly[14] to the understanding of the physics of electricity, and one regrets that he did not turn his genius to the problem of animal electricity.

Biology was invading physics and therefore was dependent on the development of instruments for detecting electrical currents and evoking them. From the discovery of the Leyden

[12]Oliver Goldsmith (1728–1774). *Survey of Experimental Philosophy.* German and Newburg, London, 1776.

[13]John Wesley (1703–1791). *The Desideratum: or Electricity made Plain and Useful by a Lover of Mankind.* Hawes, London, 1760.

[14]Joseph Priestly (1733–1804). *A Familiar Introduction to the Study of Electricity.* Dodsley, London, 1768.

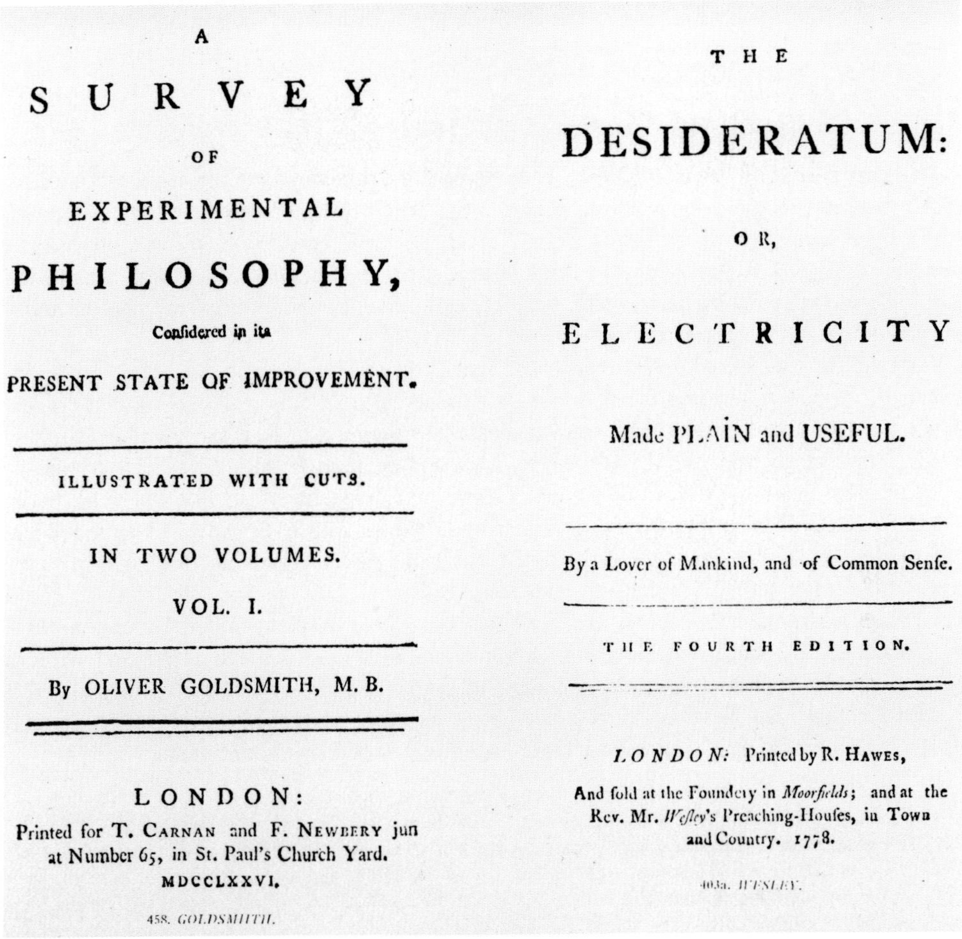

A
SURVEY
OF
EXPERIMENTAL
PHILOSOPHY,
Confidered in its
PRESENT STATE OF IMPROVEMENT.

ILLUSTRATED WITH CUTS.

IN TWO VOLUMES.
VOL. I.

By OLIVER GOLDSMITH, M. B.

LONDON:
Printed for T. CARNAN and F. NEWBERY jun
at Number 65, in St. Paul's Church Yard.
MDCCLXXVI.

458. GOLDSMITH.

THE
DESIDERATUM:
OR,
ELECTRICITY
Made PLAIN and USEFUL.

By a Lover of Mankind, and of Common Senfe.

THE FOURTH EDITION.

LONDON: Printed by R. HAWES,
And fold at the Foundery in Moorfields; and at the
Rev. Mr. Wefley's Preaching-Houfes, in Town
and Country. 1778.

403a. WESLEY.

FIG. 3. Two works by writers from markedly disparate fields who wrote to stimulate the interest of their readers, in the power of electricity. **Left:** Oliver Goldsmith (1728–1774), poet and dramatist. **Right:** John Wesley (1703–1791), religionist and preacher.

jar by Musschenbroek, immensely powerful emitters had grown, as evidenced, for example, by the instruments of van Marum,[15] now housed in the Teylers Museum in Haarlem.

The striking claims of Galvani that electricity was at the core of our muscle movements drew the majority of investigators, especially the physicists, into the study of our motor systems. But simultaneously, and largely instigated by the philosophers, was a deepening curiosity about our sensory systems. Unremembered was the proposal of Hausen[16] in the previous century that our muscles might work by electricity. He had reached his opinion when making the spectacular kinds of demonstrations so popular in his day of electrifying small boys and elicitating sparks.

The 18th century had had its Age of Enlightenment in France, and the 19th century was

[15]Martin van Marum (1750–1837).

[16]Christian August Hausen (1693–1743). Novi profectus in historia electricitatis. 4. vols. Th. Schwan, Lipsiae, 1743.

to have its Age of Reason, a name since then much disputed. The philosophies of Descartes, Locke, and Hume were replaced in Germany by that of Immanuel Kant.[17]

Johann Wolfgang von Goethe (1749–1832)

Into this period of "reason" broke a short-lived movement stemming from the intellectual revolution of the young, Sturm und Drang, which reflected the stress of the period. Short-lived though it was (it lasted from 1770 to 1778), it produced one masterpiece, *The Sorrows of Young Werther*[18] (still available today in paperback). Goethe, who was to become a major poet in his later years, was a magnet for the young scientists, and he influenced such a major neuroanatomist as Purkyně.[19]

Purkyně visited Goethe in Weimar in 1823 and, on return to Prague, started a correspondence. Both were interested in vision and the theory of colors as well as in sound waves. Their theories about these subjects are discussed in their correspondence, which still survives.[20] It is clear that their views were to draw widely apart in later years.

In the first years of Sturm und Drang, Goethe had lived in Strasbourg where he wrote *The Sorrows of Young Werther* as well as a first draft of *Faust*. But in 1775 he moved to Weimar under the patronage of Karl August, the Duke of Saxe-Weimar. For neurophysiologists today, the surprise is that this great poet, although only an amateur scientist, could have such a strong influence on experimentalists such as Alexander von Humboldt (Fig. 4). Fascinated by sensation, Goethe became absorbed in his theory of colors and wrote a long, much illustrated book on this subject entitled *The Theory of Colors*.[21] Unfortunately, he did not first read Newton,[22] and later, when criticized for this, he was unwise enough to say Newton was wrong. Passages from Newton, ignored by Goethe, read:

> This objector's (Robert Hook's) hypothesis, as to the fundamental part of it, is not against me. That fundamental supposition is, that the parts of bodies, when briskly agitated, do excite vibrations in the aether, which propagated every way from those bodies in straight lines, and cause a sensation of light by beating and dashing against the bottom of the eye, something after the manner that vibrations in the air cause a sensation of sound by beating against the organs of hearing. Now, the most free and natural application of this hypothesis to the solution of phenomena, I take to be this: that the agitated parts of bodies, according to their several sizes, figures and motions, do excite vibration in the aether of various depths of bignesses, which, being promiscuously propagated through that medium to our eyes, effect in us a sensation of light of a white colour; but if by any means those of unequal bignesses be separated from one another, the largest beget a sensation of a red colour, the least or shortest of a deep violet, and the intermediate ones of intermediate colours; much after the manner that bodies, according to their several sizes, shapes and motions, excite vibrations in the air of various bignesses, which according to bignesses, make several tones in sound: that the largest vibrations are best able to overcome the resistance of a refracting superficies, and to break through it with least refraction; whence the vibrations of several bignesses, that is the ray of several colours, which are blended together in light, must be parted from one another by refraction and so cause the phenomena of prisms, and

[17]Immanuel Kant (1724–1804). *Kritik der reinen Vernunft*. Berlin, 1781. (The Critique of Pure Reason.)

[18]Johann Wolfgang Goethe. *Die Leiden des jungen Werthers*, Weygandsche Buchhandlung, Leipzig, 1774.

[19]Jan Evangelista Purkyně (1787–1869).

[20]Some of these letters can be seen in the Stadt Museum in Prague.

[21]J.W. von Goethe. *Zur Farbenlehre*. Tübingen, 1810. (English translation: *The Theory of Colours*. John Murray, London, 1840.)

[22]Isaac Newton (1642–1727). *Optiks or a Treatise of the Reflections, Refractions, Inflections of the Colours of Light*. Book 1. London, 1704.

FIG. 4. **Left:** Johann Wolfgang Goethe (1749–1832). Charcoal drawing by David Jaffe. **Right:** The frontispiece to Alexander von Humboldt's book dedicated to Goethe. The spirit of poetry unveils the mystery of nature. (From: Speck Collection, Yale University Library.)

other refracting substances; and that it depends on the thickness of a thin transparent plate or bubble, whether a vibration shall be reflected at its further superficies, or transmitted; so that, according to the number of vibrations, interceding the two superficies, they may be reflected or transmitted for many successive thicknesses. And, since the vibrations which make blue and violet are supposed shorter than those which make red and yellow, they must be reflected at a less thickness of the plate: which is sufficient to explicate all the ordinary phenomena of those plates or bubbles, and also of all natural bodies, whose parts are like so many fragments of such plates.

Goethe was interested essentially in how we see color but was apparently unaware of Kepler's work on the retinal image.[23] Goethe boldly entitled his book *The Theory of Colours*. In this treatise, he disagreed with Newton's identification of seven colors: violet, indigo, blue, green, yellow, orange, and red. Neither Newton nor Goethe conceived of colors that the retina could not perceive (ultraviolet, infrared). The most useful tool of both men was the prism; Newton had bought his at a country fair.

Thomas Young (1773–1829)

Newton's theory was also attacked by a much better informed critic than Goethe— Thomas Young, physician and physicist, president of the Royal Society. Thomas Young was a man of many talents. Renowned for his work on the physics of color and for his research on ciliary action, he was also a great linguist and master of many languages, a skill used in deciphering the Rosetta stone. This ancient basalt stone was found in the delta of the Nile and was engraved in three languages: Greek, Coptic, and hieroglyphics. It was the last of those that Young deciphered, thus opening the understanding of many ancient

---

[23]Johann Kepler (1571–1630). *Ad vitellionem paralipomena quibus astronomia pars optica traditur.* 1604. (Supplements of Witelo Concerning the Optical Art of Astronomy.) (Kepler's manuscripts can be seen in the Pulkovo Observatory near Leningrad).

FIG. 5. The two men responsible for the Young-Helmholtz Theory of light and color. **Left:** Thomas Young (1773–1829). **Right:** Hermann Ludwig Ferdinand von Helmholtz (1821–1894).

inscriptions. Other interests included astigmatism and the structure of the ciliary muscles. It was the excitation of the retina by different colors that was Young's quarrel with Newton's theories of the previous century.

Newton had held it was the differential wavelength of light that influenced the retina and produced the sensation of color. Young reduced Newton's seven components of white light to three and proposed that our retinas are served by three types of fiber, each devoted to one frequency band. He had, of course, no physiological or anatomical basis for this concept that was in its way a harbinger of the theory of specific nerve energies to be framed later by Johannes Müller.[24, 25]

The clarification of the problem had to wait half a century for Helmholtz,[26] Müller's most famous pupil. In 1852 he wrote a paper which differentiated clearly where Goethe's ideas had gone wrong.[27]

Young's[28] views were only slightly modified by Helmholtz, resulting in the name "Young-Helmholtz Theory" (Fig. 5), though this in its turn was to be modified by the

[24]Johannes Müller (1801–1858). *Über die Phantastischen Gesichtserscheinungen.* Holscher, Koblenz, 1826. (On the Phantastic Phenomena of Vision.)

[25]Johannes Müller. *Zur vergleichenden Physiologie der Gesichtssinnes der Menschen und Thiere.* Leipzig, 1826. (On the Comparative Physiology of the Senses of Sight in Humans and Animals.)

[26]H.L.F. von Helmholtz (1821–1894). *Handbuch der Physiologischen Optik.* Vol 2. Leipzig, 1866. (Handbook of Physiological Optics.)

[27]H.L.F. von Helmholtz. Goethe's Vorahnungen kommender naturwissenschaftlicher Ideen. *Vorträge und Reden,* 2:335–361 (1892). (Goethe's Anticipation of Subsequent Scientific Ideas.)

[28]Thomas Young (1773–1829). On the theory of light and colours. *Phil. Trans. Roy. Soc.,* 92:18–21 (1802).

Scotsman James Clerk Maxwell.[29] The emphasis, in this case, was on physics rather than neurophysiology. By the ingenious superposition of different colored disks,[30] Maxwell was able to demonstrate the results of mixing red, blue, and green. For neurophysiology, the importance of the Young-Helmholtz Theory was the part it played in Johannes Müller's concept of specific nerve energies framed decades later.

In the years straddling the turn of the century, a young German scientist, who was to meet fame in another field, plunged into a long series of experiments designed to support Galvani's theories; this was Alexander von Humboldt. He was another of the young scientists who were so strongly influenced by Goethe, to whom he later dedicated his book on the geography of plants.[31]

Other prominent physiologists of the 19th century still held their admiration for the poet's interest in biology and wrote praise to him. Among them were Johannes Müller, Du Bois-Reymond[32] and Helmholtz.

BIBLIOGRAPHY

Giovanni Aldini (1762–1834)

Selected Writings

De animalis electricae theoriae ortu atque incrementis (Dissertatio) Mutinae, apud Societatem typographicam. Modena, 1792.

De animali electricitate dissertationes duae. Bologna, 1794.

Lettere del cittadino N. N. di Como al cittadino Aldini professore a Bologna intorno alla prestesa elettricita animale nelle sperienze del galvanismo. Annali di Chimia e Storia Naturale, 1798. Vol. XVI, 3–41 and 42–88.

An account of the late improvements in galvanism, with a series of curious experiments performed before the Commissioners of the French National Institute and lately repeated in the Anatomical Theatre of London. To which is added an appendix containing the author's experiments on the body of a malefactor excuted at Newgate. London, 1803. Dissertation on animal electricity; galvanic experiments on executed criminals.

Précis des expériences galvaniques faites récemment à Londres et à Calais; suivi d'un extrait d'autres expériences. Paris, 1803. Early experiments showing the effect of the electric current on the human organism.

Essai Théorique et Expérimental sur le Galvanisme. 2 vols. Fournier, Paris 1804.

General Views on the Application of Galvanism to Medical Purposes Principally in Cases of Superseded Animation. London, 1819.

Saggio di osservasioni sui mezzi atti a migliorare la costruzione e l'illuminazione dei fari, con appendice sull'illuminazione del fari col gas, Milano, dall'Imperiale Regia Stamperia, 1823. (A Short Account of Experiments Made in Italy and Recently Repeated in Geneva and Paris, for Preserving Human Life and Objects of Value from Destruction by Fire. G. Schulze, London, 1830.)

---

[29]James Clerk Maxwell (1831–1879). *The Scientific Papers of J. Clerk Maxwell,* edited by W. Niven. Cambridge, 1890.

[30]Some of these "Maxwellian" disks can be seen at the University of Pavia.

[31]Alexander Friedrich von Humboldt (1769–1859). *Ideen zu einer Geographie der Pflanzen.* 1807. (Notes on the Geography of Plants.)

[32]E. Du Bois-Reymond. *Reden.* (Collected by E. Brucke.) 2 vols. Veit, Leipzig, 1886–87.

Jean Paul Marat (1743–1793)

Selected Writings

De l'homme, ou des principes et des lois de l'influence de l'âme sur le corps et du corps sur l'âme. Amsterdam, 1775–76.

Découvertes de M. Marat. Docteur de Médecine et Médecin des Gardes du Corps de Monseigneur le Comte d'Artois sur le feu, l'électricité et la lumière, constatées par une suite d'expériences nouvelles que viennent d'être vérifées par MM. Les Commissaires de l'Académie des Sciences. Paris, 1779.

Recherches Physiques sur l'Électricité. Gloucier, Paris, 1782.

Mémoire sur l'Electricité médicale, couronné le 6 Août 1783, par l'Académie Royale des Sciences, Belles-Lettres et Arts de Rouen. Paris, 1784.

Observations de M. l'Amateur avec M. l'Abbé Sans sur la nécessité indispensable d'avoir une théorie solide et lumineuse avant d'ouvrir boutique d'Électricité médicale. En Résponse à la lettre de M. l'Abbé Sans à M. Marat sur l'Électricité Positive et Negative publiée dans le No. 16 de l'Année Littéraire. Paris, 1785.

Optique de Newton, traduction nouvelle, faite par M . . . sur la dernière Édition originale, Ornée de Vingt-unes Planches, et approuvée par l'Académie Royale de Sciences: Dédié au Roi, Par M. Beauzée, Editeur de cet Ouvrage, l'un des Quarante de l'Académie Fráncaise; de l'Académice Della Crusea; des Académies royales de Rouen, de Metz, et d'Arras; Professeur émérite de l'école royale militaire, et Secrétaire-Interprète de Monseigneur le Comte d'Artois. Paris, 1787.

Oeuvres de M. Marât. Mémoires académiques, ou nouvelles découvertes sur le lumière, relatives aux points les plus importants de l'optique; Mémoire sur la prétendue différente refragibilité des Rayons hétérogenès; Mémoire sur l'explication de l'Arc-en-ciel donné par Newton: envoyé au Concours ouvert par la Société Royale des Sciences de Montpellier en Octobre 1786; Mémoire sur les vrais causes de couleurs que presentent les lames de verre, les bulles d'eau de savon, et autres matières diaphanes extrèment mińces. Ouvrage qui a remporté le Prix de l'Académie de Sciences. Belles-lettres et Arts de Rouen, le 2 Août 1786. Paris, 1788.

Johann Wolfgang von Goethe (1749–1832)

Selected Writings

Die Leiden des jungen Werthers. Weygandsche Buchhandlung, Leipzig, 1774. (The Sorrows of Young Werther.)

Dichtung und Wahrheit Part I, Cotta, Tübingen, 1811. Part II, Cotta, Tübingen, 1812. Part III, Cotta, Tübingen, 1814. Part IV, Cotta, Tübingen, 1833. (Poetry and Truth.)

Die Wahlverwandtschaften. Cotta, Tübingen, 1809. (Elective Affinities.)

Suggested Readings

Helmholtz, Hermann Ludwig Ferdinand. *Popular Lectures on Scientific Subjects.* Translated by E. Atkinson, 1873.

Helmholtz, H. *Goethe's Vorahnungen kommender naturwissenschaftlicher Ideen.* Vorträge und Reden, 2:335–361, 1892. (Goethe's Anticipation of Subsequent Scientific Ideas.)

Kruta, Vladislav. *The Poet and the Scientist.* Prague, 1968.

Magnus, Rudolf. *Goethe als Naturforscher.* Leipzig, 1906. (*Goethe as a Scientist.* Translated by Heinz Norden. Schumann, New York, 1949.)

Sherrington, C.S. *Goethe on Nature and on Science.* Cambridge University Press, 1949.

Thomas Young (1773–1829)

Selected Writings

An Account of the Recent Discoveries in Hieroglyphical Literature and Egyptian Antiquities. J. Murray, London, 1823.

Outline of experiments and enquiries respecting sound and light. *Phil. Trans. Roy. Soc.*, 106–150 (1800).

The mechanics of the eye. (Bakerian Lecture, 1800). *Phil. Trans. Roy. Soc.*, 91:23–88 (1801).

On the phenomena of sound. *Nicholson's Journal*, 3:145–146 (1802).

On the theory of light and colours. (Bakerian Lecture, 1801). *Phil. Trans. Roy. Soc.*, 92:12–48 (1802).

An account of some cases of the production of colour, not hitherto described. *Phil. Trans. Roy. Soc.* 92:387–397 (1802).

On the velocity of sound. *Roy. Inst. J.*, 1:214–216 (1802).

A course of lectures on natural philosophy and the mechanical arts. 2 vols. Johnson, London, 1807.

Suggested Readings

Gurney, H. *Memoir of the Life of Thomas Young*. London, 1831.

Peacock and Leitch (Eds). *Miscellaneous Works of the late Thomas Young, M.D., F.R.S.* 3 vols. London, 1855.

Wood, A., Oldham, F., and Raven, C. E. *Thomas Young Natural Philosopher, 1773–1829*. Cambridge University Press, 1954.

CHAPTER II

Meeting the Challenge of Galvani

Friedrich Wilhelm Heinrich Alexander von Humboldt (1769–1859)

In 1769 two men were born who were to effect worldwide changes: Napoleon Bonaparte, in the face of Europe, and Alexander von Humboldt, in the understanding of the scientific world. The loss of German universities to Napoleon inspired Humboldt to found, with his brother, the University of Berlin.

Born in the Duchy of Brunswick, Humboldt spent his youth at Tegel, then a town outside the city of Berlin (now its airport). With his brother, Wilhelm, he was educated at Frankfurt's Academia Viedrina (not yet a university) and then went on to the University of Göttingen.

One of the great German scientists of the 19th century to interest himself in his early years in animal electricity, Alexander von Humboldt was stirred by the controversy aroused by Volta over Galvani's findings, and while quite a young man, he tested the problem himself. He demonstrated quite clearly that nerve-muscle contractions could be evoked in the absence of metals[1] (Fig. 6). He also worked on a paper on the electric fish and a few years later published on this subject.[2]

Humboldt was interested not only in the response of the muscle to stimulation but also in its degree of excitability. To this end he experimented with drugs. Although determined to refute Volta's dictum that metals must be present, Humboldt was also not entirely in accord with Galvani. For all who worked with animals, it was difficult to equate these electrical findings with their physical knowledge of electricity with its production by frictional machines, by conduction through hard wires and often with a spark. Humboldt found these

---

[1]Alexander von Humboldt. *Versuche über die gereizte Muskel und Nervenfaser, nebst Vermutungen über den chemischen Process des Lebens in der Thier-und Pflanzenwelt.* 2 vols. Dekker, Posen, 1797. (Experiments on the Excited Muscle and Nerve Fiber, with Hypothesis on the Chemical Processes of Life in the Animal and Plant World.)

[2]A. von Humboldt. *Versuch über die elektrischen Fische.* Erfurt, 1806. (Studies on the Electric Fish.)

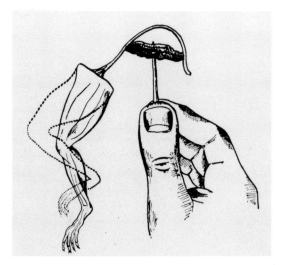 

FIG. 6. **Left:** One of Humboldt's experiments to test Galvani's claims. The isolated piece of muscle contracts as the impulse passes down between the two contacts. (From: Sirol. *Galvani et Galvanisme. L'Electricité Animale.* Vigot Frères, Paris, 1939.) **Right:** Alexander Humboldt when young, from the charcoal drawing by David Jaffe.

properties difficult to adopt for nerves and consequently came out in favor of a ''galvanic fluid'' produced by the nerve. The Académie des Sciences in Paris had established in 1796 a committee to attempt to solve this problem, but no solution came forth from their deliberations. The experimental results of Galvani were universally confirmed; however, the nature of the electrical phenomenon remained cloudy.

In his many experiments, said to be in the hundreds, Humboldt left the frog, the favorite preparation of the many experimentalists testing the results of Galvani and moved to warm-blooded animals, including himself (to whom he caused some electrical burns). At the period of Humboldt's work on the nerves and muscles, his goal, like that of Galvani's nephew, Giovanni Aldini,[3] was to disprove Volta's contention that dissimilar metals must be present for electricity to flow (Fig. 7). Volta emerged as a most formidable opponent.

After these early years in the laboratory, Humboldt became one of the most widely travelled scientists of his century—a century in which horseback and sail were the only means of travel. In 1799 he set off for the undeveloped hinterland of South America, but with a look back at his career in electricity, he stopped off at Como to visit Volta, one year before the world heard of the voltaic pile. In Asia his voyages took him to far Siberia; in the New World (where in the year of the Louisiana Purchase he visited Thomas Jefferson) he went to the West Coast, where a Pacific current is named for him as well as a county and a university in California. Among the many honors he received was membership in the American Academy of Arts and Sciences and the American Philosophical Society.

Humboldt's experiments on the electricity of nerves and muscles had ceased when he became an explorer. When he returned to Berlin at the age of 80, he found a city of turmoil—the Revolution of 1848. In the half century he had been away, the points of discord between the findings of Galvani and those of Volta had been solved, and solved largely by the group working in the university founded by his brother, the University of Berlin.

[3]Giovanni Aldini (1762–1834). *Essai Théorique et Expérimental de Galvanisme.* Vol. 1. Fornier, Paris, 1804.

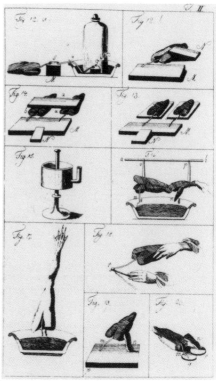

FIG. 7.   Left: Alexander von Humboldt (1769–1859) at the age of 37. **Right:** Some of his many experiments undertaken to support Galvani's findings. (From: *Versuche über die gereizte Muskel-und-Nervenfaser, oder Galvanismus, nebst Vermuthungen über den chemischen Process des Lebens in der Thier- und Pflanzenwelt.* Decker and Rottman, Posen, Berlin, 1797). (Experiments on Stimulated Muscle and Nerve Fibers, or Galvanism, along with Conjectures on the Chemical Process of Life in the Animal and Plant World.)

His brother, Wilhelm, was also outstanding in his time, though somewhat eclipsed by his famous brother. In 1810 Wilhelm Humboldt founded the University of Berlin, now named for him: the Humboldt University. This became urgently needed after the victory of Napoleon's troops and the Peace of Tilsit in 1807, a treaty by which Prussia lost all its lands west of the Elbe. This meant the loss of all its universities other than those of Frankfurt and of Königsberg far north on the Baltic Sea.

The building in which the University of Berlin opened was the former Prince Henry Palace built in the 18th century. It seems strange that Berlin had no university as it was the seat of the Academy of Sciences and (in medicine) of the famous Charité, the hospital founded in 1710. On each side of the door of this great university (now in East Berlin) stand statues of the Humboldt brothers designed by the sculptor Begas. The fine gilded equestrian statue of Frederick the Great, who died in 1786, has been restored to its plinth in front of the university in the well-known thoroughfare of Unter den Linden. This has resumed its original name after a brief period as Stalin Allée.

The University of Berlin opened on October 6, 1810 with four faculties: theology, law, medicine, and philosophy. In the opening classes there was one student in theology, three in law, one in medicine, and one pharmacist. Within less than half a century, the university was to become the site of the most intense study of bioelectricity in the 19th century, spear-headed by Du Bois-Reymond.

Alexander von Humboldt died in 1859 at the age of 90, having left the field of neuro-physiology nearly a half century before.

Galvani's Claims Meet the World of Physics

By the beginning of the 19th century, the field opened by Galvani[4, 5, 6] had reached the textbooks and here we already find the adoption of the word "galvanic" to cover almost everything electrical as it does to this day. In the most famous textbook[7] of the first decade of the 19th century, the excerpt reads:

Galvanisme. Un professeur d'anatomie à l'université de Bologne, Galvani, faisait un jour des expériences sur l'électricité. Dans son laboratoire, et non loin de la machine, se trou-voient des grenouilles écorchées, dont les membres entroient en convulsion chaque fois que l'on soutiroit une étincelle. Surpris de ce phenomene, Galvani en fit le sujet de ses recher-ches, et reconnut que des métaux appliqués aux nerfs et aux muscles de ces animaux, déter-minoient des contractions fortes et rapides, lorsqu'on les disposoit d'une certaine manière. Il a donné le nom de l'électricité animale à cet ordre de nouveaux phénomènes, d'après l'analogie qu'il crut apercevoir entre ces effets et ceux que produit l'électricité. Cette dé-couverte fut annoncée; plusieurs savants, et principalement ceux d'Italie, parmi lesquels on distingue Volte, s'empresserent d'ajouter aux travaux de l'inventeur.

Alessandro Giuseppe Antonio Anastasio Volta (1745–1827)

The first important challenge to Galvani came from Volta. Alessandro Volta was born in Como on the shore of Lake Garda where he was schooled at the Jesuit College and later at a seminary. He never entered the university as a student or tried for a degree. With his devotion to physics, Volta taught his subject to students in the public schools of his home-town of Como with such success that in 1778 he was appointed to a professorship of Exper-imental Physics at the University of Pavia, a position he held for 41 years (Fig. 8).

Volta had crystallized many of his views on physical phenomena while still young, as we know from his early letters. Even from the age of 16, he had already made his goal in science the phenomenon of electricity and especially its instrumentation. A rich friend of his, Giulio Gattoni, had set up a laboratory in his home, and it was there that the young Volta began his practical experiments. In this period two types of electricity were held to exist, atmospheric and natural, and then a third was being claimed—animal electricity. Volta revealed in private correspondence with the Abbé Vassalli that he had begun experi-ments to check on Galvani's claims in the same year that those were published.

One of Volta's earliest letters, written in 1793 to the Abbé Carminati, states that he had repeated Galvani's experiment on the frog and had confirmed the result. His letter praised Galvani, but he was to change his mind. His enthusiasm for animal electricity barely lasted out the year of Galvani's *Commentarius*. From the journal published by Luigi Brugnatelli,

---

[4]L. Galvani (1737–1798). De viribus electricitatis in motu musculari. Commentarius. *De Bononiensi scienti-arum et Artium Instituto atque Academia Commentarii,* 7:363–418, (1791).

[5]L. Galvani. *Dell'Uso e dell'Attivitá dell Arco Conduttore nelle Contrazioni dei Muscoli,* con Supplento (pub-lished anonymously). D'Aquino, Bologna, 1794.

[6]See the first volume of this work for a fuller account of Galvani's own work: M.A.B. Brazier, *A History of Neurophysiology in the 17th and 18th Centuries: From Concept to Experiment.* Raven Press, New York, 1984.

[7]A. Richerand. *Nouveaux Eléments de Physiologie.* 2 vols. Chapart, Caille et Ravier, Paris, 1807.

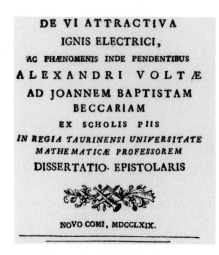

DE VI ATTRACTIVA
IGNIS ELECTRICI,
AC PHÆNOMENIS INDE PENDENTIBUS
ALEXANDRI VOLTÆ
AD JOANNEM BAPTISTAM
BECCARIAM
EX SCHOLIS PIIS
IN REGIA TAURINENSI UNIVERSITATE
MATHEMATICÆ PROFESSOREM
DISSERTATIO· EPISTOLARIS

NOVO COMI, MDCCLXIX.

FIG. 8. Left: Volta's first publication on electricity written at the age of 24 and published in his home town of Como. **Right:** A memorial tablet made in his honor.

it is known that Volta began this work as early as March 1792, with results which he reported as follows:

> What can ever be accomplished that is worth doing, if things are not reduced to degrees and measures, particularly in physics? How can we evaluate causes without determining, not only the qualities, but also the quantities and strength of the effects themselves?

Volta regarded this observation as proof of the fact that electricity (as opposed to Galvani's opinion) is: "negative on the nerve side, that is within the interior of muscles, where the nerve enters, and positive on the outer surface."[8]

Before the end of the previous century, Volta had rushed to attack Galvani's explanation of his results. As a Foreign Member of the Royal Society, he chose to send his attack for publication in the *Philosophical Transactions*. His two long letters,[9] as they were termed, were read to the Royal Society on January 3, 1793, only a few months after the appearance of Galvani's *Commentarius*. At great length, and writing in French, Volta raised no less than 51 arguments against Galvani's interpretations, and he backed these up with descriptions of approximately 17 experiments he had done himself. These descriptions show his interest in the current that flows between different metals—the idea that was to develop later into his famous pile. In his 14th comment he wrote:

> Il est vrai qu'on ne reussit pas, à beaucoup près, si bien de cette manière que de l'autre, et qu'il faut, dans ce cas, avoir recours à un artifice, dont nous aurons occasion de parler plus au long, et qui consiste à employer deux métaux differents; artifice qui n'est pas absolument necessaire lorsqu'on expérimente suivant le procédé de GALVANI, decrit ci-dessus, ... du moins tant que la vitalité dans l'animal, ou dans ses membres coupés, se soutient en pleine vigueur; mais enfin, puisque avec des armures de differents métaux appli-

[8]This letter appeared in *Annali di Chimica e Storia naturale,* by Luigi Brugnatelli, 1794, Vol. 6, 142–166, and in *Giornale Fisico-medico,* 1794, Anno VII, Vol. 3, 97–121. It is the second of the five letters that Volta sent to the Abbé Anton Maria Vassalli and that, as a whole, constitutes the Nuova Memoria sull'elettricita animale.

[9]Account of some discoveries made by Mr. Galvani of Bologna; with experiments and observations on them. In: *Two Letters from Mr. Alexander Volta,* F.R.S. Professor of Natural Philosophy in the University of Pavia, to Mr. Tiberius Cavallo, R.R.S. *Phil. Trans. Roy. Soc.*, Part 1, pp. 12–44, 1793.

FIG. 9. **Left:** Alessandro Volta (1745–1827). **Right:** His friend and correspondent Martin van Marum (1759–1837), the great Dutch developer of electrical apparatus. (Courtesy of the Teyler's Museum in Haarlem.)

quées, soit aux nerfs seuls, soit aux seuls muscles, on vient à bout d'exciter les contractions dans ceux-ci, et les mouvements des membres, on doit conclure que s'il y a des cas (ce qui pourroit bien encore paroitre douteux) où la pretendue décharge entre nerf et muscle . . . est cause des mouvements musculaires, il y a bien aussi des circonstances, et plus fréquentes, ou l'on obtien les mêmes mouvements, par un tout autre jeu, par une tout autre circulation, du fluide électrique. . . .

Galvani, who had himself experimented with dissimilar metals, could not accept Volta's interpretation. Though he died before the end of the century, having followed his classic *Commentarius*[10] with an anonymously published treatise,[11] his nephew, Aldini, also joined in the fray. Until his death in 1827, Volta was unrelenting in his interpretation, bending it to meet the rising number of experiments from other centers that favored animal electricity.

The lines of thinking followed by Volta are known because he reported his work and the development of his hypotheses in letters to various scholars, and these are now available to us. From his Jesuit upbringing, he corresponded with the Abbés Beccaria, Carminati, and Vassalli, but he also consulted the great Dutch physicist Martin van Marum (Fig. 9), whose collection of electrical instruments is housed in the Teyler's Museum in Haarlem. Van Marum[12, 13] made a giant version of Volta's pile, capable of delivering half a million volts. In 1976 a working copy of this toured the United States.

---

[10]Luigi Galvani (1737–1798). De viribus electricitatis in motu musculari. Commentarius. *De Bónoniensi Sciertiarum et Artium Instituto atque Academia Commentarii,* 7:363–418 (1791).

[11]L. Galvani. Dell'uso e dell'attività dell'arco conduttore nelle contrazioni dei muscoli. (Anon.) Bologna, 1794.

[12]Martin van Marum (1759–1837). La description d'une machine électrique, construite d'une manière nouvelle et simple, et qui reunit plusieurs avantages sur la construction ordinaire. (Lettre à M. Jean Ingenhousz.) Haarlem, 1791.

[13]Martin van Marum. Lettre à M. Berthollet contenant la déscription d'un gazometre, constuit d'une manière différente de celui de Lavoisier & Meusnier et d'un appareil pour faire très exactement l'expérience de la composition de l'eau, par combustion continuelle, avec plus de facilité et moins de frais. Haarlem, 1791.

Against the fact that muscular contractions can be evoked even by using only a single metal, Volta argued that the current of electric fluid may in this case originate "from some accidental, even imperceptible' difference" between the two ends of the arc. Against Aldini's contrivance using a mercury bath in order to obtain a perfectly homogeneous arc, he answered that "the demanding Chemist, challenged to find a difference among parts of the same mercury, rectified after the strictest canons of art, unfortunately does find a difference, and a great one, between its inner parts and its surface, which quickly loses its polish and begins to suffer slight calcination. . . ."[14]

When it came to checking Galvani's claims experimentally, Volta was clearly inexperienced in the dissection of the frog, but he had personally examined the marine torpedo. It was the structure of the electric organ of this fish which, strangely for him, contained no metals. His classic description of his pile constructed of two dissimilar metals was sent to the Royal Society of London in the opening year of the 19th century and published in the *Philosophical Transactions* the same year.[15]

From this great controversy, two fundamental truths were to emerge: that the agent for nervous transmission was electrical and that a current flows between dissimilar metals. The first led to the understanding of the nervous system, the second to the operation of the electric battery. Galvani's findings were confirmed by everyone who copied his experiments (including Volta), but it was their interpretations of their results that were in error.

Volta's friend Vassalli[16, 17, 18, 19] had also experimented with electricity and, like Aldini before him, attempted the electrification of cadavers. Valli was to claim that Vassalli anticipated many of Galvani's observations. He assumed animal electricity and the nervous fluid were one and the same.

Volta's life, in a still divided Italy, spanned the period of the rise of Napoleon and his invasion of the northern provinces, but Volta received praise from the Emperor as well as imperial honors. The photographic reproduction of an elaborate painting of Volta receiving these honors from the Emperor, to whom he is presenting what is said to be his first pile, can be seen at the Académie des Sciences[20] in Paris. There is also Volta's original letter of July 2, 1801, describing his invention of the electric pile.

Volta died in Como in 1827 having received many honors, not only from his own country but from many European centers, still a convinced opponent of electrophysiology in an era that in Italy had produced some of the greatest contributors to this field—Caldani, Fontana, and Galvani. After his death, a temple was built in his honor at Como (Fig. 10).

---

[14]A. Volta. *Le Opere di Alessandro Volta.* Edizione Nazionale sotto gli auspici della Reale Accademia dei Lincei e del Reale Instituto Lombardo di Scienze e Lettere. Vol. 1. U. Hoepli, Milano, 1918.

[15]A. Volta. On the electricity excited by the mere contact of conducting substances of different kinds. (Letter to Sir Joseph Banks, March 29, 1800.) *Phil. Trans. Roy. Soc.*, 90:403–431 (1800).

[16]Emilio Giulio Vassalli (1761–1825). Rapport présenté à la classe des sciences exacts de l'Académie de Turin le 27 thermidor, sur les expériences galvaniques faites les 22 and 26 du même mois sur la tête et le tronc de trois hommes peu de temps après leur décapitation. Turin, 1802. (Report presented at the class of the exact sciences of the Academy of Turin on the 27th thermidor on galvanic experiments made on the 22nd and 26th of the same month on the head and the trunk of three men shortly after their decapitation.)

[17]E. G. Vassalli. Lettre d'un ami contenant un précis des expériences de Louis Galvani, inseré dans la Bibl. de Turin de l'année 1792, Mars, Vol. 1. p. 261. *Observations sur la physique,* 41:66–71 (1792). (Letter of a friend giving a summary of the experiments of Luigi Galvani, deposited in the library of Turin in the year 1792, March, Vol. 1, p. 261. *Observations on Physics,* 41:66–71, 1792.)

[18]E. G. Vassalli. *Expériences sur la torpille,* 1789.

[19]E. G. Vassalli. Lettre à J. C. de la Metherie sur le Galvanisme et l'origine de l'Électricité animale. *Journal de Physique,* Germinal, 1789, p. 336.

[20]For a reproduction of this portrait, see Figure 99 in the first volume of this work: M.A.B. Brazier, *A History of Neurophysiology in the 17th and 18th Centuries: From Concept to Experiment.* Raven Press, New York, 1984.

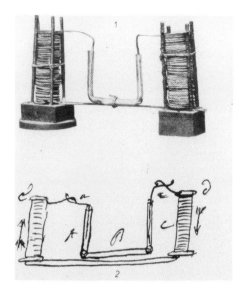

FIG. 10. **Left:** A sketch by Volta of a pile and an example from the temple to Volta at Como. **Right:** Two fine examples made of copper and zinc. (Courtesy of the Museo Nationale della Scienza e della Technica, Milan).

Intended to form a repository for his scientific instruments and others of this period, it was tragically burned to the ground with considerable looting taking place during the disaster. A few years later, some odd instruments began to turn up in the pawn shops; among them was a frictional machine believed to have been that used by Galvani. (This instrument has been said to be the one that came to the collection of the Wellcome Historical Museum in London.)

As just related, all who repeated Galvani's experiments found the same result and, in spite of Volta's unrelenting refusal to accept the biological source of the electricity, there were some who did. One of the earliest was Valli.

Eusebio Valli (1725–1816)

Eusebio Valli was born in Tuscany near Pisa and was educated at its university, moving in 1789 to Pavia, where he was exposed to the teaching of Volta. Valli was a physician and closer to biology than the physicist Volta. The experiments he saw in Volta's laboratory drew him immediately to the conviction that the source of the electricity was the animal tissue.[21] At first, he discussed his experiments in correspondence with friends,[22] but by 1793 he published a major account (Fig. 11) in London.[23] Published[24] also just before the turn of the century, their impact was outstandingly influential in combatting Volta's adverse views that became so well known in the following decade. Valli included in the account

[21]Eusebio Valli (1755–1816). Sur l'électricité animale. Observations sur la physique par l'Abbé Roziers, 41:66 (1792).

[22]E. Valli. Lettre à Prosper Balbo, contenant un précis des expèriences de Galvani. *Journal de Physique,* 1792.

[23]E. Valli. *Experiments on Animal Electricity with their Application to Physiology.* J. Johnson, London, 1793.

[24]E. Valli. Die Vermutung einer Identität von ''Nervenkraft'' und ''thierischer Elektrizität'' war schon in der sog. vorgalvanischen Epoche aufgetaucht. 1793. (The idea of the identical nature of ''nerve power'' and ''animal electricity'' had already appeared before Galvani.)

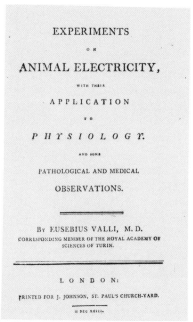

EXPERIMENTS

ON

ANIMAL ELECTRICITY,

WITH THEIR

APPLICATION

TO

PHYSIOLOGY.

AND SOME

PATHOLOGICAL AND MEDICAL

OBSERVATIONS.

———

BY EUSEBIUS VALLI, M. D.

CORRESPONDING MEMBER OF THE ROYAL ACADEMY OF
SCIENCES OF TURIN.

———

L O N D O N :

PRINTED FOR J. JOHNSON, ST. PAUL'S CHURCH-YARD.

M DCC XCIII.

FIG. 11. **Left:** Eusebio Valli, a physician and one of the first experimenters to support the claims of Galvani. (Portrait courtesy of the Istituto Lombardo, Accademia di Scienze e Lettere, Milan.) **Right:** Title page of Valli's book that he chose to publish in England.

of his experiments some in which no metals were present. Sections taken from his series of experiments, many of which were reported in letters, are now preserved at Pavia. He also corresponded with the Abbé Nollet of Paris whose interesting reply is in the Museo Nazionale della Scienza e della Technica, Milan.

Valli had a very definite concept of the electrical currents in muscles; he held that the muscle was positive on the outside and negative in the interior and that the period of inactivity after a contraction was due to the electricity being exhausted so that the muscle was inactive. As a physician, he related these findings to electrical observations of muscle movement. Valli was clearly an early supporter of Galvani, for in a series of letters[25] he described many experiments on frogs. Of his experiments, Valli wrote:

> I get hold of an already unsheathed frog, prepared as usual, then I apply my fingers, wetted with blood, to the crural nerves, or rather I touch them with a thigh recently removed from another frog's body. The animal, of whose electricity I now become a conductor, jerks slightly and for a short time. If, as often happens, it does not jerk, then I can evoke contractions by applying to the spinal cord or nerves the tip of my tongue or my moistened lips. . . .
>
> Sometimes the electricity has been lying dormant in the animal's organs from the very beginning of the operation. It can be caused to circulate by a second's application of the metal stimulator. To this end, a rusted and moistened piece of iron is more active than zinc, gold, or silver, taken separately. A slight irritation or simply the gentle heat of my mouth gave me effects perfectly comparable to those of metals . . .

———

[25]Now in the Library of the University of Pavia. Most have been published. The first eight letters were published in *Observations sur la Physique sur l'Histoire Naturelle et sur les Arts*. Vol. 41, 1792. The ninth appeared in the same periodical, Vol. 42, 1793. The tenth was lost.

> After moistening nerves and muscles with saliva I often noticed that contractions became stronger if they had been weak, and appeared again when they had been lacking.
>
> Mr. Fowler had already noticed, that metals need moisture to exert their power on the nervous fluid, a power which he calls by the new name of influence. In my experiments, water could not successfully replace saliva.[26]

Valli was an army physician and like another ardent explorer of the nervous system (Ramon y Cajal), he served a term in Cuba (half a century before Jesse Lazear and Walter Reed)[27] where he engrossed himself in the battle against yellow fever, using himself to test whether or not a much diluted injection could combat the disease. The experiment was a failure, and he died in Havana in 1826.

The surge from the great Italian schools changed the understanding of the nervous system so fundamentally that no longer would there be two distinctly different sciences: that of physics with its hard-wired apparatus and the world of the soft fragile nerves that decayed when the organism died. These two apparently disparate worlds shared the same energizer, electricity.

Christoph-Heinrich Pfaff (1773–1852)

Volta's refusal to accept this common bond drew others from physics into the astounding acceptance of the role of electricity in the animal body. Not from Ben Franklin in the New World but from Germany one finds serious examination of the evidence. For example, Christoph-Heinrich Pfaff[28, 29]was one of a family interested in physics.[30, 31]

Distinguished son of a distinguished father, Pfaff headed for a career in chemistry. Later, he was to attain a professorship in Kiel, having written his doctoral thesis on the so-called animal electricity. A great traveller, he met all the prominent scientists of his time, including Volta, who was then in Paris enjoying to the full Napoleon's interest and praise. This led to Pfaff being put on a commission to test Volta's claims. Pfaff wrote no less than 40 essays on what had become known as ''galvanism,'' although his major works stem from his career in chemistry.

Johann Wilhelm Ritter (1776–1810)

Another German scientist to enter the field was the physicist Johann Wilhelm Ritter.[32] A member of the Bavarian Academy in Munich, he was attracted to the claims for animal electricity, toward which he made a more reasonable approach than that of Volta. Ritter

---

[26]Printed in Mantova, nella Stamperia di Giuseppe Braglia, 1794.

[27]W. Reed, J. Carroll, A. Agramonte, and J. W. Lazear. The etiology of yellow fever. *Phil. Med. J.*, 6:790–796 (1900).

[28]Christoph-Heinrich Pfaff. *Dissertationes de electricitate sic dicta animali*. Stuttgart, 1793.

[29]C-H. Pfaff. Über thierische Elektrizität und Irritabilität. Leipzig, 1795.

[30]Johann Wilhelm Andreas Pfaff (1774–1835). *Übersicht über den Voltaismus und die wichtigsten Saetze zu Begruendung einer Theorie desselben*. Stuttgart, 1804.

[31]J. W. A. Pfaff. Die Umkehrung der voltaischen Pole durch Herrn Pohl, oder die durch seine Philosophie geheilte 25 jaehrige Blindheit der Naturforscher. Nuremberg, 1827. (Theoretical Consideration regarding the Voltaic Pile.)

[32]Johann Wilhelm Ritter. Beweis, dass ein beständiger Galvanismus den Lebensprozess im Thierreiche begleitet. Weimar, 1798. (Evidence that a Constant Galvanism Accompanies the Life-Process in the Animal Kingdom.)

had become very close to anticipating Volta's invention of the pile. He had demonstrated that an electric current could "divide" water into hydrogen and oxygen, taking them to opposite poles. This result seemed mysterious at the time for electrolytes and free ions were not understood, water being thought to be one of the elements and presumably indivisible. He did not, however, think of adding zinc and copper cells in a series and, when he learned of Volta's success, blamed himself for not foreseeing the obvious development from his own experiments. So admiring was he of Volta's discovery that he dedicated his own book to Volta and Alexander von Humboldt and made translations of Volta's works into the German language.

Ritter had not only come close to Volta's field, he was one of the first to accept Galvani's findings. As early as 1797, only six years after the world heard of Galvani's claims, he gave a lecture in support of it at Jena, where he was a student. He published this the following year. It is of interest that one finds lingering traces of *Natürphilosophie* in his writings, though he was not an open espouser of this philosophy.

In the controversy that broke out after the publication of Galvani's findings, we find Ritter taking a middle position. Clearly, from his own work he knew Volta's concepts of metallic action were correct, but he also supported Galvani's proposal that electricity lay within our bodies. In fact, he went further than Galvani, saying every animal cell consists of "a system of infinitely small, infinitely numerous galvanic chains."

Interest in this post-Galvani period was most intensely focused on the effect of electricity on the muscle and nerve, but Ritter decided to examine also the effect of chemicals on the excitability of muscle to a nerve stimulus and had some success with an oxygenated sodium-chloride solution. The danger of the destructive power on organic tissues of too strong a solution became obvious. Ritter did not fail to experiment on himself with electric currents to the point of pain, about which he gave some rather vivid descriptions.

Some of these experiments were on stimulation[33, 34] of his ear, since he wanted to investigate the action of current on sense organs. He described several kinds of sounds heard following this stimulation by his silver-copper battery. It is noticeable that during this decade following Galvani's death, the scientists were using the terms "galvanism" and "galvanic currents," as they do today nearly 190 years later.

Ritter was born in Silesia in 1776, a year of resurgence in both Europe and the New World; he died in 1810 before the issue of nerve transmission was finally accepted. It was to be nearly another 40 years before claims to have found a final proof were made by Emil Du Bois-Reymond, who leaned heavily on Ritter's physics for his own claims.

BIBLIOGRAPHY

Alexander von Humboldt (1769–1859)

Selected Writings

Über die gereizte Muskelfaser. Aus einem Brief an Herrn Hofrath Blumenbach von Herrn Oberbergrath. F. A. v. Humboldt. *Grens, N.J. de Physik.*, 115, 1795. (On Stimulated Muscle Fiber. From a Letter to Counselor Blumenbach from Inspector F. S. v. Humboldt.)

---

[33]J. W. Ritter. *Beiträge zur näheren Kenntniss des Galvanismus und der Resultate seiner Untersuchung.* Vol. I in 4 parts; Vol. 2 in 5 parts. Jena, 1800–1805. (Contributions to the more exact recognition of galvanism and the results of its investigation.)

[34]J. W. Ritter. *Physisch-Chemische Abhandlungen in chronologischer Folge.* Vols. I, II, III. Leipzig, 1806. (Physical-chemical Works in Chronological Order.)

Versuche über die gereizte Muskel-und-Nervenfaser, oder Galvanismus, nebst Vermuthungen über den chemischen Process des Lebens in der Thier- und Pflanzenwelt. Decker and Rottman, Posen and Berlin, 1797. (Experiments on Stimulated Muscle and Nerve Fibers, or Galvanism, along with Conjectures on the Chemical Process of Life in the Animal and Plant World.) (French translation by J. F. N. Jadelot. *Experiences sur le Galvanisme.* Paris, 1799). Two volumes, 1797. Posen and Berlin, 1797.

Suite des experiences sur l'irritation de la fibre nerveuse et musculaire. *J. de Physique de Chimie,* 46:198, 310, 465 (1798).

Versuch über die elektrischen Fische. Erfurt, 1806. (Investigation of Electric Fish.)

Kosmos. Entwurf einer physischen Weltbeschreibung. 5 vols. Stuttgart and Tübingen, 1845–62. (Plan for a Physical Description of the World.)

Suggested Readings

Botting, D. *Humboldt and the Cosmos.* Harper & Row, New York, 1973.

Nelken, Halina. *Alexander von Humboldt. Bildnisse und Künstler, eine dokumentierte Ikonographie.* Reimer, Berlin, 1980. (Alexander von Humboldt. Pictures and Artists, a Documented Iconography.)

Rothschuh, K.E. *Geschichte der Physiologie.* Berlin and Heidelberg, Krieger, 1953. (History of Physiology.)

Rothschuh, K.E. Alexander v. Humboldt und die Physiologie seiner Zeit. Sudhoffs Arch. Beih., 43:97 (1959). (Alexander v. Humboldt and the Physiology of His Time.)

Rothschuh, K.E. Die neurophysiologischen Beiträge von Galvani und Volta. In: *Essays on the History of Italian Neurology,* edited by L. Belloni. Milan, pp. 117–130, 1963. (The Neurophysiological Contributions of Galvani and Volta.)

Alessandro Volta (1745–1827)

Selected Writings

De vi attractiva ignis electrici, ac phaenomenis inde pendentibus ad Joanem Bapt. Beccarium dissertatio epistolaris. Novo-Comi. Typis Octovii Staurenghi, Como, 1769.

Del modo di render sensibilissima la pir debole elettricita sia naturale, sia artificiale. *Phil. Trans. Roy. Soc.,* 71:237–238 (1782). (Of the method of rendering very sensible the weakest natural or artificial electricity.)

Letter to Cavallo. Written May 1793. *Gren's Journal,* 2:159.

Account of some discoveries made by Mr. Galvani, of Bologna; with experiments and observations of them. In two letters from Mr. Alexander Volta to Mr. Tiberius Cavallo. *Phil. Trans. Roy. Soc.,* 83:10–44 (1793).

Schreiben an den Herrn Abt Anton Maria Vasali ueber die Thierische Elektrizitaet als eine Fortsetzung der Schriften desselben ueber die thierische Elektrizitaet, herausgegeben von Dr. Johann Mayer. Prag Calve, Prague, 1796. (Letters to Anton Maria Vasali on animal electricity as a continuation of his writings on animal electricity, edited by Dr. Johann Mayer.)

Letter to Sir Joseph Banks, March 29, 1800. On electricity excited by the mere contact of conducting substances of different kinds. *Phil. Trans. Roy. Soc.,* 90:403–431 (1800).

Schriften über Elektricität und Galvanismus; aus dem Italienischen und Franzoesischen übersetzt von C.F. Nasse. Vol. 1. Halle, 1803. (Writings on Electricity and Galvanism; translated from the Italian and French by C.F. Nasse.)

Neueste Versuche über Galvanismus; Beschreibung eines Neuen Galvanometers und andere kleine Abhandlungen über diesen Gegenstand. Vienna, 1803. (Latest Experiments on Galvanism; Description of a New Galvanometer and other Small Writings on this Subject.)

L'identita del fluido elettrico col cosi detto fluido galvanico vittoriosamente dimostrate, con nuove esperienze ed osservazioni. Memoria communicata al signore Pietro Configliachi. E aggiunto il catalogo delle sue opere stampate sino a tutto l'anno 1813. Pavia, 1814.

Collezione dell'opere dell Cavaliere conte Alessandro Volta. 3 vols. Piatti, Florence, 1816.

Le Opere di Alessandro Volta. Edizione Nazionale sotto gli auspici della Accademia dei Lincei e dell Instituto Lombardo di Scienze e Lattere. Vol. I. U. Hopeli, Milan, 1918. (*The Works of Alexander Volta.* National publication under the auspices of the Academy of the Lincei and the Lombardian Institute of Science and Letters. Vol. 1, 1918.)

A bibliography of the works of Volta will be found in: Configliachi, Pietro. "L'Identita del fluido elettrico col cose detto fluido Galvanico. Garelli, Pavia, 1814. A collection of Volta's works including the letters referred to was made in 1816; Collezione dell'Opere Alessandro Volta. Piatti, Florence, 1816 and again in 1918.

Suggested Readings

Cohen, I. B. *Volta, Alexxandro (1745–1827) Introduction.* To: *Galvani's Commentary.* English translation by M. G. Foley. Burndy Library, Norwalk (Conn.), 1954.

Dibner, B. *Galvani-Volta. A Controversy that Led to the Discovery of Useful Electricity.* Burndy Library, Norwalk (Conn.), 1952.

Geddes, L. A., and Hoff, H. E. The discovery of bioelectricity and current electricity. The Galvani-Volta controversy. *IEEE Spectrum,* (12):8, 38–46 (1971).

Hoff, H. E. Galvani and the Pre-Galvanian Electrophysiologists. *Ann Sci,* 1:157–172 (1936).

Heilborn, J. L. Volta's path to the battery. In: *Selected Topics in the History of Electrochemistry,* edited by G. Dubpernell and J. H. Westbrook. The Electrochemical Society, Princeton, N.J., 1978.

Polvani, G. *Alessandro Volta.* Domus Galilaeana, Pisa, 1942.

Pupilli, G. C. and Fadiga, E. The origins of electrophysiology. *J. World Hist.,* VII/2:141–589 (1963).

Rothschuh, K. E. Die neurophysiologischen Beiträge von Galvani und Volta. In: *Essays on the History of Italian Neurology,* edited by L. Belloni, Milan, 1963.

Rothschuh, K. E. *Bull Hist Phys,* 72:149 (1973).

Wetzels, W. D. *Johann, Wilhelm Ritter: Physik im Wirkungsfeld der Deutschen Romantik.* Walter de Gruyter, Berlin, New York, 1973. (Physics in the Field of Influence of German Romanticism.)

Emilio Giulio Vassalli (1761–1826)

Selected Writings

Lettre d'un ami . . . contenant un précis des expériences de Louis Galvani, inseré dans la Bibl. de Turin de l'année 1792, mars, vol. 1. p. 261. *Observations sur la Physique,* 41, 1792.

Rapport presenté à la classe des sciences exactes de l'Académie de Turin le 26 thermidor, sur les expériences galvaniques faites les 22 et 26 du meme mois sur la tête et le tronc de trois hommes peu de temps après leur decapitation. Turin, 1802.

Schreiben an den Herrn Abt Anton Maria Vasalli ueber die Thierische Elektrizitaet als eine Fortsetzung der Schriften desselben ueber die thierische Elektrizitaet, herausgegeben von Dr. Johann Mayer. Prag Calve, Prague, 1796. (Letter to Abbot Anton Maria Vasalli on animal electricity, as a continuation of his writings on animal electricity, edited by Dr. Johann Mayes.)

Fusebio Valli (1755–1816)

Selected Writings

Sur l'électricité animale. Observations sur la physique par l'Abbé Roziers, 41:66 (1792). (On animal electricity. Observations on physics by Abbot Roziers.)

Experiments on Animal Electricity with Their Application to Physiology, and Some Pathological and Medical Observations. J. Johnson, London, 1793. (Animal electricity and the nervous fluid are assumed to be one and the same.)

Die Vermutung einer Identität von ''Nervenkraft'' und ''thierischer Elektrizität'' war schon in der sog. Vorgalvanischen Epoche aufgetaucht. 1793. (The idea of the identical nature of ''nerve energy'' and ''animal electricity'' had already emerged in the time before Galvani.)

Johann Wilhelm Andreas Pfaff (1774–1825)

Selected Writings

Üebersicht üeber den Voltaismus und die wichtigsten Saetze zu Begruendung einer Theorie desselben. Stuttgart, 1804. (Laws and theory of the voltaic cell.)

Die Umkehrung der voltaischen Pole durch Herrn Pole, oder die durch seine Philosophie geheilte 25 jaehrige Blindheit der Naturforscher. Nuremberg, 1827. (Theoretical consideration regarding the Voltaic Pile.)

Johann Wilhelm Ritter (1776–1810)

Selected Writings

Beweis, duss ein beständiger Galvanismus den Lebensprozess im Thierreich begleitet. Weimar, 1798. (Evidence that a Constant Galvanism Accompanies the Life-Process in the Animal Kingdom.)

Beiträge zur näheren Kenntniss des Galvanismus und der Resultate seiner Untersuchung. Vol. 1 in 4 parts; Vol. 2 in 5 parts. Jena, 1800–1805. (Contributions to the More Exact Recognition of Galvanism and the Results of its Investigation.)

Das elektrische System der Körper, ein Versuch. Leipzig, 1805. (The Electrical System of the Body, an Experiment.)

Physisch-Chemische Abhandlungen in chronologischer Folge. 3 vols. Leipzig, 1806. (Physical-Chemical Works in Chronological Order.)

CHAPTER III

The Italian School Comes to the Support of Galvani

Leopold Nobili (1784–1835)

Leopold Nobili was born in 1784 in Trassilico in Garrangnono. Starting out in the army, he trained in the military school in which he later rose to be teacher of physics. In the year 1812, famous for Napoleon's assault on Moscow and later for Tchaikovsky's great overture, Nobili fought in the campaign and was taken prisoner by the Russian army. When released, he returned to live in Reggio Emilia and devoted himself solely to research. Caught, as was his pupil Carlo Matteucci, in the intense disturbances of the time, he moved first to France and then to Tuscany. There, he caught the eye of Leopold II, the Grand Duke of Tuscany, who appointed him Professor of Physics in the Musei Fisica e Storia Nazionale in Florence. Matteucci came under his influence at this time. Known among physicists mainly for his work on magnetic fields (long before these were found in the brain), Nobili's contributions to animal electricity were based on his work with frogs.

When Nobili began this work with frogs, he approached the problem from the physicist's point of view. He repeated one of the experiments others had made, but instead of being satisfied by an observable muscle twitch, he detected the "animal electricity" by a physical instrument. This was the astatic galvanometer that he himself had designed. In using an astatic galvanometer, Nobili was adapting the brilliant discovery by Oersted of electromagnetism.

Hans Christian Oersted (Fig. 12) was Professor Extraordinary at the University of Copenhagen. Any relationship between electricity and magnetism had been denied by such giants of the times as André Marie Ampère and Thomas Young. But Oersted, with an *a priori* hunch that they were related in a certain way, had so much faith in his idea that he tested it for the first time in a lecture to students. The experiment worked. He passed a current through a single loop of wire above a magnetic needle. When a current flowed, the

FIG. 12. **Left:** A contemporary sketch depicting Oersted giving his successful demonstration of electro-magnetism. (Louis Figuier. *Les Merveilles de Science.* Furne, Jouvet, Paris, 1867. Vol. 1, p. 743). **Right:** Bust in Copenhagen of Hans Christian Oersted (1777–1851).

needle moved. This is the only example in the history of science of a major scientific discovery being made during a classroom lecture.[1, 2, 3, 4]

This great discovery led to the development of instruments with multiple windings and of moving coil galvanometers, though it was still a few years before they became more sensitive than the frogs' legs. This advance was helped by the invention of Nobili.[5, 6] His contribution was the further design of the astatic galvanometer[7] in which two coils of wire, wound in opposite directions, cancelled the effect of the earth's own magnetism (Fig. 13). Always, his goal was to produce a detector of electricity more sensitive than the frog. But Nobili mistakenly interpreted the phenomenon as a thermal current due to cooling. The astatic galvanometer which he used in his experiments on frogs was for a specific purpose, namely, to rid his physical measurements of the influence of the terrestrial magnetic field. Many of Nobili's experiments were performed with a physician, Marianini, and (rather inevitably) they decided to test whether or not the electrical stimulation could help patients with paralysis of different kinds. Nobili made no reference to Aldini's attempts 30 years

[1]Hans Christian Oersted (1777–1851). Experimenta circa effectum conflictus electrici in Acum magneticam. *J. F. Chem. Phys.*, 29:275–281 (1820).

[2]H. C. Oersted. *La découverte de l'électromagnétisme faite au 1820 par H. C. Oersted,* edited by A. Larsen. Copenhagen, 1820.

[3]H. C. Oersted. Neuere electro-magnetische Versuche, *J. Chem. Phys.*, 29:364–369 (1820).

[4]H. C. Oersted. Galvanic magnetism. *Phil. Mag.*, 56:394 (1820).

[5]Leopold Nobili. Comparaison entre les deux galvanomètres les plus sensibles, la grenouille et le multiplicateur à deux aiguilles, suivies de quelques resultat nouveaux. *Ann. Chim. Phys.*, 38:225–245 (1828).

[6]L. Nobili (1784–1835). Descrizione di un nuovo galvanometro. Read before the Réale Accademia di Scienze, Lettere et d'Arti of Modena of May 13, 1825. Printed in *Memorie et Instrumenti del Cav. Prof. Leopoldo Nobili.* Vol. 1. D. Passigli et Socj, Florence, 1834. 1–6.

[7]A large collection of Nobili's instruments, including the galvanometer, can be seen in the Museo di Storia delle Scienza, Florence.

FIG. 13. Left: Nobili's astatic galvanometer, which consisted of a series of magnetic needles set into motion by the current flowing in a coil. A compass was inserted in the base. (Courtesy of the Museo di Storia delle Scienze, Florence). **Right:** Leopold Nobili (1784–1835). Professor of Physics in Florence.

before when he published his results in the *Annales de Chimie*. He reported rather cautiously as follows:

> L'électricité a eu, dans le traitement de plusieurs maladies, une si grande vogue, qu'elle a peut-être été employée de toutes les manières possibles. Il n'est donc point question ici de présenter des méthodes nouvelles; il s'agit seulement de diriger la practique médicale d'après des principes certains au lieu de continuer à employer l'électricité d'une manière totalement empirique, comme on l'a fait jusqu'ici.[8]

It was about this time that his pupil Carlo Matteucci became interested in animal electricity.

Carlo Matteucci (1811–1865)

Carlo Matteucci was born in Forli in Venezia in 1811 in a divided Italy, and his life spanned the Risorgimento and saw the unification of Italy in 1861. His training and research covered many fields. He first studied mathematics at the University of Bologna; he then went to Paris, trained there with Arago,[9] and formed a love for France, whose language he was later to use for many of his publications. Returning to Pavia, he worked in meteorology and in agriculture until 1834, when he moved to Florence. In Florence he came under the influence of the physicist Nobili[10] and of Amici,[11] whose beautiful designs

---

[8]L. Nobili. Analyse expérimentale et théorique des phénomènes physiologiques produits par l'électricité sur la grenouille; avec un appendice sur la nature du tétanos et de la paralysie, et sur les moyens de traiter ces deux maladies par l'électricité. *Ann. Chim. Phys.*, 44:60–94 (1830).

[9]Dominique François Arago (1786–1853).

[10]Leopold Nobili (1784–1835).

[11]Giovanni Battista Amici (1786–1863).

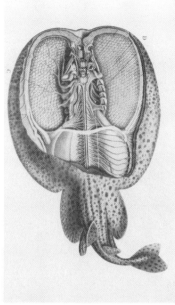

FIG. 14. **Left:** Carlo Matteucci (1811–1865). Portrait preserved at the University of Pisa. **Right:** One of Matteucci's illustrations. (From: *Traité des Phénomènes Electro-physiologiques des Animaux.* Fortin et Masson, Paris, 1844.)

of microscopes keep his memory alive. A fine example can be seen in the Josephenum Museum in Vienna and another at the Museo di Storia e delle Scienze in Florence.

Matteucci became interested in animal electricity, and the first form he investigated was the electric fish.[12, 13] This work he pursued at this hometown of Forli in 1836, while he was still a very young man. But financial needs led to his taking a post in pharmacy at the hospital in Ravenna. This was, however, only a temporary appointment, for in 1840 he was called by the Grand Duke of Tuscany to the Chair of Physics at the University of Pisa, and it is with this great university that his name is usually associated (Fig. 14). Nobili had called the electric currents "corrente della rana." Matteucci was to repeat these experiments and introduce the term "muscle currents," i.e., electricity of the active muscle as distinct from the demarcation potential of the nerve.

Matteucci demonstrated demarcation potentials in muscle but failed to find them in nerve. "Nous croyons," he wrote, "être autorisés à conclure qu'il n'existe aucune trace de courants électriques dans les nerfs des animaux vivants, appreciable à l'aide des instruments que l'on possède aujourd'hui."

Matteucci first published in 1830, and he devoted most of his productive years to the study of the muscle current of the frog (Fig. 15), using the same name for them as suggested by Nobili, "la corrente propria della rana." He also wrote much in French, using the expression "courant propre," which was mistranslated into English, where it was referred to in all the journals of the time as "the proper current of the frog." Matteucci demonstrated the fallacies of Volta's and Nobili's theories but was himself hampered by a con-

[12]C. Matteucci. Experiences sur la torpille. *C. R. Acad. Sci.* 3:430–431 (1836).

[13]C. Matteucci. Recherches physiques, chimiques et physiologiques sur la torpille. *C. R. Acad. Sci.,* 5:788–792 (1837).

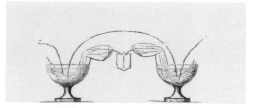
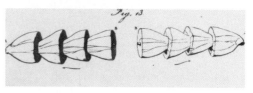

FIG. 15.   Left: Carlo Matteucci. From an old yellowing photograph preserved at the Scuola Normale Superiore in Pisa. (Courtesy of the late Professor G. Moruzzi). **Right:** Three of Matteucci's many experiments to prove the existence of "muscle currents." At the *top* is his repeat of Nobili's experiment. *Center* and *below* is his demonstration that piles, imitating Volta's, can be formed from muscle tissue. These were to evoke scorn from Du Bois-Reymond. (From: *Traité des Phénomènes Électro-physiologiques des Animaux.* Fortin et Masson, Paris, 1844.)

cept of "nerve force" that he appeared to have regarded as a specific form of energy. Matteucci found his frog's nerve-muscle preparation so sensitive to small currents that the muscle would contract if the nerve were laid across another contracting muscle. This is the "rheoscopic" frog of the physiology textbooks, although Matteucci called it the "galvanoscopic frog." He used it to detect current in the muscles of other animals. He deposited a sealed letter with Monsieur Dumas,[14] the President of the French Academy, registering his claim to have recorded muscle currents in warm-blood animals.

The bridge from the electricity of the electric fish to that of the muscle was crossed by Matteucci in his publication of 1838,[15] and from his wording it is clear that he foresaw that the same process would be found in the nerve.[16, 17] To quote his own words:

> Toute action extérieure ou irritation excerceé sur le corps de la torpille vivante, et qui determine la décharge électrique, est transmisé par les nerfs du point irrité au quatrième lobe du cerveau. Toute irritation excerceé sur ce quatrième lobe, ou sur les nerfs qui en sortent et qui vont à l'organe, est suivié par une décharge electrique sans aucune espèce de contraction. La liaison qui entre le quatrième lobe et les nerfs qui en partent d'une part, et la

[14]C. Matteucci. Experiences sur les phénomènes de la contraction induite. Lettre à M. Dumas. *Ann. Chim. Phys.,* 15:64–70 (1845).

[15]C. Matteucci. Sur le courant électrique ou propre de la grenouille; seconde mémoire sur l'électricité animale, faisant suite à celui sur la torpille *Ann. Chim. Phys.,* 2ème serie, 67:93–106 (1838).

[16]C. Matteucci. Sur le courant électrique ou propre de la grenouille. *Biblio. Univ. Genève,* 15:156–168 (1838).

[17]C. Matteucci. *Essai sur les Phénomènes Électriques des Animaux.* Carilian, Goeury et Dalmot, Paris, 1840.

substance de l'organe de l'autre, est precisement la même que elle qui se trouve entre un nerf quelconque et les muscles dans lesquels il est ramifié.

He goes on to claim: "This is the greatest analogy that we have between the unknown force in nerves and that of electricity." He had no doubt that it was the agent in both the fish and the frog.

Matteucci had, in 1838, worked on what he called the "courant propre" of the injured muscle. What we recognize now to be a demarcation potential. He seems to have discovered this current of injury almost by accident, for he wrote:

Un autre cause qui modifie grandement le courant propre de la grenouille, c'est son état tétanique. Il arrive très-souvent avec des individus vivaces, qu'en les préparant rapidement on les voit étendre leurs jambes et les roidir de tel sorte qu'il devient impossible de les plier; on peut aussi, avec une solution de strichnine ou de l'extrait de noix vomique, déterminer en peu de secondes la convulsion tétanique. L'influence du tétanos est telle que le courant propre manque toujours lorsque la grenouille en est attaquée. Nous n'avons plus de contractions, ni de signes au galvanomètre. Si l'animal e été tué par le poison, on ne réussit plus à en obtenir, mais si, au contraire, le tétanos a été produit par l'irritation qu'on a donnée à la grenouille en la préparant, une fois que les convulsions son passées, les signes du courant propre apparaissent encore.

He is here warning the experimentalists that the condition of the animal affects the results obtained and that the frogs chosen for the experiment should be very fresh. Care must be taken in preparing them lest they go into tetanic convulsions, in which case it is impossible to record the "courant propre."

It is essentially to Matteucci that we owe the clarification of the current that flows between a cut surface and the uninjured surface of a muscle or nerve. Today this is known as the demarcation potential. In Matteucci's day, however, the fact that there is a resting potential of the membrane was not known. His own description in his *Traité* reads:

On prend pour cela, la grenouille preparée que j'aì appelée grenouille galvanoscopique; ensuite on coupe d'une manière quelconque le muscle d'un animal vivant, et on introduit, dans la blessure, le nerf de la grenouille galvanoscopique. En se bornant à cela, il arrive souvent que la grenouille se contracte. Si l'on fait l'expérience avec soin, on découvre facilement, qu'afin de reussir il faut toucher, avec deux points différents du filament nerveux, deux points differents de la masse musculaire. C'est ainsi qu'en touchant avec le bout du nerf de la grenouille galvanoscopique, le fond de la blessure la grenouille se contracte constamment. Ceci prouve évidemment que c'est bien un courant électrique qui circule dans le nerf, puisqu'il faut former un arc dan lequel ce même nerf est compris.[18]

For the physicists, the crucial test was not the observation that a muscle twitched but the detection by a galvanometer of the current flow. Matteucci used a galvanometer placing one electrode on the intact surface and one on the cut portion. He made this critical test in 1838[19] and described it more fully in 1840,[20] but it did not appear in print until 1842,[21] shortly after Du Bois-Reymond had achieved the same result. This led to the acrimonious claim for priority.

The demarcation potential was what Matteucci called "le courant propre." He then

[18]C. Matteucci. *Traité des Phénomènes Électro-physiologiques des Animaux suivi d'Études Anatomiques sur le Système Nerveux et sur l'Organe Électrique de la Torpille.* Fortin et Masson, Paris, 1844.

[19]C. Matteucci. Sur le courant électrique ou propre de la grenouille. Second memoire sur l'électricité animale, faisant suite à celui sur la torpille. *Ann. Chim. Phys.,* 2ème serie, 67:93–106 (1838).

[20]C. Matteucci. Essai sur les Phénomènes Électriques des Animaux. Carilian, Goeury et Dalmot, Paris, 1840.

[21]C. Matteucci. Deuxième mémoire sur le courant électrique propre de la grenouille et sur celui des animaux à sang chaud (1): *Ann. Chim. Phys.,* 3ème serie. 6:301–339 (1842).

moved on to look for it in birds and mammals—pigeons, rabbits, and goats. He placed one electrode in a wound and the other on the surface of a muscle. Matteucci's own description written in 1842 reads:

> J'ai obtenu encore un courant bien distinct et quelquefois de 20 à 30 degrés, en faisant une blessure dans le muscle de la poitrine ou de la cuisse d'un animal vivant (pigeon, lapin, brebis), en plongeant une des lames dans l'intérieur de la blessure, et enposant l'autre sur la surface mise à nu du muscle blessé. Le courant était constamment dirigé, dans l'animal, de l'intérieur de la blessure à la surface extérieure du muscle. La constante direction de ces courants et les signes bien distincts que j'avais à mon galvanomètre auraient du m'assurer qu'on ne devait les attribuer à aucune imperfection de l'expérience.''

One of Matteucci's most striking demonstrations of this flow of current from intact surface to injured tissue is a series of chunks of muscles placed end to end. This was one of the experiments Du Bois-Reymond was to challenge in his published letter to Bence Jones in 1858, a letter in which he complained of much of Matteucci's reports that failed to acknowledge his own priority.

For the laboratory scientist, a great contribution of Matteucci's was his construction of a kymograph (Fig. 16). He described this instrument in one of his seven reports to the Royal Society.[22] His description reads:

> The apparatus principally consists of a solid brass support AB fixed upon a wooden stand, in which slide two pieces of metal C, D, capable of being fixed in different places by means of pressure screws (vis de pression). The piece of metal C is furnished with a vice E, in which is to be held the morsel of spinal marrow of the prepared frog, and is fastened there by three screws. The other piece F, of a forked shape, is provided with a hole in each extremity of the fork, in which a very fine wire G is fixed or regulated. At one end of the wire is a hook to which is affixed the claw of the frog, the other end of the wire is attached to a silken thread which winds round the little pulley I. Upon this pulley another thread of silk is wound, in contrary sense to the former, and to this is attached a small leaden weight 0. The axis of the pulley is furnished with a kind of double index PQ, in the form of a semicircle. The axis is fixed upon two pivots, which admit of being more or less approximated. One of these pivots is the centre of a circle RS, which bears a division. A long ivory needle TV is attached to this pivot; it is very light, and turns with the slightest possible touch. The use of this ivory index is obvious. In effect, when this index is brought in contact with the semicircular one PQ, wihch is attached to the axis of the pulley, and the pulley is put in motion, the movement is communicated to the ivory index, and this latter will stop at the point at which it arrives in its gyration even when the pulley is brought back to its former position by the little weight. It must be allowed that without such an index as the one described, it would have been impossible to have judged of the extent of the movement of the pulley produced by the contraction on account of its short duration. The weight I have been in the habit of using is 0.600 gramme, sufficient to allow of the limb returning to its position after the contractions have ceased; a heavier weight than this would stretch the nerve too much.

Contemporaries of Matteucci's were Garibaldi, with his campaigns, and Mazzini, the apostle of the Risorgimento. Both had been born in Napoleon's France: Garibaldi in Nice and Mazzini in Genoa. But the fall of the Emperor established the House of Piedmont. Matteucci himself, in addition to his intensive work in animal electricity, was also a prominent figure in the Risorgimento. A great liberal and a great patriot, he attempted to coordinate efforts of European liberals when the Revolution of 1848 broke out. When the two

---

[22]Electrophysiological researchers 4th Memoir. The physiological action of the electric current. *Phil. Trans Roy. Soc.*, 136:483–490 (1846).

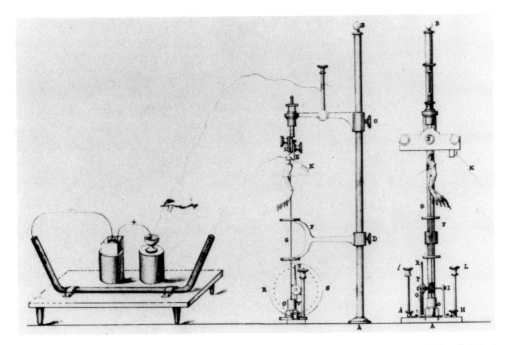

FIG. 16. The kymograph designed by Matteucci for recording the contraction of a frog's leg (labelled G). See text for explanation as given in *Phil. Trans. Roy. Soc.,* 136 483–490 (1846).

Italies were united in 1861, he became a Senator. Matteucci was one of the early Ministers of Public Information in Italy. He died in 1868.

BIBLIOGRAPHY

Leopold Nobili (1784–1835)

Selected Writings

Introduzione alla Meccanica della Materia. Paolo Emilio Giusti, Milan, 1819.

Nuovo Trattato d'Ottica o sia la Scienza della luce Dimostrata coi Puri Principi di Meccanica. Paolo Emilio Giusti, Milan, 1820.

Nuovi Trattati Sopra il Calorico, l'Elettricita e il Magnetismo. Eredi Soliana Tipi Reali, Modena, 1822.

Comparaison entre les deux galvanometres les plus sensibles, la grenouille et le multiplicateur a deux aiguilles, suivies de quelques résultats nouveaux. *Ann. Chim. Phys.,* 38:225–245 (1828).

Analyse expérimentale et théorique des phénomènes physiologiques produits par l'électricité sur la grenouille; avec un appendice sur la nature du tetanos et de la paralysie, et sur les moyens de traiter ces deux maladies par l'électricité. *Ann. Chim. Phys.,* 44:60–94 (1830).

Descrizione di un nuovo galvanometro. Read before the Reale Accademia di Scienze, Lettere et d'Arti of Modena on May 13, 1825. Printed in: *Memorie ed Instrumenti del Cav. Prof. Leopoldo Nobili.* Vol. 1. D. Passigli et Soc., Florence, 1834.

Memorie ed Osservazioni colla Descrizione ed Analisis de'suoi apparati ed Instrumenti. Florence, 1834.

Nuovi Trattati Sopra il Calorico l'Elettricita e il Magnetismo. Modena, 1838.

Carlo Matteucci (1811–1865)

Selected Writings

Action de la pile sur les substances animales vivantes. *Ann. Chim. Phys.*, 2ème serie, 43:256–258 (1830).

Mémoire sur l'éléctricité animale. *Ann. Chim Phys.*, 2ème serie, 56:439–443 (1834).

Expériences sur la torpille. *C. R. Acad. Sci.*, 3:430–431 (1836).

Sur le courant éléctrique ou propre de la grenouille; second memoire sur l'éléctricité animale, faisant suite a celui sur la torpille. *Ann. Chim. Phys.*, 2ème serie, 68:93–106 (1838).

Sur le courant éléctrique ou propre de la grenouille. *Biblio. Univ. Genève*, 15:156–168 (1838).

Sur les phénomènes éléctriques de la torpille. *Biblio. Univ. Geneve*, 17:373–378 (1838).

Essai sur les Phénomenes Éléctriques des Animaux. Carilian, Goeury et Dalmot, Paris, 1840.

Deuxieme mémoire sur le courant éléctrique propre de la grenouille et sur celui des animaux a sang chaud. *Ann. Chim. Phys.*, 3ème serie, 6:301–339 (1842).

Expériences rapportées dans un paquet cacheté deposé par M. Dumas au nom de M. Matteucci, et dont l'auteur, present a la séance, desire aujourd'hui l'ouverture. *C. R. Acad. Sci.*, 15:797–798 (1842).

Sur un phénomène physiologique produit par les muscles en contraction. *Ann. Chim. Phys.*, 3ème serie, 6:339–343 (1842).

Mémoire sur la mésure de la force nerveuse dévelopée par un courant éléctrique. *C. R. Acad. Sci.*, 19:563–565 (1844).

Traité des Phénomenes Éléctro-physiologiques des Animaux suivi d'Études Anatomiques sur le Système Nerveux et sur l'Orange Éléctrique de la Tropille. Fortin et Masson, Paris, 1844.

Expériences sur les phénomènes de la contraction induite. Lettre à M. Dumas. *Ann. Chim. Phys.*, 15:64–70 (1845).

Recherches physiques, chimiques et physiologiques sur la torpille. *C. R. Acad. Sci.*, 5:788–792 (1847).

Leçons sur les Phénomènes Physiques des Corps Vivants. Edition française publiée, avec additions considerables sur la deuxième édition italienne. Fortin et Masson, Paris, 1847.

Lectures on the physical phenomena of living things. Longman, Brown, Green and Longmans, London, 1847.

Electro-physiologica researches. Fifth series. Part I. Upon induced contractions. *Phil. Trans. Roy. Soc.*, 137:231–237 (1847).

Electro physiological researches on induced contraction. Ninth series. *Phil. Trans. Roy. Soc.*, 140:645–649 (1850).

Lettre de Charles Matteucci à Mr. H. Bence Jones, F. R. S. & c. Editeur d'une Brochure intitulée "On Animal Electricity." Extrait de Découvertes de Mr. Du Bois-Reymond. Imprimerie Le Monnier, Florence, 1853.

Corso di electro-fisiologia in sei lezioni date in Torino. Castellazzo e Vercellino, Torino, 1861.

Suggested Readings for Carlo Matteucci

Bernabeo, Raffaele. *Carlo Matteucci (1811–1868); profilo della vita e dell'opera (con 10 documenti inediti).* Universita delgi Studi di Ferrara, Ferrara, 1972.

Moruzzi, Giuseppe. The electrophysiological work of Carlo Matteucci. In: Essays on the History of Italian Neurology. Proc. Int. Symp. Hist. Neurol., 1963.

Moruzzi, Giuseppe. *L'Opera elettrofisiologica di Carlo Matteucci. Il contributo di Carlo Matteucci alla creazione del modello fisico del nervo.* Universita degli Studi di Ferrara, Ferrara, 1973.

CHAPTER IV

Exploration of the Spinal Cord in the 19th Century

Jǔrí Prochaska (1749–1820)

With the growing interest in the behavior of muscles, the question of their control by the animal body as a whole brought attention to what anatomists had clarified in the spinal cord. Once it had been generally accepted that the long fibers of the spinal cord served as conduction routes to and from the brain, another possibility emerged, namely, that some incoming sensations might pass within the cord and result in a movement without necessarily having travelled to the brain. This contravened the established teaching of many. A challenge came, for example, from Prochaska,[1, 2] whose writings were published for the general scientific community in Latin and nearly a century later were translated into English by Thomas Laycock[3] for the Sydenham Society and into Prochaska's native language in 1854.

In this era, all of those who studied the nervous system were concerned with the constitution of the "vis nervosa," so named by Albrecht von Haller to denote outflow from the brain to the muscles. Prochaska (Fig. 17) used the same term but in a rather different sense—the capability of the nerves to receive impressions and then transmit them. He realized that this might take place unconsciously and therefore without participation by the brain. He described this transmission as a "reflection" taking place in what he called the "sensorium commune," a term used by many earlier workers, none of whom agreed on its anatomical site. The unconscious transmissions from sensory to motor that Prochaska had in mind were mostly respiratory and cardiac responses, although he noted the with-

[1]J. Prochaska. *De Structura Nervorum*, Vondobonnae (Graeffer), Prague, 1779.

[2]J. Prochaska. *De Funcionibus Nervosi Commentatio*. Wolfgange Goerle, Prague, 1784.

[3]T. Laycock (1812–1876). *Dissertation on the Functions of the Nervous System by George Prochaska*, Sydenham Society Translation, London, 1851.

 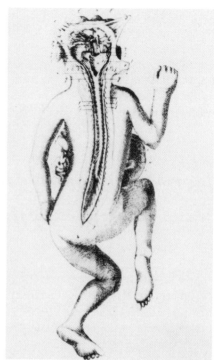

FIG. 17.   **Left:** Jŭrĭ Prochaska (1749–1820) Professor of Anatomy, Physiology and Ophthalmology in Prague and later in Vienna. **Right:** Prochaska's drawing of the spinal cord in an ancephalic monster. (From: J. Prochaska. *De Structura Nervorum,* Vondobonnae (Graeffer), Prague, 1780.)

drawal of a limb by a decapitated animal and concluded that part of the "sensorium commune" must therefore lie in the medulla spinalis.

The behavior of muscles below a spinal transection had sparked a famous controversy. Haller[4] held that the contraction of a muscle when irritated was an intrinsic response that did not need sensory transmission of the impression to the "soul." This tenet led to the well-known controversy with Robert Whytt.[5]

Haller and Whytt essentially concerned themselves with the muscles and the nerves rather than the spinal cord, but the concept that some representation of consciousness existed within the cord was to persist to the end of the 19th century. Its chief protagonist was E. F. W. Pflüger,[6,7] the founder of the famous *Pflüger's Archiv.* It was not until the fine dissection and illustrations of F. Vicq d'Azyr[8] that not only were the columns more clearly

[4]A. von Haller (1708–1777). De partibus corporis humani senibilibus et irritabilibus. *Comm. Soc. Reg. Sci. Göttingen,* 2:114–158 (1752).

[5]R. Whytt (1714–1766). *An Essay on the Vital and Other Involuntary Motions of Animals,* Hamilton, Balfour and Neill, Edinburgh, 1751.

[6]E. F. W. Pflüger (1829–1910). Die sensorischen Functionen des Rückenmarks der Wirbelthiere nebst einer neuen Lehre über die Leitungsgesetze der Reflexionen. Hirschwald, Berlin, 1853. (The Sensory Functions of the Spinal Cord of Vertebrates, Together with a New Theory of the Laws of Transmission of Reflexes.)

[7]For fuller account of this dispute, see the first volume of this study: M. A. B. Brazier. *A History of Neurophysiology in the 17th and 18th Century: From Concept to Experiment.* Raven Press, New York, 1984.

[8]F. Vicq. d'Azyr (1748–1794). Recherches sur la structure du cerveau, du cervelet, de la moelle longee, de la moelle épinière; et sur l'origine des nerfs de l'homme et des animaux. *Hist. Acad. Roy. Soc.,* 495–622 (1781).

identified, but attention was also given to the distribution of the gray matter. C. F. Bellingeri[9] had thought the origin of the sensory nerves to be in the gray matter and that of the nerves for movement in the white matter.

In the 19th century, an influential book on this subject was published by R. D. Grainger,[10] a London physician. He wrote: "The following axioms appear to be susceptible of satisfactory proof: 1. That the source of all power in the nervous system is in the gray matter; 2. That the white fibers are merely conductors."

Luigi Rolando (1773–1831)

The cellular contents of the spinal gray matter (as declared by Luigi Rolando in 1824[11]) became clearly identifiable when the microscope entered the armamentarium of the anatomist. One of the first to use it for the study of the spinal cord was B. Stilling.[12, 13] In hardened but unstained sections, he identified fibers crisscrossing in the gray independently of the direction-oriented white fiber columns. These formed a mesh or plexus among which he identified nerve cells of different sizes.

Stilling, however, did not use stains and was soon challenged by Albert von Kölliker,[14] whose goal was to show that nerve fibers derived from nerve cells. In the course of this pursuit, he examined both brain and spinal cord, recognizing several processes originating in ventral horn cells, which he called "the motor ganglia."

Robert Remak (1815–1865) and Marshall Hall (1790–1857)

Kölliker claimed his finding to be the first observation of the continuity of nerve and fiber though, in fact, Robert Remak in 1838[15] had given clear illustrations from the spinal cords of oxen. But one of the greatest contributors to the clarification of spinal reflexes was Marshall Hall, who studied cold-blooded animals such as the turtle and frog (Fig. 18). He chose these lower vertebrates because they had been known for centuries to move after decapitation. Alexander Stuart's demonstrations to the Royal Society in 1739 of the movements of a frog after pithing were disturbing evidence that the brain was not essential for movements of the limbs.

Hall, an English physician who had set up a laboratory in his own home, conducted experiments that he hoped would elucidate the clinical signs he observed in his patients. He is responsible for much of the early knowledge of the segmental reflex, and he introduced the term "reflex arc." However, he realized that sensory impulses coming into the medulla spinalis had more far-reaching effects in the nervous system than merely the seg-

[9]C. F. Bellingeri (1789–1848). *De Medula Spinali Nerisque ex ea Prodiunfibus.* Turin, 1823.

[10]R. D. Grainger (1801–1865). *Observations on the Structure and Function of the Spinal Cord.* Highly, London, 1837.

[11]L. Rolando. *Dizionario Periodico di Medicina,* Turin, 1824, p. 338.

[12]B. Stilling (1810–1876). *Untersuchungen über den Bau und die Verrichtungen des Gehirns.* Mauka, Jena, 1846.

[13]B. Stilling. *Untersuchungen über die Textur des Rückenmarks.* Wigand, Leipzig, 1842.

[14]Albert von Kölliker (1817–1905). Neurologische Bemerkungen. *Zeit. Wiss. Zool.,* 1:135–163 (1849).

[15]R. Remak. *Observationses Anatomicae et Microscopicae de Systematis Nervosis Structuri.* Reimer, Berlin, 1838.

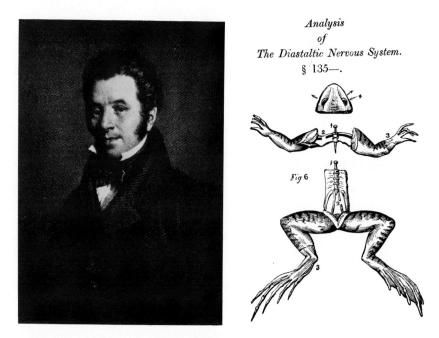

Analysis
of
The Diastaltic Nervous System.
§ 135—.

Fig 6

FIG. 18.   Left: Marshall Hall (1790–1857). **Right:** One of Hall's experiments to demonstrate the three parts of the reflex arc. The arc was broken by any of the following procedures: (a) skinning the extremity (at 3) (the "esodic" nerves); (b) sectioning of the "brachial or the lumbar or femoral nerve leading to the point irritated" (i.e., the "exodic" nerve at 2); or (c) removing the spinal marrow (the "spinal centre"). (From: M. Hall, *Synopsis of the Diastaltic Nervous System,* being outlines of the Croonian Lectures delivered at the Royal College of Physicians in April 1850.)

mental motor response. "But the operation," he wrote, "of the reflex function is by no means confined to parts corresponding to distinct portions of the medulla. The irritation of a given part may, on the contrary, induce contraction in a part very remote."

Marshall Hall placed the location of the central section of the three-part arc in the spinal marrow, with nervous pathways presumed to ascend and descend alongside the major columns but acting independently of them. He called these "excitomotory nerves." Like most of his contemporaries, Hall was still bothered by the specter of the will, for he (anticipating Sherrington) had discovered that he could get exaggerated reflexes by removing the brain. He did not wish to be drawn into the controversy raised by Descartes decades before by which these movements were described for man as "la machine de son corps." To protect himself, he wrote that his results on decapitated animals were "all beautiful and demonstrative of the wisdom of Him who fashioneth all things after his own will."[16]

Marshall Hall's experiments and the claims he made for them raised a storm of arguments, some of them questioning the findings and some of them accusing him of plagiarism. It was not until 1844, when the great Müller gave his work recognition in his famous *Handbuch,* that Hall's work was accepted.

The interaction within the spinal cord was far from clear, and the teachings of Claude

---

[16]Marshall Hall. On the reflex function of the medulla oblongata and medulla spinalis. *Phil. Trans. Roy. Soc.,* 123:635–665 (1833).

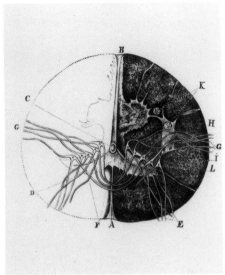

FIG. 19. Left: Claude Bernard (1813–1878). **Right:** Connections between anterior and posterior roots of the spinal cord as taught (erroneuously) by Claude Bernard. (From: *Leçons sur Physiologie et Pathologie du Système Nerveux.* Baillière, Paris, 1858.)

Bernard (Fig. 19) did little to clarify this question. His famous textbook,[17] published in 1858, taught students that there were direct connections not only between the dorsal and ventral roots but also between cells in the two sides of the spinal cord. With the eventual coming of the neuron doctrine and the theory of the synapse, this concept was to die. It is of interest, however, that before the neuron doctrine was developed, there was a textbook which suggested that interaction of nerve cells could result not only in excitation but in inhibition. This concept of inhibition is found in the lectures given to medical students at Guy's Hospital by Pye-Smith[18] and published in 1885 (Fig. 20).

Santiago Ramón y Cajal (1852–1934) and Camillo Golgi (1844–1926)

The generally held and rather simplistic view of the spinal columns running uninterruptedly up and down the cord received a radical reassessment with the discovery of collaterals and thus replaced Hall's concept of an "excitomotory" system. Ramón y Cajal,[19] using a myelin stain, found what he described as innumerable fine fibers, more or less horizontal,

[17]Claude Bernard (1813–1878). *Leçons sur la Physiologie et la Pathologie du Système Nerveux.* Ballière, Paris, 1858.

[18]P. H. Pye-Smith. *Syllabus of a Course of Lectures on Physiology Delivered at Guy's Hospital.* Churchill, London, 1885.

[19]S. Ramón y Cajal. Contribución al estudio de la structura de la medula espinal. *Rev. trim. Histol norm. pathol.,* 3 and 4 (1889).

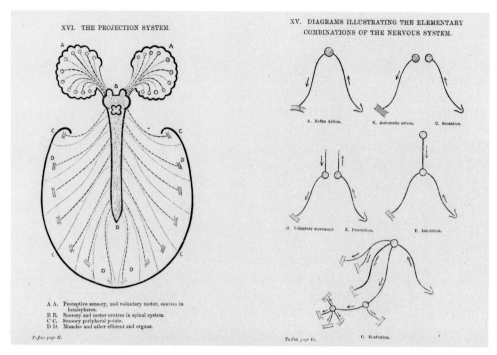

FIG. 20.   Some examples of the connections of the spinal cord as taught in the 19th century. **Left:** The projection system. A = receptive areas in the brain. B = in the spinal cord. C = sensory projections. D = - motor projections. **Right:** Connections of the nervous system illustrating excitation, inhibition, and reflex action. (From: P. H. Pye-Smith. *Syllabus of a Course of Lectures on Physiology Delivered at Guy's Hospital. Churchill, London, 1885.)*

which left the columns at all points to lose themselves in the gray susbstance (Fig. 21). He proceeded to use Golgi's silver chromate stain and was able to demonstrate that these were truly bifurcations of the longitudinal fibers. He published this in 1889 and only later was he challenged for priority in their discovery by Golgi,[20] who had described them nearly 10 years earlier. Previous workers such as Gerlach, Kölliker, and Waldeyer had seen these horizontal fibers but had presumed them to be axis cylinders of nerve cells separate from the fibers of the columns. The discovery that spinal nerves could bifurcate in this way was so novel and imposed such a basic reconsideration of both the anatomy and function of the spinal cord that Ramón y Cajal found anatomists reluctant to accept his finding. Kölliker, so he tells us, was the first savant to confirm his identification of these fibers as collaterals, and Ramón y Cajal suggested that they form the anatomical construct of a reflex, namely, long sensory collaterals making contact with cells of motor nerves, a proposal quickly taken up by Kölliker and other anatomists of the time.

These anatomical clarifications were to become of the greatest importance for neurophysiologists when the reports of François Magendie became known, for these concerned the problem of the spinal roots.[21]

---

[20]C. Golgi. *Studi Istologici sul Midollo Spinale.* Communication made to the Third Italian Congress of Psychiatry held in 1880 at Reggio (in Emilia), 1880.

[21]F. Magendie. Expériences sur les fonctions des racines des nerfs rachidiens. J. Physiol. Exp. Pathol., 3:276–279 (1822).

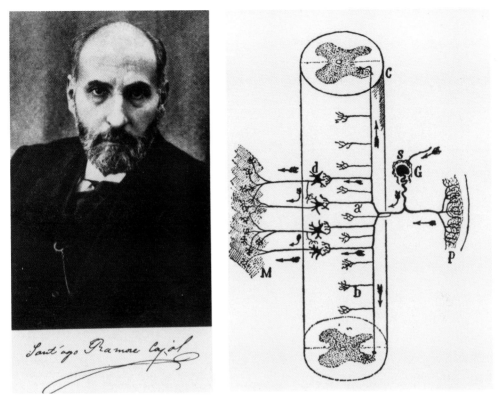

FIG. 21.   Left: Santiago Ramón y Cajal (1852–1934). **Right:** His schema for a spinal cord reflex using collaterals to connect the sensory imput from the dorsal root ganglion to the motor cells in the ventral horn. (From: S. Ramón y Cajal. *Histologie du Système nerveux de l'Homme et des Vertébrés.* 2 vols. Azonlay, Paris, 1909.)

BIBLIOGRAPHY

Suggested Readings for the Spinal Cord

Bell, C. *The Nervous System of the Human Body as Explained in a Series of Papers Read before the Royal Society of London.* Longman, London, 1830.

Brazier, M. A. B. The development of ideas about the spinal cord in the eighteenth and nineteenth centuries. In: Windle, W. E., *The Spinal Cord and its Reaction to Traumatic Injury.* Dekker, New York, 1986.

Cranefield, P. E. *The Way In and the Way Out.* Futura, New York, 1974.

Gutmann, E. Jǔrí Prochaska a reflexni theorie. *Czecholov. Fisiol.,* 1:1–8 (1952).

Kruta, M. V. Prochaska's and Purkinje's contributions to neurophysiology. In: *Von Boerhaave bis Berger,* edited by K. Rothschuh. Fisher, Stuttgart, 1964.

Liddell, E. G. T. *The Discovery of Reflexes.* Clarendon Press, Oxford, 1960.

Neuberger, M. Die Physiologue Georg Prochaska ein Vorlaufer Purkinje's. *Vienna med. Woch,* 1155–1157 (1937). (The Physiologist Georg Prochaska as a Forerunner of Purkinje.)

Stilling, B. *Untersuchungen über die Textur des Rückenmarks.* Wigand, Leipzig, 1842. (Investigations on the Texture of the Spinal Cord.)

Windle, W. F. (Ed.) *The Spinal Cord and its Reaction to Traumatic Injury.* Dekker, New York, 1986.

CHAPTER V

The Great Era of Experimental Physiology Opens

François Magendie (1783–1855)

A great change took place in experimental research when a technique was developed not only for anesthetizing the animal during surgery but also for artificial respiration. Initiation to this technique was owed to Julien LeGallois in 1812. At the beginning of the century, LeGallois had introduced a primitive method of artifical respiration while working mostly with rabbits. He used a syringe to introduce air into the lungs and controlled the inflow by manipulation of an opening in a side tube (Fig. 22). His method was rapidly adopted by those eager to solve the problem of neuromuscular action. His book[1] was published in both French and in English.

François Magendie, who was to become the outstanding experimental physiologist of 19th-century France, grew up under the lingering traces of vitalism, which he at first accepted. But, eventually, he was to see (and lead) the release of natural science from its bonds. Important as this release was for all physiology, its most critical impact was for the nervous system and specifically for knowledge of the brain of man.

Magendie was a Bordelais, born in 1783 into a medical family that was ardently republican in the 18th-century meaning of that word in France. When the revolution broke out, Magendie's family moved to Paris where he lived the rest of his life. He spent his early student days in anatomy, obtained his medical degree in 1808, and entered a medical world involved in the restructuring era of Napoleon's empire. Maintaining his clinical interests all his life, he had appointments at the famous old Hôtel-Dieu while holding the position of Professor of Experimental Physiology and Pathology at the Collège de France, the renamed Collège du Roi of pre-revolutionary days.

---

[1]J. J. D. LeGallois (1770–1840). *Expériences sur le Principe de la Vie, notament sur celui des Mouvements du Coeur, et le Siège de ce Principe*. D'Hautel, Paris, 1812.

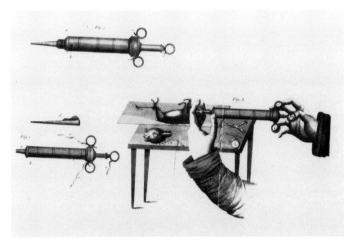

FIG. 22. Left: Technique for artificial respiration as designed by LeGallois. His own description reads (in part): "Figure represents a decapitated rabbit, kept alive by pulmonary inflation performed with the small syringe to which the silver tube is adapted. Control of inflow is made by the operator's thumb over the hole. (b in the diagram)." **Right:** Title page of the book by J.J.D. LeGallois on experiments on the life process.

The investigators who entered the study of the nervous system after the end of the 18th century inherited a wealth of established knowledge and advanced technology without which the progress that today's student takes for granted could not have been made. Also taken for granted by the modern student is a materialistic approach to this research, a trend developed in large part from the leadership of Magendie in France. Central to this problem were the relationship of mind to body and the existence of free will and voluntary movement. In the previous century, the science of the nervous system had reached different levels in the various countries of Europe. In Germany a retrogression developed owing to the extremely influential figure Georg Ernst Stahl[2] at Halle who, in opposition to the growing materialistic approach, reintroduced an immaterial anima which he held to be the sole activating principle of the body parts. Since the search for an immaterial agent lies outside the scope of science, this doctrine virtually extinguished experimental enquiry. This was particularly disturbing to the philosopher Leibniz.

Biological scientists were just beginning to free themselves from the tenet that man's behavior was under command not from an organ (the brain) but from an abstraction (the soul). But soon they were to become mired in the now time-worn controversy of body versus mind. Magendie did not hesitate to make his position clear. He wrote: "Le cerveau est l'organe matériel de la pensée: une foule de faits et d'expériences le prouvent." Remembering the climate of his own student days and the view held by his teachers, he spoke out unequivocally when he came to discuss the intellect in the section on the brain in his *Précis Elémentaire:*[3]

> L'intelligence de l'homme se compose de phénomènes tellement différents de tout ce que presente d'ailleurs la nature, qu'on les raporte à un être particulier que l'on regarde comme un émanation divine et dont le premier attribut est immortalité. . . .

[2]G. E. Stahl (1660–1734). *Theoria Medica Vera*. Halle, 1708.

[3]François Magendie. *Précis Elémentaire de Physiologie*. Meguignon-Marvis, Paris. Vol. 1, 1816; Vol. 2, 1817; 2nd ed., 1825; 3rd ed., 1834; 4th ed., 1836.

> Le physiologiste reçoit de la religion cette croyance consolatrice, mais la sévérité de langage ou de logique que comporte maintenant la science, exige que nous traitions de l'intelligence humaine comme si elle était le résultat de l'action d'un organe.

In terms of today's neurophysiology, Magendie's foresight of the role of the brain is impressive. Writing ostensibly of the intellect, he said:

> Quelque soient le nombre et la diversité des phénomènes que appartiennent à l'intelligence de l'homme, quelque differents qu'ils paraissent des autres phénomènes de la vie, et quoiqu'ils soient évidemment sous la dépendance de l'âme, il est indispensable de les considérer comme le résultat de l'action du cerveau et de ne les distinguer ainsi en aucune manière des autres phénomènes qui dépendent des action d'organe. En effet, les fonctions du cerveau sont absolument soumises aux mêmes lois générales que les autres fonctions. . . .

Magendie was writing in a period when interest was growing substantially among those whose concern was principally the brain. He had, behind him, the climate of free thought that had grown with the Idéalogues, the brilliant group of thinkers centered around the salon of Mme Helvetius in Auteuil. In the previous century, the leaders (and notably Condillac) had still been tied to a philosophy regarding sensations such as pleasure and pain as transactions of the soul.

In Magendie's time, it was one of Condillac's[4-6] most ardent followers, De Stutt de Tracy,[7] anxious at all costs to gain acceptance for a more materialistic view, who took the lead among the Idéologues in modifying the model to bring it into the domain of the physical world. Among those profoundly influenced by the writings of De Tracy was Magendie. He urged the readers of his *Précis Elémentaire* to study the works of Condillac, Cabanis,[8] and, above all, De Stutt de Tracy's book *Eléments d'Idéologie*, a book which introduced a term for the science of ideas before the word ''psychology'' had been coined.

Magendie was not only a deep thinker but an active and talented experimenter in the laboratory. In his research on the brain and especially on the cranial nerves, he used unanesthetized animals, a procedure that brought much disaproval, especially from scientists in England. But he argued that no other way could supply the information needed for uncovering the sensory systems. For example, to explore the effect of sectioning the fifth cranial nerve, he cut this nerve in a live rabbit, giving a very clear sketch of the procedure (Fig. 23). His technique was so impressive that his pupil Claude Bernard[9] later included it in his famous textbook.

Magendie, in the observations he made on decerebrate animals,[10] anticipated Sherrington by an accurate and detailed description of decerebrate rigidity in rabbits. This was in the days before the discovery of ether anesthesia,[11] and Magendie was severely criticized for his use of vivisection. The question of experimentation on animals was exacerbated by the

[4]Étienne Bonnot de Condillac, Abbéde Condillac (1715–1780). *Essai sur l'Origine des Connaissances Humaines. Ouvrages où Reduit à un seul Principe tout ce qui concerne l'Entendment Humain.* 1746.

[5]E. B. de Condillac. *Traité des Sensations (Extrait Raisonne* added to the 2nd edition, 1778). 1754.

[6]E. B. de Condillac. *Traité des Animaux.* Amsterdam, 1755.

[7]A. L. C. De Stutt de Tracy (1754–1836). *Eléments d'Idéologie.* Paris, 1804.

[8]P. J. G. Cabanis (1757–1808). *Rapports du Physique et du Moral de l'Homme.* Paris, 1802.

[9]Claude Bernard (1813–1878). *Leçons sur la Physiologie et la Pathologie du Système Nerveux.* Baillière, Paris, 1858.

[10]F. Magendie. Sur la siege au mouvement et du sentiment dans la moelle épinière. *J. Physiol. Expér. Path.,* 3:153–157 (1823).

[11]William Thomas Greene Morton (1819–1868). *On the Physiological Effects of Sulphuric Ether and its Superiority to Chloroform.* Clapp, Boston, 1850.

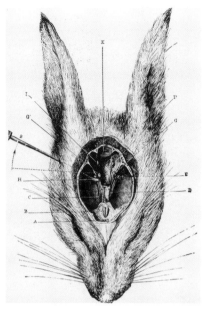

FIG. 23. **Left:** François Magendie (1783–1855). Unsigned portrait from the Collège de France, thought to be by Paul Guerin. **Right:** Magendie's technique for sectioning the fifth nerve in the living rabbit. The dissection is to demonstrate the insertion of the instrument. On the rabbit's right, the probe is seen entering the base of the skull and reaching the trunk of the fifth nerve at H. On the animal's left, the end of the instrument is seen at E and the sectioned nerve at G. (From: C. Bernard. *Leçons sur la Physiologie et la Pathologie du Système Nerveux.* Baillière, Paris, 1858.)

publication in 1859 of Charles Darwin's *Origin of Species,*[12] which brought the animals closer to man.

So prolific was Magendie that he founded his own journal, *Journal de Physiologie Expérimentale et Pathologique du Système Nerveux,* and it is in the pages of this successful journal, which first appeared in 1821, that his many reports were published. He was a generous man; when, after writing of his work on the spinal fluid, he found that this same discovery had been made years before by Domenico Cotugno, he not only made the acknowledgment in print but reproduced in his journal Cotugno's original description from its obscure source.

In 1839 Magendie published his famous textbook *Leçons sur les Fonctions du Système Nerveux,* which witnesses to the long experimental trail that led him away from vitalism. This followed his popular *Précis Elémentaire de Physiologie* (Fig. 24) that he published four years earlier and in which he had taken his stand against Rolando's view[13] of cerebellar activity, insisting that the role of the cerebellum was to maintain equilibrium.[14] He reached this conclusion from observations of the disturbed gait of a duck from which he had removed the cerebellum unilaterally.

It was no doubt Magendie's distress at meeting the antagonism evoked by his experi-

[12]Charles Darwin (1809–1882). *On the Origin of Species by Means of Natural Selection.* Murray, London, 1859.

[13]Luigi Rolando (1773–1831). Osservazioni sul cervelletto. *Mem. Reale. Accad. Sci.,* 29:163 (1825).

[14]F. Magendie. Mémoire sur un liquide que se trouve dans le crâne et le canal vertébral de l'homme et des animaux mammifères. *J. Physiol. Expér. Pathol.,* 5:27–37 (1825).

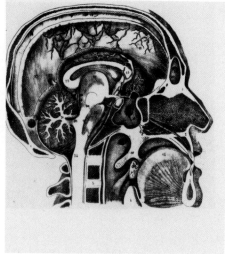

FIG. 24. Left: François Magendie (1783–1855) at midlife. **Right:** His illustration of the interior of the human brain. (From: *Précis Elémentaire de Physiologie.* Maquignon-Marvis, Paris, 1816–17.)

ments with unanesthetized animals that led to his immediate interest in the first records of ether anesthesia. The demonstration of ether anesthesia by Morton (in man) took place at the Massachusetts General Hospital in Boston in 1846, and in the following year we find Magendie writing:

> I should hope that no one supposes that I had the intention of provoking hilarity. On the contrary, I regard the consequences of ether intoxication as extremely serious. I should not be very happy if my wife or my daughter had been the subject of scenes similar to those of which I have been a witness, in which chaste and modest girls have been transformed in a few minutes into bacchanals. Might not the taking of ether lead to results similar to those produced by the hashish and opium of the Orient?

He was writing from the experience he had had as a physician at the Charité.

In regard to the spinal roots, Magendie wrote:

> For a long time, I had wished to perform the experiment of cutting the anterior and posterior roots of the nerves arising from the spinal cord of an animal . . . I had then before my eyes the posterior roots of the lumbar and sacral pairs, and, raising them up successively on the blade of a pair of small scissors, I cut them on one side . . . I reunited the wound by means of a suture through the skin and observed the animal. I thought at first that the member corresponding to the cut nerves was entirely paralyzed. It was insensitive to pricks and to the strongest compression; it also appeared to me to be immovable; but soon, to my great surprise, I saw it move perceptively, although sensibility was always entirely absent. A second and third experiment gave me exactly the same result. I began to regard it as probable that the posterior roots of the spinal nerves might have different functions from the anterior roots, and that they were particularly designed for sensibility . . . As in the preceding experiments, I made the section on only one side . . . One can imagine with what curiosity I followed the effects of this section. The results were not doubtful; the member was completely immovable and flaccid, although it preserved an unequivocal sensibility. Finally, that nothing might be neglected, I cut at the same time the anterior and posterior roots; there was a complete loss of sensation and motion. . . .[15]

[15]F. Magendie. Expériences sur les fonctions des racines des nerfs qui nassent de la moelle épinière. *J. Physiol. Exper. Path.,* 2:366–371 (1822).

Magendie's researches covered many areas of physiology—the urinary system, the digestive system, cardiac activity, the importance of blood sugar—but, in the context of this account, it is his experiments on the nervous system that are outstanding. Intent on establishing whether nerves were motor or sensory (not only in the spinal cord but in the brain), he made experiments on the cranial nerves, in particular on the fifth and seventh, the dual roles of which were a puzzle. The nerves that had both sensory and motor branches proved the most difficult. Magendie at first thought that the fifth nerve was sensory and nutrient to the face and that the seventh nerve was entirely motor, since cutting it caused facial paralysis without relieving neuralgia.

Charles Bell (1774–1842)

In 1821 Charles Bell,[16] dissecting the nerves of the face, noticed that the fibers of the seventh nerve went to the muscle whereas those of the fifth entered the skin. He suspected that they served different functions and, being himself an anatomist rather than an experimentalist, asked his close collaborator and brother-in-law John Shaw to make a study of the effect of sensations on these nerves. Using an unusual experimental animal, the donkey, Shaw was able to demonstrate paralysis in the one case and loss of reaction to touch in the other; neither he nor Bell, whose fine drawings illustrate his findings, recognized the mixed nature of the nerve. After this beginning, several workers added their contributions to the further clarification of the cranial nerves, prominent among those being Mayo,[17] who taught the course in anatomy and physiology at King's College, London.

It was the reports of Magendie on his researches into sensory and motor systems that caught the attention of Charles Bell and started a quarrel for priority that continued until the death of Bell in 1842, an ignominious quarrel that led to other physiologists taking sides. Bell was not a neurophysiologist but an anatomist of excellence who received a Knighthood toward the end of his career. Altercation arose only in 1822 when Magendie reported his first definitive experiment on this point.[18] The classic entry in the proceedings of the Académie des Sciences in 1847 at which this famous experiment was reported reads in translation: "M. Magendie reports the discovery he has recently made, that if the posterior roots of the spinal nerves are cut, only the sensation of those nerves is abolished, and if the anterior roots are cut, only the movements they cause are lost."

He was apparently unaware of Bell's work 11 years before on the function of the spinal roots, in which stimulation of the anterior root in an unconscious animal evoked contraction, and section of the posterior root had no effect. The latter root Bell thought was both motor and sensory. According to John Shaw, Bell was unwilling to work on living animals unless they were stunned into unconsciousness, thus handicapping his ability to study sensation. In France, where the antivivisection movement was less strong than in England, Magendie repeatedly used the living, conscious animal. He was denounced for cruelty when he went to give demonstrations in England and was later attacked on this issue in his own country.

Only later did altercation become acrimonious, the flame being fed by their respective supporters as much as by the protagonists themselves. What emerges are the striking char-

[16]Charles Bell. On the nerves: giving an account of some experiments on their structure and functions which lead to a new arrangement of the system. *Phil. Trans. Roy. Soc.*, 111:398–414 (1821).

[17]Herbert Mayo (1792–1852). *Anatomical and Physiological Connections.* Underwood, London, 1822 and 1823.

[18]F. Magendie. Expériences sur les fonctions des racines des nerfs rachidiens. *J. Physiol. Exp. Pathol.*, 2:276–279; 366–371 (1822).

 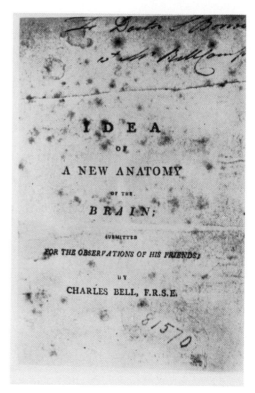

FIG. 25. Left: Charles Bell (1774–1842). **Right:** The privately distributed pamphlet that formed Bell's claim to be the first to establish the functions of the motor and sensory roots of the spinal cord. Only four copies are now known. (Courtesy of the Sterling Library of Yale University.)

acters—not only of the two main protagonists but also of their friends and vilifiers. Pierre Flourens, who in an earlier campaign to destroy Francis Joseph Gall had nearly destroyed the search for cortical localization, brought his attack to Magendie.

In fact, in the case of the spinal roots, much of the dispute took place in retrospect, for it was in 1847 that Flourens,[19] before the whole Académie des Sciences, aggravated Magendie by giving the priority to Bell for his pamphlet[20] privately circulated in 1811, *Idea of a New Anatomy of the Brain* (Fig. 25). What emerges far more clearly than a scientific interest in nerve roots is Flourens' antagonism to Magendie. The antagonism was deep-set and long-lasting; Flourens had opposed him for the Académie and had persisted in ridiculing his character even after his death.

On the other side of the Channel, Bell, until his death in 1842, pressed his claim, reading back into his earlier writings a meaning that was, in fact, by no means clear in the originals. He even introduced insertions to strengthen his case. Once again, it is the character as well as the differing scientific drives of these men that come out so clearly: Bell, essentially the breeder of ideas and the anatomical dissector, and Magendie, the scrupulous empiricist. Bell, the first to die, had continued all his life to champion, with questionable arguments, his own claim, incomplete as it was. We are told again by Flourens, that Bell, after

---

[19]P. J. M. Flourens (1794–1847). C. R. Acad. Sci., 24:253–320 (1847).

[20]Charles Bell. *Idea of a New Anatomy of the Brain.* Strahan and Preston, London, 1811.

FIG. 26. Left: The nerves of the face as dissected by Charles Bell. **Right:** Sketch of Bell teaching the anatomy of the skull.

reporting his work to the Royal Society, remarked: "My discovery will place me by the side of Harvey." Supportive evidence that Bell ever made this remark is missing.

Today's student, with modern knowledge of the roots, may question this assessment as too simplistically equating, as it does, direction with function. Consider the modern discovery of some sensory fibers in the anterior root, a finding[21] that supports Bell's surmise so ridiculed by historians, the dorsal root reflex, and consider the muscle spindle efferent system which modulates the sensory inflow from the intrafusal fibers. The latter is under control from higher centers and is one of several examples of centrifugal control of input to the nervous system exerted at the level of the receptor—what scientists today call a "feedback loop" but what Bell envisaged as a "circle of nerves," though the substrate for his idea was not found for over a century. Writing in 1824, Bell[22] stated: "There must be a sense of the condition of the muscle communicated to the brain as well as energy proceeding from the brain towards the muscles" (Fig. 26). How right he was.

Magendie's excursions into ablation of portions of the brain[23] (the cerebellum, the olfactory lobes, and even the cerebral hemispheres) led him to the observation of the foramen that carries his name. He died in 1855; a medallion of his profile was made by David d'An-

[21]L. Leksell. The action potential and excitatory effects of the small ventral root-fibers to skeletal muscle. *Acta Physiol. Scand.,* 10(Suppl. 31):1–84 (1945).

[22]Charles Bell. *An Exposition of the Natural System of the Nerves of the Human Body with a Republication of the Papers Delivered to the Royal Society, on the Subject of Nerves.* Spottiswoode, London, 1824.

[23]F. Magendie. Mémoire physiologique sur le cerveau. *J. Physiol. Exper. Pathol.,* 8:211–229 (1828).

 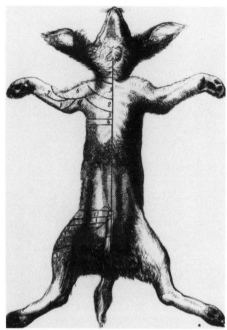

FIG. 27. **Left:** Ludwig Türck (1810–1868). Physician at the Allgemeine Krankenhaus, Vienna. His research focused on the innervation of the skin. **Right:** One of his experiments on a dog illustrating the spinal roots serving the neck, the thorax, and the lumbar region and demonstrating some sharing. (From: Ludwig, *Lehrbuch der Physiologie,* Leipzig, 1858.)

gers, where it can be seen in that town's museum. His career had spanned the Revolution and the restoration of the monarchy in 1830. He had received many honors—membership in the Académie des Sciences in 1821 to be followed by membership in many foreign societies (English, Danish, Polish) and by many prizes.

In addition to the two main protagonists in the arguments about the function of the spinal roots, there had been others working less publicly on the same question. One of these was Ludwig Türck. He had a broad training in Paris in the 1840s with Magendie, Flourens, and Longet; he then went to Vienna to the Allgemeine Krankenhaus as Professor of Pathology. It was there, in the laboratories of Brücke and Ludwig,[24] that he did most of his work on the nervous system.[25] He is best known for tracing the nerves from the neck, the chest, and the lumbar region among which he found some sharing (Fig. 27).

Claude Bernard (1813–1878)

The favorite pupil of Magendie, and one to attain worldwide acclaim, was Claude Bernard. He was born in St. Julien in the Rhone Valley in a France experiencing Napoleon's continuing retreat from Moscow—to the Vistula, to the Oder, to the Elbe—and the alliances that then formed against him (Russia, England, Prussia, Sweden, and Spain). But by

[24]Ludwig Türck (1810–1868). In: Ludwig's *Handbuch der Physiologie.* Leipzig, 1858. (Handbook of Physiology.)

[25]L. Türck. Ueber secondäre Erkrankungen einzelner Rückenmarksstränge und ihre Fortsetzungen zum Gehirne. *S. B. Akad. Wiss.,* 6:288 (1851). (On Secondary Diseases of Individual Spinal Cords and their Extension to the Brain.)

FIG. 28. **Left:** Claude Bernard (1813–1878). **Right:** Copy of the document admitting him for study under Magendie.

the time Bernard was adult, he was living in a different France, one in which scientists could once again interact with those of other countries. Unlike the development of most world renowned scientists, Bernard's way was extremely varied. He worked several years for a pharmacist, but his interest then appears to have been in the writing of plays—in which he had some success. He was 21 before he turned with any seriousness toward medicine. He enrolled in medical school in Paris in 1834. His good fortune came when later he was able to work under Magendie (Fig. 28) at the old Hôtel-Dieu on the Île-de-la-Cité. And it was through Magendie's influence that Bernard was appointed in 1841 ''préparateur'' at the Collège de France, the distinguished center of learning founded as the Collège Royale in 1530 by Francis I.

At the age of 30, and while still with Magendie, Bernard published his first paper,[25a] an anatomical study of the chorda tympani. While Bernard was working under Magendie's influence, it was inevitable that he should become involved with the controversy over the nature of the spinal roots. In addition to the attacks by Bell, Magendie had had to face those of Longet[26] and thus inevitably Bernard had to work on the problem, work which resulted in a paper.[27] Bernard was slow to find his major drive. He was 30 years old before undertaking any laboratory work and was the same age when getting his medical degree, though he had no intention of practicing medicine. He continued to work in Magendie's

[25a]Claude Bernard. Recherches anatomiques et physiologiques sur la corde du tympan our servir à l'hémiplégie faciale. *Ann. Méd. Physiol.*, 1: 408–439 (1843).

[26]F. A. Longet. Fait physiologique relative aux racines des nerfs rachidiens. *C. R. Acad. Sci.*, 8: 861–883 (1839).

[27]C. Bernard. Recherches expérimentales sur les fonctions de nerf spinal, ou accessoire de Willis étudié spécialement dans ses rapports avec le pneumo-gastrique. *Arch. Gén. Méd.* 4:307–424 and 5:51–96 (1844).

laboratory, although they never published jointly. In 1855 Magendie died; Bernard inherited his Chair and was also appointed to the newly created Chair of General Physiology at the Sorbonne. From this time on, it is clear that Bernard's outstanding gift was as a teacher, though reports of experimental work, mainly on the nervous supply to the digestive system continued to come from his hand for another seven years. The first volume of his most famous book, *Leçons de Physiologie expérimentale appliquiés à la Médecines*,[28] was published in 1855, the second volume in the following year. It is for this work that his name still stands high in the world of physiology. Ten years later, he published another book,[29] which was more philosophical in nature and revealed his interest in Goethe. In this he embarked on an elaborate examination of how laboratory observations are achieved. The book gives a thoughtful overview of his more scientific papers and includes a long section on disputes in physiology, including his own with Longet about the spinal roots.

The drive behind the major laboratory researches of Bernard was to understand the "milieu intérieur." For the neurophysiologist, the most meaningful were his experiments on digestion, the role of the liver, and the vasomotor system. His thesis[30] for his medical degree comes in the first of these groups and was followed by one in which he showed that gastric digestion was due to enzymes rather than to acid.[31]

For the nervous system's role in digestion, Bernard examined the pancreas and established the action of pancreatic juice in the digestion of neutral fats and the transformation of starch into sugar.[32] Those aspects of the studies that considered the action of the nerves included experiments on the innervation of the salivary gland. His major neurophysiological contribution was the unravelling of the vasomotor control of the digestive system.[33, 34] Not only had this been started by Stilling but similar experimental results had been found by Ludwig who, in his textbook published in 1861,[35] wrote (in translation):

> The famous Parisian academic has for many years been describing experiments carried out long before him by Dr. Rahn in my laboratory. Since Dr. Bernard, as he has already shown, possesses a fine sense of literary ownership, then his silence as to the time authorship of these experiments can only be due to his ignorance of them.

Bernard later ranked Ludwig's experiments on saliva secretion on par with his own important contributions to this area of physiology.

Bernard found that section of the cervical sympathic caused vasodilation and an increase of skin temperature of the homolateral side (Fig. 29). Later, he showed that spinal transec-

---

[28]2 vols. Baillière, Paris, 1855–56.

[29]C. Bernard. *Introduction a l'étude de la Médecine expérimentale*. Baillière, Paris, 1865. (English translation by H. Copley Greene. Dover, New York, 1957).

[30]C. Bernard. *Du sac gastrique et de son rôle dans la nutrition*. Rignoux, Paris, 1844.

[31]C. Bernard. Sur l'independence de l'élément moteur et de l'élément sensitif dans les phénomènes du système nerveux. *Mem. Soc. Biol.*, 1:15 (1849).

[32]C. Bernard. Influence de systèm nerveux sur production du sucre dans l'économie animal. *Mém. Soc. Biol.*, 1:49 (1849).

[33]C. Bernard. De l'influence du système nerveux grand sympathique sur la chaleur animale. *Comp. Rand. Acad. Sci.*, 34:472–475 (1852). (Bernard was not the first to propose a vasomotor control by sympathetic nerves. This had been suggested by Heinrich Stilling (1810–1879). *Untersuchungen über die Spinalirritation*. Wiehelm, Leipzig, 1840.)

[34]C. Bernard. Recherches expérimentales sur le grand sympathique spécialment sur l'influence que la section ce nerf exerce sur la chaleur animal. *Mém. Soc. Biol.*, 5:77 (1853).

[35]C. Ludwig, *Lehrbuch der Physiologie des Menschen*. (Die zweite, neu bearbeitete Auflage, 2 Bnde. Leipzig und Heidelberg, 1858, 1861.) (Textbook on the Physiology of Man.)

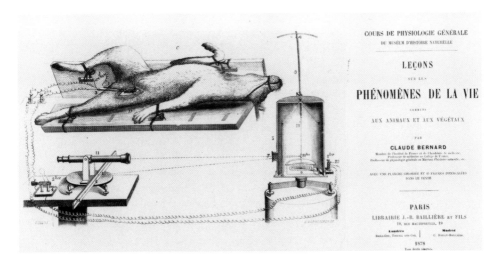

FIG. 29. Left: Bernard's apparatus for measuring animal heat. The current from the thermo element (1.2.) went to the d'Arsonval Galvanometer (at 10). **Right:** One of Bernard's many famous publications.

tion at the cervical level caused a spectacular fall in blood pressure.[36] He was on the threshold of defining the self-regulatory control of venous tone by the vasomotor centers in the medulla, a mechanism that a few years later was clarified by Goltz of Strasbourg.

What grew in Claude Bernard's mind as he pursued his experiments was that there must be some form of self-regulation within the animal body. Another example of Bernard's experimental approach was the question of sugar in the blood and where it came from. The ruling theory was that it came only from the food, that it was essentially a nutrient to the lungs, and that it was destroyed in the lungs by combustion.

By experiments with animals on sugar-free diets, Bernard discovered that there was always sugar in the blood that he withdrew from their hearts. He found there was even more sugar in the portal vein from the liver. From a long series of experiments that cannot be detailed here, he proved that sugar was actually formed in the liver. Out of these experiments on the physiological role of glycogenesis in the liver, he moved on to discover the role of the sympathetic control, which, if he interfered with it by puncture in the floor of the fourth ventricle, made the animal diabetic.[37]

When hearing the name of Claude Bernard, physiologists today think immediately of the "milieu intérieur," though in fact he did not arrive at this concept until some years after he had ceased to work in the laboratory.[38, 39]

In 1865 poor health led Bernard to drop his work, and he withdrew to the family home in St. Julien for two years. He returned to Paris in 1867, but three years later he went back for the very serious reason of the Prussian invasion of Paris in 1870. The Franco-German war lost Alsace to Prussia and heralded the formation by Bismarck of the German Empire.

[36]C. Bernard. Chiens rendus diabètiques. *C. R. Soc. Biol.*, 1:33–36 (1849).

[37]C. Bernard. Sur les effets de la portion de la section encéphalique du grand sympathetique. *Mem. Soc. Biol.*, 4:168–170 (1852).

[38]C. Bernard. Influence de grand sympathetique sur la sensibilité et la calorifucation. *C. R. Soc. Biol.*, 3:163–164 (1852).

[39]C. Bernard. De l'influence des deux ordres de nerfs qui detérmine les variations du conleur du sang varieux dans les organes glandulares. *C. R. Soc. Biol.*, 47:245–253, 393–400 (1858).

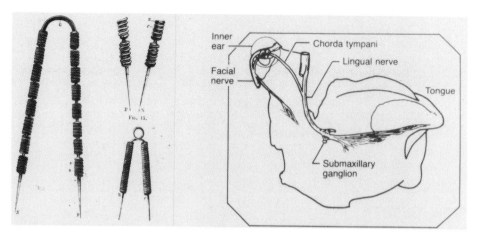

FIG. 30. Left: The electrodes used by Claude Bernard. These were originally designed by Du Bois-Reymond and became of general use in all the contemporary electrophysiological laboratories. **Right:** Claude Bernard's dissection of the chorda tympani showing the facial nerve going to the submaxillary ganglion and the chorda tympani to the tongue. (From: *Leçons sur la Physiologie et Pathologe du Système Nerveux.* Ballière, Paris, 1858.)

The year 1870 had seen the disastrous defeat of Napoleon II's army at Sedan, the fall of the Second Empire, and the passing into exile of the Emperor to England, where he died in 1873 after a long illness. Bernard's laboratory researches were over and his health failing. However, he continued to publish many didactic "leçons."

Bernard would find himself out-ot-tune with science in the 20th century, for he was strongly opposed to statements or figures established only by statistics, which he equated with conjecture.[40] This was part of his desire to make medicine an experimental science rather than a collection of observations. Direct experimental observation was to replace hypothesis (Fig. 30). "Statistics," he wrote "can create conjecture only."

There was another development that was making inroads into physiology and this was Darwinism. Bernard did not meet this immediately and was essentially noncommital as were many other physiologists. First came the *Origin of Species*[41] in 1859 to be followed by the more disturbing *Descent of Man.*[42] Later, Bernard faced these challenges in his collection of writings called *Leçons sur la Phenomène de la Vie commune aux Animaux et aux Végéraux.*[43] He may have been cautious because of the strong opposition made public by Cuvier, the powerful Permanent Secretary of the Académie des Sciences whose opposition was based on his study of fossil bones.[44]

In 1868 came another change. Bernard moved from the Sorbonne to the Museé de l'Historie Naturelle, and his personal laboratory work came to an end. He died in 1878. He was honored by a statue in the grounds of the Collège de France. This was torn down during "les evénements de Mai" in 1968 but has been restored to its place. After his death a great

[40]C. Bernard. *Introduction a l'Étude de la Médecine.* Baillière, Paris, 1865. (English translation by H. C. Greene. Dover, New York, 1857.)

[41]C. Darwin (1809–1883). *On the Origin of Species by Means of Natural Selection.* Murray, London, 1859.

[42]C. Darwin. *The Descent of Man and Selection in Relation to Sex.* Murray, London, 1871.

[43]C. Bernard. *Leçons sur la Phénomène de la Vie commune aux Animaux et des Végéraux.* Baillière, Paris, 1878.

[44]Georges Cuvier (1769–1832). Recherches sur les ossemments fossilles des quadripèdes. Deterville, Paris, 1812.

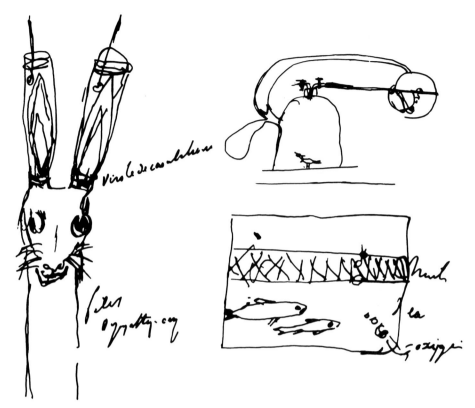

FIG. 31. Sketches by Claude Bernard of experiments to be made. **Left:** Technique for studying blood flow. **Right:** Suggested experiment on a bird under a bell jar to measure the amount of carbon dioxide expired (*upper*). To measure the absorption by fish of oxygen administered in measured quantity (*lower right*) and prevented from evaporation by a layer of oil. (From: C. Bernard. *Cahier Rouge*. J. Delhorne Gallimard, Paris, 1942.)

number of his papers were found at the Collège, and these have provided a wealth of material for scholars, resulting in many biographies of great interest. Among the papers were the notes he kept for 10 years in a red book. These were edited and published after his death as the *Cahier Rouge*.[45] A most interesting small book, it did not contain scientific findings so much as Bernard's plans for examining ideas that had come to him and included suggestions for testing them—a most revealing account of the thinking process in one of the most prominent figures in physiology (Fig. 31).

A great teacher, he had attracted to his laboratories many who later became distinguished in their own right, for example, Wilhelm His, Elie Cyon, and Ivan Michaelovich Sechenov. Others from Russia who came to study with Claude Bernard were Yakubovich, the histologist, and Ovsyanikov and Torkhanov, physiologists. Bernard's popularity in Russia is evidenced by Dostoyevsky's mention of him (by name) several times in his famous book *The Brothers Karamazov*.[46] A museum in Bernard's honor has been established in Paris[47]

[45]C. Bernard. *Cahier Rouge*. Delhoure J. Gallimard, Paris, 1942. (English translation by H. E. Hoff. Schenkman, Cambridge, 1967.)

[46]In the Russian text of *The Brothers Karamazov*, Bernard's name is referred to by Dostoyevsky on pages 101, 111, 203, 208, and 273.

[47]Musée des Universités, Paris.

and contains some of his apparatus. His home in St. Julien is also preserved as a museum.

The debt that neurophysiologists owe to Claude Bernard is due to his incessant curiosity about the innervation of the "milieu intérieur." He did not introduce this concept until 1859 when we find it mentioned in his book *Leçons sur les Propriétés Physiologiques*.[48] Oddly, this very famous book was not entirely written by himself but is a collection of the notes taken by his students of his lectures and was authorized by him for publication under his name. The concept at this stage was merely descriptive and favored the blood as the integrating mechanism. The explicit concept of the "milieu intérieur" is not very fully developed in Bernard's writings until 1864, many years after his period of experimental work. One finds it in the famous book *Introduction à l'Étude de la Médecine Expérimentale*.

In translation, the key phrase is ". . . all the phenomena of a living body are in such reciprocal harmony, one with another, that it seems impossible to separate any part without at once disturbing the whole organism." He rejected as the agent for this "reciprocal harmony" the earlier concept of "a vital force."

In a later book, *Leçons sur les Phénomènes de la Vie commune aux Animaux et aux Végétaux*, published in the year of his death, he wrote:

> The stability of the internal environment is the condition for free and independent life: the mechanism which permits it is that which ensures the maintenance in the internal environment of all conditions necessary for the life of the elements.

He goes on to say: "The stability of the environment implies the perfection of an organism such that the external variations are compensated and balanced at every moment." For Bernard, neurophysiology was not an end in itself but the role it played in the life of the "milieu antérieur."

Johannes Müller (1801–1858)

Johannes Müller was born in Coblenz in the Rhineland in the year 1801, the very year his homeland was lost to France by the Treaty of Luneville. After his early education at Bonn until he was 22, he went for the first time for one year to Berlin, the center where he was to become so prominent as a teacher. At Bonn he had been drawn to the *Natürphilosphie* of Schelling and the last fading stages of vitalism. Müller, however, began to move away from this concept, perhaps largely because of his brilliant pupil Du Bois-Reymond, who himself became the leader of the materialist position. Müller's own studies were mainly in the glandular system and the development of the genitals from which field he moved on to examining blood and lymph. During this period he travelled to centers in Holland and France, but in 1833 the call came to come to Berlin as Professor of Anatomy and Physiology. There he founded his famous *Handbuch der Physiologie;* however, his interests at this time were not in the nervous system but in the biology of such animals as snails. In 1848 Berlin experienced the revolution that shook Europe and caused so many to emigrate. It was also the year of the Communist Manifesto.

Müller (Fig. 32) is widely remembered for his theory of specific nerve energy[49] but it should be noted that this use of the word "energy" is not that used in physics.[50] His own words (in translation) read:

---

[48]C. Bernard. *Leçons sur les Propriétés Physiologiques et les Altérations Pathologiques des Liquedes de l'Organisme*. 2 vols. Baillière, Paris, 1859.

[49]J. Müller. *Zur vergleichenden Physiologie des Gesichtssinnes des Menschen und der Thiere*. Gnobloch, Leipzig, 1826. (On the Comparative Physiology of the Sense of Sight of Humans and Animals.)

[50]The words Müller used were: Spezifische Sinnesenergien.

FIG. 32. Left: Johannes Müller at the age of 25. **Right:** Müller's drawings of the optic chiasm of a two-year-old child. His dissections were designed to differentiate those fibers from the eye that crossed from those that did not. (From: J. Müller. *Zur Vergleichenden Physiologie des Gesichtssinnes des Menschen und der Thiere.* Gnobloch, Leipzig, 1826.)

If the nerves are mere passible conductors of the impressions of light, sonorous vibrations, and odours, how does it happen that the nerve which perceives odours is sensible to this kind of impressions only, and to no others, while by another nerve odours are not perceived; that the nerve which is sensible to the matter of light, or the luminous oscillations, is insensible to the vibrations of sonorous bodies; that the auditory nerve is not sensible to light, nor the nerve of taste to odours; while, to the common sensitive nerve, the vibrations of bodies give the sensation, not of sound, but merely of tremors? These considerations have induced physiologists to ascribe to the individual nerves of the senses a special sensibility to certain impressions, by which they are supposed to be rendered conductors of certain qualities of bodies, and not of others.

This concept of Müller's came under fire from many sides but most outstandingly from Lötze.[51] He strongly challenged Müller's generalization that any kind of stimulus in a given sensory pathway provokes only one sensation in a given sensory modality. He later modified this view to include inner psychological stimuli. But Müller's theory of the specificity of nerves, even those coming from the periphery, was shown experimentally to be untenable. These were the experiments of A. Vulpian,[52] Professor at the Faculté de Médecine in Paris and his assistant J. M. Philipeaux[53] at the National History Museum in work published in 1853.

[51]R. H. Lotze. *Allgemeine Pathologie und Therapie als mechanische Natur-Wissenschaften.* Weidmann'sche Buchhandlung, Leipzig, 1842.

[52]J. M. Philipeaux and A. Vulpian. Recherches expérimentales sur la reunion bout à bout de nerfs de fonctions différentes, *J. Physiol. Homme Animaux,* 6:421–455 (1853). (Experiments on the End-to-End Union of Nerves having Different Functions.)

[53]J. M. Philipeaux and A. Vulpian. Recherches expérimentales sur la reunion de fonctions différentes. Deuxième série d'expériences. *J. Physiol. Homme Animaux,* 6:474–516 (1853). (Experiments on the Union of Different Functions.)

FIG. 33.   Left: Johannes Müller (1801–1858). **Right:** Title page of one of his first publications on vision. The title page reads: "On the fantastic phenomena of sight. A physiological investigation with a physiological document on dreams by Aristotle. Dedicated to philosophers and doctors. Dr. Johannes Müller. Associate professor of medicine at Bonn University, practicing doctor in that city, member of the Royal Academy of Naturalists."

These investigators, working on the innervation of the tongue, cut the lingual nerve and the hypoglossal and cross-joined them. They were able to demonstrate that stimulation of the peripheral branch of the hypoglossal nerve activated the tongue. Others were to explore the cross-ligaturing of nerves, but it was the work of Vulpian that essentially destroyed the theory of "specific nerve energy."

In 1826 Müller wrote a book[54] on the comparative physiology of the sense of vision (Fig. 33). In the opening of this book, he stated very clearly the philosophy that ruled all his experimental work. He declared as follows:

> In proposing to speak on the close connection of philosophy and physiology, I have undertaken no less a task than to show that a doctrine which uses a great amount of empirical knowledge in its composition is a *science,* and not simply a logical connection of empirical facts ordered according to categories of reason; I have to show how and in what way physiology, excluding all other forms of its existence, can become a science. I said excluding all other forms of its existence; because in the presentation of my point of view I would like to exclude any other concept which one might have if one heard talk of the common features of philosophy and physiology.

After a long philosophical introduction, Müller reported on his studies of vision in lower animals. He studied (and illustrated) species with movable, converging eyes adaptable to the distance of the object viewed and to ambient light. He examined the optic nerves and the optic chiasms in a variety of species—vertebrates, including frogs and fish, and inverte-

[54]J. Müller. *Über die phantastischen Gesichtserscheinungen.* F. Hölscher, Coblenz, 1826. (On Fantastic Phenomena of Vision).

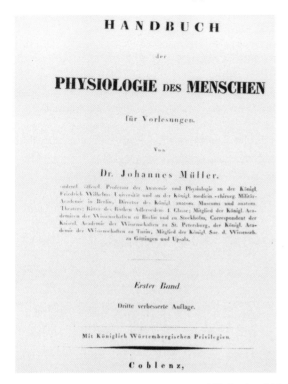

FIG. 34. **Left:** The famous *Handbook* of Müller in which he develops more fully his theory of specific nerve energy. **Right:** Alfred Vulpian (1826–1887) who disproved this concept.

brates, such as the snail. He illustrated his findings (for example, of the optic chiasm) in many species.

Müller's contribution was essentially as a teacher. That was his principal drive. This was evidenced not only in the classroom but also in the *Handbuch* (Fig. 34) and in the journal he founded: *Archiv für Anatomie und Physiologie und Wissenschaftliche Medizin,* popularly known as Müller's *Archiv.* This journal formed the major outlet also for scientists from Czarist Russia and for such prominent English scientists as Thomas Henry Huxley.[55]

Müller was still deeply concerned with the part played by the "soul" in his explorations of the senses and especially in his research on vision, a work in which his tie to Goethe is clear. He wrote: "The soul is only a special form of life among many other vital manifestations which are subjects for physiological investigation. Thus, I am also convinced that research in physiology will deal with psychological matters in the last analysis." Later, in his *Handbuch,* Müller expressed similar views about the relationship between life and soul. But one sees a gradually changing approach in the developing series of animal reports Müller wrote on the progress of science. He published these in his *Archiv* from 1834 to 1838. In his later reviews, he asked for physical and chemical proofs in physiology. Thus, in his later years, he evinced a material approach with few traces left of vitalism. Müller died in 1858 after a long illness.

---

[55]Thomas H. Huxley (1825–1895). Über die Sexualorgane der Diphydae und Physophoridae. *Archiv. für Anat. Physiol. Wiss. Med.,* 380–384, 1851. (On the Sexual Organs of the Diphydae and Physophoridae.)

BIBLIOGRAPHY

François Magendie (1783–1855)
Selected Writings

Essai sur les usages du voile du palais, avec quelques propositions sur la fracture du cartilage des côtes. (Presented and defended at the École de Médecine.) Paris, 1808.

Examen de l'action de quelque vegetaux sur la moelle épinière. Read 24 April 1809, Paris, 1809.

Institut de France, Académie des Sciences, Próces verbaux, 4:208–210 (1808–1811).

Quelque idées generales sur les phénomènes particularies aux corps vivantes. *Bull. Sci. Med.*, 5:145–170 (1809).

Institut de France, Académie des Sciences Procès verbaux, 5:205–208, 174–179; 7:109–111 (1812–1815).

Présic Elémentaire de Physiologie. Mequignon-Marvis, Paris, Vol. 1, 1816, Vol. 2, 1817, 2nd ed., 1825, 3rd ed., 1834.

Extrait d'un mémoire sur les vaisseaux lymphatique des oiseaux. Societé Philmatique de Paris. *Bull Sci. Ser. 3*, 6:89–92 (1819).

Mémoire sur le mechanisme de l'absorption chez les animaux à sang rouge et chand. *J. Phys. Pathol.*, 1:1–17 (1821).

Histoire d'une maladie singulière du système nerveux. *J. Physiol. Exp. Pathol.* 2:99–104 (1822).

Expériences sur les fonctions des racines des nerfs qui naissent de la moelle épinère. *J. Physiol. Exper. Path.*, 2:276–279; 366–371 (1822).

Expériences sur les fonctions des racines des nerfs rachidiens. *J. Physiol. Exp. Pathol.*, 3:276–279, 366–371 (1822).

Sur le siège du mouvement et de sentiment dans la moelle épinière. *J. Physiol. Exp. Pathol.*, 3:153–157 (1823).

Remarques sur une destruction d'une grande partie de moelle épinière. *J. Physiol. Exp. Pathol.*, 3:186 (1823).

Le nerf olfactif est-il l'organe de l'odorat? Expèriences sur cette question. *J. Physiol. Exp. Pathol.*, 4:169–175 (1824).

Mémoire sur des fonctions de quelque parties du système nerveux. *J. Physiol. Exp. Pathol.*, 4:399–407 (1824).

De l'influence de la clinquième paire des nerfs sur la nutrition et les fonctions de l'oeil. *J. Physiol. Exp. Pathol.*, 4:176–302 (1824).

Mémoire sur un liquide qui se trouve dans le crâne et le canal vértébral de l'homme et des animaux mammifère. *J. Physiol. Exp. Pathol.*, 5:27–37 (1825).

Précis Elémentaire de Physiologie. Paris. Vol. 1, 1816; Vol. 2, 1817. English translation by E. Milligan. John Carfrae & Son. Edinburgh, 1826.

Mémoire sur le liquide qui se trouve dans le crâne et l'épine de l'homme et des animaux vértébres. *J. Physiol. Exp. Pathol.*, 7:1–17; 17–29; 66–82 (1827).

Quelque réflexions sur la dissertation de Cotugno "De Ischiade Nervosa" contenue dans le Thesaurus Dissertationum de Sandifort. Tome II. p. 411, Rotterman, 1769, *J. Physiol. Exp. Pathol.*, 7:85–96 (1827).

Sur l'insensibilitié de la retine de l'homme. *J. Physiol., Exp. Pathol.*, 5:37 (1825).

Seconde mémoire sur le liquide qui se trouve dans le crâne et l'épine de l'homme, et des animaux vértébres. *J. Physiol. Exp. Pathol.*, 3:1–29 (1826).

Notice sur l'heureuse application de galvanisme aux nerfs de l'oeil. *Arch Gen. Med.*, 2:3–12 (1826).

Extrait de la dissertation de Cotugno de Ischiade Nervosa contenue dans le Thesarus Dissertationum de Sandifort; ave quelques reflexions. *J. Physiol. Exp. Pathol.*, 7:85–96 (1827).

Mémoire physiologique sur le cerveau. *J. Physiol. Exp. Pathol.*, 8:211–229 (1828).

Troisième et dernière partie du seconde mémoire sur le liquide qui se trouve dans le crâne et l'épine de l'homme et des animaux vértébres. *J. Physiol. Exp. Pathol.*, 7:66–82 (1827).

Lectures on Experimental Physiology. *London Med. Gaz.*, 1 (1828).

Note on Bell's attitude regarding Bell-Magendie controversy. *J. Physiol. Exp. Pathol.*, 10:1–3 (1830).

Résultats de quelque nouvelles experiences sur les nerfs sensitifs et sur les nerfs moteurs. *C. R. Acad. Sci.*, 8:787–788 (1836).

Communication relative à une guerison obtenue par des courants électrique portés directement sur la corde de tympan; restitution des sens de góut et de l'ouie abolis par suite d'une commotion cerebrale. Deductions tirée de ce fait quant à l'origine de nerf de tympan. *C. R. Acad. Sci.*, 2:447–452 (1836).

Traitement de certaines affections nerveuses par l'électropunture des nerfs. *C. R. Acad. Sci.*, 5:855–856 (1837).

Quelque nouvelles expériences sur les fonctions du systeme nerveux. *C. R. Acad. Sci,* 8:865–867 (1839).

Leçons sur les Fonctions et les Maladies du Système Nerveux Professées au Collège de France. 2 vols. Ebard, Paris, 1839–1841.

Recherches physiologiques et cliniques sur le liquide cephalorachidien ou cerebro-spinal. Mequignon-Marvis, Paris, 1842.

De l'influence des nerfs rachidiens sur les mouvements du coeur. *C. R. Acad. Sci,* 25:875–879, 926–928 (1847).

Suggested Readings

Cranefield, P. *The Way In and the Way Out.* Futura, New York, 1974.

Davson, P.M. *A Biography of Francois Magendie.* Brooklyn, N.Y., 1908.

Deloryers, L. *François Magendie (1783–1855).* Presses Universitaires de Bruxelles, 1970.

Lesch, J.E. *Science and Medicine in France.* Harvard Univ. Press, Cambridge, 1984.

Olmstead, J.M.D. *François Magendie.* Schumann, New York, 1944.

Temkin, O. This philosophical background of Magendie's physiology. *Bull. Hist. Med.,* 20:10–35 (1946).

Charles Bell (1774–1842)

Selected Writings

Idea of a New Anatomy of the Brain; submitted for the observations of his friends in 1811 by Strahan and Preston, London; not published but privately circlulated.

Idea of a new anatomy of the brain . . . *The Baltimore Medical and Philosophical Lycaeum, 1:*303–318 (actually published in or after March 1812).

A Series of Engravings explaining the Course of the Nerves. With an Address to Young Physicians of the Study of the Nerves. Second edition. Longman, London, 1816.

On the nerves; giving an account of some experiments of their structure and functions, which lead to a new arrangement of the system. *Phil. Trans. R. Soc., 111:*398–424 (1821).

On the nerves which associate the muscles of the chest, in the actions of breathing, speaking, and expression. Being a continuation of the paper on the structure and functions of the nerves. *Phil. Trans. R. Soc., 112*:284–312 (1822).

The Nervous System of the Human Body. Longman, London, 1824. (2nd ed., 1830, 3rd ed., 1844).

An Exposition of the Natural System of the Nerves of the Human Body with a Republication of the Papers Delivered to the Royal Society on the Subject of Nerves. Spottiswood, London, 1824.

On the nervous circle which connects the voluntary muscles with the brain. *Phil. Trans. R. Soc.,* 2:172 (1826).

Lectures on the physiology of the brain and nervous system. *Ryan's Med. Surg. J.,* 1:682, 752 (1832).

The Nervous System of the Human Body as Explained in a Series of Papers Read Before the Royal Society of London. Black, Edinburgh, 1836.

Suggested Reading

Letters of Charles Bell Selected from His Correspondence with His Brother George Joseph Bell. John Murray, London, 1870.

Claude Bernard (1813–1878)

Selected Writings

Disposition des fibres musculaires dans la veine cave inférieure du cheval. *Mem Soc. Biol.,* 1:33 (1849).

Chiens rendus diabétiques. *Mem. Soc. Biol.,* 1:60 (1849).

Autopsie d'un diabétique. *Mem. Soc. Biol.,* 1:80 (1849).

De l'influence du système nerveux grand sympathique sur la chaleur animale. *C. R. Acad. Sci.,* 34:472–475 (1852).

Sur les effets de la portion de la section encéphalique du grand sympathique. *Mem. Soc. Biol.,* 4:168–170 (1852).

Recherches expérimentales sur le grand sympathique spécialement sur l'influence que la section ce nerf exerce sur la chaleur animale. *Mem. Soc. Biol.,* 5:77–107 (1853).

Leçons de physiologie expérimentale appliquée à la médecine. Paris, Ballière, 2 vols., 1855–56.

De l'influence qu'exercent différents nerfs sur la sécrétion de la salive. *Mem. Soc. Biol.,* 9:85–86 (1857).

Leçons sur la physiologie et la pathologie du système nerveux. Paris, Baillère, 2 vols., 1858.

De l'influence de deux ordres de nerfs qui déterminent les variations de couleurs du sang vineux dans les organes glandulaires. *C. R. Acad. Sci.,* 47:245–253 (1858).

Sur l'action des nerfs, sur la circulation et la sécrétion des glandes. *Mem. Soc. Biol.,* 11:49–51 (1859).

Recherches expérimentales sur les nerfs vasculaires et calorifiques du grand sympathique. *J. Physiol.* (Paris), 5:383–418 (1862); *C. R. Acad. Sci.,* 55:228–236; 305–312; 341–350 (1862).

Suggested Readings for Claude Bernard

Foster, Michael. *Claude Bernard.* Fisher Union, London, 1899.

Ganguilhem, G. *Études d'Histoire et Philosophie des Sciences.* Verin, Paris, 1968.

Grmek, M. D. *Claude Bernard, Cahier de Notes. 1850–1860.* Presenté et commenté par Mirkio Drazen Grmek. Editions Gallimard, Paris, 1965.

Grmek, M. D. *Raisonnement expérimental et recherches toxicologique chez Claude Bernard*. Droz, Geneva, 1972.

Holmes, F.L. *Claude Bernard and Animal Chemistry*. Harvard Univ. Press, Cambridge, 1974.

Kay, A. S. Claude Bernard. *Dictionary of Scientific Biography*. 2:24–35 (1970).

Olmsted, J. M. D. *Claude Bernard, Physiologist*. Harper & Row, New York, 1938.

Olmsted, J. M. D. *Claude Bernard and the Experimental Method in Medicine*. Schumann, New York, 1952.

Alfred Vulpian (1826–1877) and J.M. Philipeaux

Selected Writings

Philipeaux, J.M., et Vulpian, A. Recherches expérimentales sur la réunion bout à bout nerfs de fonctions differents. *J. Physiol. l'Homme Animaux*, 6:421–455 (1853).

Philipeaux, J.M., et Vulpian, A. Recherches expérimentales sur la réunion de fonctions differents. Deuxième serie d'expériences. *J. Physiol. l'Homme Animaux*, 6:474–516 (1853).

Vulpian, A. *Leçons sur la physiologie générale et comparée du système nerveux faites au Musée d'Historie Naturelle*. Baillière, Paris, 1866.

Vulpian, A. *Leçons sur l'appareil vaso moteur (Physiologie et pathologie) faites à la Faculté de Médecine de Paris*. 2 vols. Baillière, Paris, 1874–75.

Vulpian, A. *Maladies du système nerveux; Leçons professées à la Faculté de Médecine*. 2 vols. Doin, Paris, 1879–89.

Johannes Müller (1801–1858)

Selected Writings

De respiratione foetus commentatio physiologica, in Academia Borussica Rhenana praemio ornata. Cum Tabula seri incisa. Knoblock, Leipzig, 1823.

Über das Bedürfniss der Physiologie nach einer philosophischen Naturbetrachtung. *Antrittsvorlesung*, 27–28 (1824). (On Physiology's Need for a Philosophical View of Nature.)

Über die Entwickelung der Eier im Eierstock bei den Gespenstheuschrecken. *Nova Acta phys.-med. Acad. Caes. Leopold nat. curios.*, 12:553–672 (1825). (On the Development of Eggs in the Ovaries of Stick Insects.)

Zur vergleichenden Physiologie des Gesichtssinnes des Menschen und der Thiere. Gnobloch, Leipzig, 1826. (On the comparative physiology of the sense of sight of humans and animals.)

Über die phantastischen Gesichtserscheinungen. F. Hölscher, Coblenz, 1826. (On Fantastic Phenomena of Vision.)

De glandularum secernentium structurea penitiori. L. Vossii, Lipsiae, 1830.

Bestätigung des Bell'schen Lehrsatzes. *Notiz. Natur Heilk*. 30:113–122 Weimar (1831). (Confirmation of Bell's Doctrine.)

Über die Existenz von vier getrennten, regelmässig pulsirenden Herzen, welche mit dem lymphatischen System in Verbindung stehen, bei einigen Amphibien. *Arch. Anat. Physiol. Wiss. Med.*, 296–300 (1834). (On the Existence of Four Separated, Regularly Beating Hearts, which are Connected to the Lymphatic System, in Some Amphibians.)

Über die äusseren Geschlechtstheile der Buschmänninnen. *Archiv. Anat. Physiol. Wiss. Med.*, 319–345 (1834). (On the Outer Sexual Organs of Bush Females.)

Handbuch der Physiologie des Menschen. J. Hölscher, Coblenz, 1834–1840. (Handbook of Human Physiology.) Translated by W. Baly as *Elements of Physiology*. Baylor and Walton, London, 1842.

Entdeckung der bei der Erection des männlichen Gliedes wirksamen Arterien bei dem Menschen und den Thieren. *Arch. Anat. Physiol. Wiss. Med.*, 202–213 (1835). (The discovery of Arteries Operative during the Erection of the Penis in Man and Animals.)

Untersuchung eines Schildkrötenharns. *Arch. Anat. Physiol. Wiss. Med.*, 214–219 (1835). (Investigation of the Urine of a Turtle.)

Versuche über die künstliche Verdauung des geronnenen Eiweisses. *Arch. Anat. Physiol. Wiss. Med.*, 66 89 (1836.) (Experiments into Artificial Digestion of Coagulated Albumen.)

Über Knorpel und Knochen. *Ann. Pharm.*, 21:277–282 (1837). (On Cartilage and Bone.)

Gedächtnisrede auf Carl Asmund Rudolphi. *Abh. Preuss. Akad. Wiss.* Berlin, 23 (1837).

Über den feineren Bau und die Formen der krankhaften Geschwülste. G. Reimer, Berlin, 1838. (On the Finer Structure and Forms of Malignant Tumors.)

Über die Compensation der physischen Kräfte am menschlichen Stimmorgan, mit Bemerkungen über die Stimme der Säugethiere, Vögel und Amphibien. A Hirschwald, Berlin, 1839. (On the Compensation of Physical Forces in the Human Vocal Chords, with Remarks on the Voices of Mammals, Birds and Amphibians.)

Über die Lymphherzen der Schildkröten. Druckerei d. k. Akad. Berlin, 1840. (On the Lumph Hearts of Tortoises.)

Anatomische Studien über die Echinodermen. *Arch. Anat. Physiol. Wiss. Med.*, 117–155 (1850). (Anatomical studies on the echinoderms.)

Berichtigung und Nachtrag zu den anatomischen Studien über die Echinodermen. *Arch. Anat. Physiol. Wiss. Med.*, 225–233 (1850). (Correction and continuation of the anatomical studies on the echinoderms.)

Fortsetzung der Untersuchungen über die Metamorphose der Echinodermen. *Arch. Anat. Physiol. Wiss. Med.*, 452–484 (1850). (Continuation of the investigations on the Metamorphosis of the Echinoderms.)

Über eine den Sipunculiden verwandte Wurmlarve. *Arch. Anat. Physiol. Wiss. Med.*, 439–451 (1850). (On a worm larva related to the sipunculids.)

Über eine eigenthümliche Wurmlarve, aus der Classe der Turbellarien und aus der Familie der Planarien. *Arch. Anat. Physiol. Wiss. Med.*, 485–507 (1850). (On a peculiar worm larva, from the class of turbellarians and the family of planarians.)

Suggested Readings

Haberling, W. *Johannes Müller: Das Leben des rheinischer Natürforschers.* Akad. Verlag, Leipzig, 1924. (The Life of the Scientist from the Rhineland.)

Paly, *Elements of Physiology* (a translation of Müller's *Handbuch)*, 2 vols. Taylor & Walton, London, 1838–1842.

Hernstein, R., and Boring, E.G., eds. *A Source Book in the History of Psychology.* Harvard University Press, 1965.

Diamond, S., ed. *The Roots of Psychology, a Sourcebook in the History of Ideas.* Basic Books, New York, 1974.

Koller, Gottfried. *Johannes Müller: Das Leben des Biologen, 1801–1858.* Wissen. Verlag, Stuttgart, 1958. (The Life of the Biologist.)

Rothschuh, K. *History of Physiology.* Krieger Publishing Co., New York, 1973.

Steudel, Johannes. *Von Boerhaave bis Berger.* Fischer, 1964. (From Boerhaave to Berger.)

Woodward, William R. Hermann Lotze's Critique of Johannes Müller's Doctrine of Specific Sense, *Medical History,* vol. 2, 1975.

CHAPTER VI

The Great German Schools

Hermann Ludwig Ferdinand von Helmholtz (1821–1894)

In the first half of the 19th century, the neuron theory had not yet been formulated, so that the relationship of nerve to gray matter was still obscure. But in 1842 a 21-year-old student, in an inaugural thesis,[1] demonstrated for the first time that axons were processes of nerve cell bodies. He demonstrated this discovery in the invertebrate nerve, using the crab. The name of this young student was Hermann von Helmholtz, who later became one of the outstanding physiologists of all time.

Helmholtz was born in Potsdam, the German capital before it moved to Berlin. He studied in Berlin and remained essentially a Berliner at heart. As a student of Müller's, Helmholtz worked on invertebrates, and was to return to this kind of work after taking a full medical training at the Institute for Medicine and Surgery, which was named for Kaiser Friedrich Wilhelm. On qualifying, he served a residency at the famous old Charité hospital, and then he spent two years as an army surgeon. When his service was over, Helmholtz returned to his research. Later, his professional life took him to Königsberg on the Baltic Coast of Prussia where in time he rose to be Professor in its university, a position he held until 1871, when he moved back to Berlin. In Helmholtz's day, Königsberg was in the farthest eastern point of the Kingdom of Prussia. After the First and Second World Wars, it is now in the Lithuanian Soviet Republic and is named Kaliningrad, where the State University preserves one example of Helmholtz's ingeniously designed instruments.

The number of Helmholtz's major contributions is impressive. Perhaps the one affecting all science was his establishment of the Conservation of Energy,[2] a condition as universal in our world in the 19th century as $E = mc^2$ is in the 20th. The recognition of this rule

---

[1]H. von Helmholtz. De fabrica systematis nervosi evertebratorum (thesis). Berlin, 1842. (Structure of the Nervous System of Invertebrates.)

[2]H. von Helmholtz. *Über die Erlangung der Kraft*. Berlin, 1847. (On the Conservation of Energy.) The word Helmholtz used was "Kraft," directly translated as "Force." But the reading of his works shows that his concept was the Conservation of Energy.

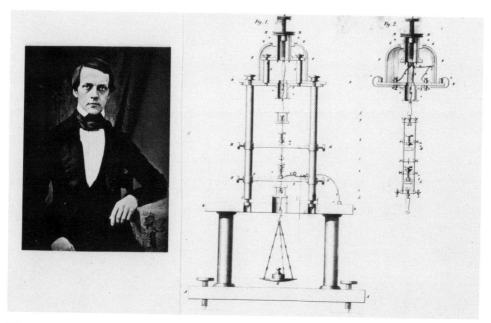

FIG. 35. Left: Hermann von Helmholtz (1821–1894) at the period of his life devoted to research on nerve conduction (From a daguerreotype made in 1848). **Right:** Diagrams of his instrumentation published by him in 1850. (From: *Arch. Anat. Physiol. Wiss. Med.*, 276–365, 1850). Two years later Helmholtz published another decription with full instructions for constructing this apparatus. (From: *Arch. Anat. Physiol. Wiss. Med.*, 199–216, 1852.)

struck a blow to vitalism. He was 26 and still living in Berlin when he published his essay on the Conservation of Energy. He read his theory first to the Physical Society of Berlin in 1847. He reported and lectured on this "rule" on many occasions in London.[3, 4] In a lecture in the 1870s, he wrote (in translation):

> If a certain quantity of mechanical work is lost, an equivalent quantity of heat or of chemical force, is gained and, conversely, when heat is lost we gain an equivalent quantity of chemical or mechanical force; and again, when chemical force disappears, we gain an equivalent of heat or work; so that, in all these interchanges between various inorganic natural forces, the working force may indeed disappear in one form, but then it reappears exactly equivalent in some other form; it is thus neither increased nor diminished, but always remains in exactly the same quantity. . . . The same law holds good also for processes in organic nature, so far as the facts have been tested.

The law of the Conservation of Energy applies to all science, including neurophysiology.

Many of Helmholtz's contributions come closer to the problems of the nerve and muscle (Fig. 35). He worked in this field and was the first to succeed in measuring the rate of conduction in the nerve.[5] His pursuit of this measurement was in spite of his teacher (Johannes Müller), who still held a less mechanistic belief. Confident of his findings, Helm-

---

[3]H. von Helmholtz. The application of the law of conservation of force to organic nature. *Proc. R. Inst.*, 3:347–357, 1861. Also in: *Arch. Anat. Physiol. Wiss. Med.*, 565–580 (1861).

[4]H. von Helmholtz. Lectures on the conservation of energy, delivered at the Royal Institution, London. *Med. Times Gaz.*, 1:385–388, 415–418, 443–474, 499–501, 527–530 (1864).

[5]H. von Helmholtz. Über die Fortpflanzungsgeschwindigkeit der Nervenreizung. *Arch. Anat. Physiol. Wiss. Med.*, 71–73 (1850). (On the Speed of Propagation of Nerve Stimulation.)

holtz sent reports not only to Du Bois-Reymond for the Physical Society in Berlin but to Müller and to the French and Berlin Academies of Science.[6]

Helmholtz was quite modest when describing his achievement in measuring the speed of nerve conduction. Writing from Königsberg to the French Academy of Science, he said he did not find it difficult to measure the time between irritation of the spinal root in the frog and contraction of the gastrocnemius muscle: "J'ai trouvé qu'il faut à l'irritation nerveuse pour arriver du plexus sciatique au muscle gastrocnemien d'une grenouille, un espace de temps qui n'est pas trop difficile d'évaluer." To record the moment of stimulation was simple. To catch the moment of muscle contraction, he attached a weight to the gastrocnemius which, when lifted by the responding muscle, made contact with the circuit. In other words, he was not recording directly from two points on the nerve itself but from the moment of stimulation of the spinal root to the moment of effective muscular contraction. Direct recording of the actual potential of the nerve itself still lay in the future. Helmholtz published a drawing of this experimental design in Müller's *Archiv* in 1850.[7, 8] Helmholtz was able to convince himself that the use of two circuits—one broken by the instant of stimulation and the other by closure when the weight was lifted—did not introduce significant error.[9]

Helmholtz was aware that this measurement would depend on the state of excitability of the muscle and on the weight it was expected to raise but that neither of these variations influenced conduction velocity of the nerve itself. He attempted leading instead from the point of entry of the nerve to the muscle, but this introduced too many vagaries when contraction took place. Therefore, to deduce the conduction time in the nerve itself (Fig. 36), he made a pair of experiments on the same preparation but with the stimulus applied at different distances from the muscle. Thus, any difference between the time at which the weight was lifted (and activated the second circuit) could be assigned to conduction time in that length of the nerve. He reported that the weight was lifted "un peu plus tard" when the stimulus was applied to the spinal root rather than further down—near to where the nerve entered the muscle. A second finding was that even when the nerve was stimulated as close as possible to its entry into the muscle, there was a delay. Helmholtz was thus able to measure also the neuromuscular delay, or latent period.[10] To measure such small intervals of time was a great achievement in the mid-19th century.

Helmholtz used a galvanometer, but none of these original records has survived. Helmholtz calculated the speed of conduction in the motor nerve of the frog to be 27.25 meters per second (m/sec.).

He then proceeded to show that cooling the nerve lengthened the conduction time.[11] Helm-

[6]H. von Helmholtz. Note sur la vitesse de propagation de l'agent nerveux dans les nerfs rachediens. *C. R. Acad. Sci.*, 30:204–206 (1851).

[7]H. von Helmholtz. Über die Fortpflanzungsgeschwindigkeit der Nervenreizung. *Arch. Anat. Physiol. Wiss. Med.*, 71–73 (1850). (On the Speed of Conduction of Nerve Stimulation.)

[8]H. von Helmholtz. Messungen über den zeitlichen Verlauf der Zuckung animalischer Muskeln und die Fortpflanzungsgeschwindigkeit der Reizung in den Nerven. *Arch. Anat. Physiol. Wiss. Med.*, 276–364 (1850). (Measurement of the Chronological Progress of Contractions in Animal Muscles and the Spread of Conduction of Nerve Stimulation.)

[9]H. von Helmholtz. "Mais je me suis assuré que la difference de temps entre l'ouverture et la closure de deux circuits restant de beaucoup inférieure à une dixième de la duré qui s'agissant d'évaluer." *C. R. Acad. Sci.*, 30:204–206 (1850).

[10]H. von Helmholtz. Messungen über Fortpflanzungsgeschwindigkeit der Reizung in den Nerven. Zweite Reihe. *Arch. Anat. Physiol. Wiss. Med.*, 844–861 (1852). (Measurement of the Speed of Propagation of the Nerve Stimulus.)

[11]H. von Helmholtz. On the later views of the conduction of electricity and magnetism. Annual Report of the Smithsonian Institution for 1873. Washington, D.C., 1874, pp. 247–253.

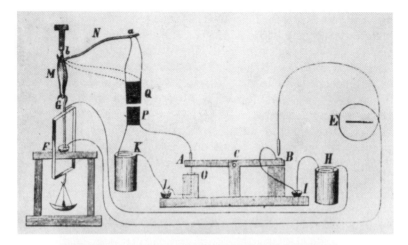

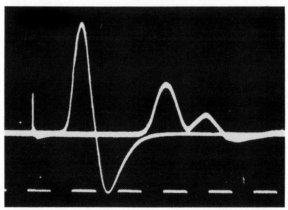

FIG. 36. Above: Helmholtz's apparatus for measuring the velocity of nerve conduction. The nerve is stimulated at "a" and the muscle (M) contracts, signalling this to E. (Reproduced from Julius Budge. *Lehrbuch der speciallen Physiologie des Menschen.* Vol. 8. Leipzig, 1862.) **Below:** The change in the action potential of the sciatic nerve of the frog when stimulated at different distances from the muscle. The difference between the nearest and farthest points of stimulation was 50 mm. The action potential when stimulation is close to the muscle is diphasic. When the stimulation is distant, the action potential is nearly monophasic. Distant stimulation produces two waves, corresponding to a rapid and a slower group of fibers. (From W. Blasius. In: *Von Boerhaave bis Berger.* Fischer, Stuttgart, 1964.)

holtz's own comment on the rapidity of conduction he found was how fortunate we were that there was little delay in messages reaching our brains, commenting that we are more fortunate than the whale in the distances the impulse has to travel.

In a second report[12] to the French Academy, he went into more detail about the time lost between the arrival of the nerve impulse and the reaction to it. He called his research on this lost time "la recherche du temps perdu," a phrase that was to ring out in another field.[13]

In 1864 a rather clumsy attempt[14] was made by another German scientist, R. Schelske,

[12]H. von Helmholtz. Deuxième note sur la vitesse de la propagation de l'agent nerveux. *C. R. Acad. Sci.,* 33:262–265 (1851).

[13]M. Proust. *A la Recherche du Temps Perdu.* Paris, 1913–1927.

[14]Rudolf Schelske. Neue Messungen der Fortpflanzungsgeschwindigkeit des Reizes in den menschlichen Nerven. *Arch. Ana. Physiol. Wiss. Med.,* 151–174 (1864). (New Measurements of the Speed of Transmission of Stimulation in the Human Nerves.)

who measured the interval between administering an electric shock to the foot of a man and noting an electrical change in the recording from the groin. In one subject he found a speed of transmission of 31 m/sec. In another it was 25 m/sec. When stimulating the foot and recording from the neck, he found a speed of 32.6 m/sec.

For the neurophysiologist, another great contribution was Helmholtz's work on the eye. Intrigued by color, he researched rods and cones, having himself designed an ophthalmoscope[15] to study the retina. He made measurements of the cornea of the human eye[16] and made calculations about accommodation[17] and on the physics of stereoscopic vision.[18]

In Helmholtz's theory of color:

> Each primary sensation of red, green and violet is excited in some degree by almost every ray of the spectrum, but the maxima of excitation occur at different places, while the strength of stimulation in each case diminishes in both directions from the maximal point. Thus when the three sensations are equally excited, white light is the result; green is caused by a very weak violet stimulation, a stronger red, and still stronger green stimulation. At each end of the spectrum we have only the simple sensations of red and violet, and all the intermediate color sensations are compounds of varying proportions of the three primaries.

In summary, he held that:

> (1) Red excites strongly the fibers sensitive to red, and feebly the other two—sensation, red.
> (2) Yellow excites moderately the fibers sensitive to red and green, feebly the violet—sensation, yellow.
> (3) Green excites strongly the green, feebly the other two—sensation, green.
> (4) Blue excites moderately the fibers sensitive to green and violet, and feebly the red—sensation, blue.
> (5) Violet excites strongly the fibers sensitive to violet, and feebly the other two—sensation, violet.
> (6) When the excitation is nearly equal for the three kinds of fibers, then the sensation is white.[19]

From his interest in the sensory systems of the eye, which he had pursued in Königsberg, Helmholtz moved in 1836 to Bonn and to the auditory system (Fig. 37). His interest in acoustics led him to the study of tones and eventually to an analytical study of music written from the groundwork of both physics and physiology. In striving to find the physics basic to hearing, he dissected the ossicles of the ear and published a small report on them (translated into English by the New Sydenham Society).[20] He published four papers on acoustics in the 1860s. Helmholtz showed that the distinctive quality of timbre of a complicated sound is determined by its components, which give a particular sound its distinctive auditory character. Thus, sinusoidal sound waves are not simple.

---

[15]H. von Helmholtz. Beschreibung eines Augenspiegels zur Untersuchung der Netzhaut im lebenden Auge. *Arch. Anat. Physiol. Wiss. Med.,* 229–260 (1851). (Description of an Ophthalmoscope for the Investigation of the Retina of the Living Eye.)

[16]H. von Helmholtz. On the normal motions of the human eye in relation to binocular vision. *Arch. Anat. Physiol. Wiss. Med.,* 25–43 (1864).

[17]H. von Helmholtz. Über die Accommodation des Auges. *Arch. Anat. Physiol. Wiss. Med.,* 283–345 (1855). (On Accommodation in the Eye.)

[18]H. von Helmholtz. *Handbuch der physiologischen Optik.* 3 vols. 1856, 1860, 1866. Voss. Leipzig, 1867. (Third edition translated into English by James P. C. Southall and published by the Optical Society of America in Menasha, Wisconsin, in 3 vols., 1909–1911.)

[19]From the translation by J. G. McKendrick. *Hermann Ludwig Ferdinand von Helmholtz.* Longmann Green, London, 1899.

[20]H. von Helmholtz. *The Mechanism of the Ossicles and the Membrana Tympani.* New Sydenham Society, London, 1874. Translation by A. J. Ellis.

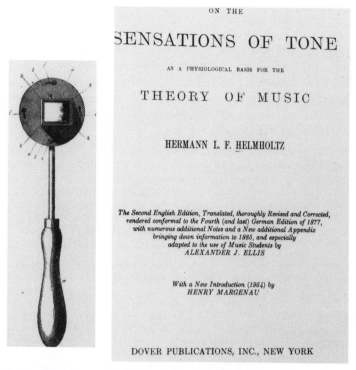

ON THE

SENSATIONS OF TONE

AS A PHYSIOLOGICAL BASIS FOR THE

THEORY OF MUSIC

HERMANN L. F. HELMHOLTZ

*The Second English Edition, Translated, thoroughly Revised and Corrected,
rendered conformal to the Fourth (and last) German Edition of 1877,
with numerous additional Notes and a New additional Appendix
bringing down information to 1885, and especially
adapted to the use of Music Students by*
ALEXANDER J. ELLIS

With a New Introduction (1954) by
HENRY MARGENAU

DOVER PUBLICATIONS, INC., NEW YORK

FIG. 37. **Left:** The ophthalmoscope designed by Helmholtz. (From: *Arch. Anat. Physiol. Wiss. Med.,* 261–270, 1852.) **Right:** The English translation of the 4th edition of *Die Lehre von den Tonempfindungen als physiologische Grundlage für die Theorie der Musik.* Braunschweig, 1863.

Helmholtz made many experiments using tuning forks, and it is clear that he was himself endowed with an accurate sense of pitch. He was fortunate to have the patronage of the King of Bavaria, Maximillian, who gave him a special collection of tuning forks. His prodigious work in this field was collected in his famous book[21] translated into English with the title *Sensations of Tone.* This included his various researches on vowel sounds, the tones of violin strings and organ pipes, and even Persian music with its scale so different from that used in Europe. Tradition has it that he himself played the piano with an especial enjoyment of Bach's fugues. Helmholtz summed up his theory of tones as follows:

> Whenever the vibrations of the air or of other elastic bodies, are set in motion at the same time by two generating simple tones (so powerful that they can no longer be considered infinitely small), mathematical theory shows that vibrations of the air must arise which have the same vibrational numbers as the combination of tones.

Helmholtz's interest in tones included his experiments showing that the contraction of a muscle itself produced a "tone." In a report to the Academy in 1864,[22] he wrote:

> The well-known, but often doubted muscle sound can be heard very clearly and under circumstances in which there is definitely no rubbing of the ear or of the stethoscope against

[21]H. von Helmholtz. *Die Lehre von den Tonempfindungen als physiologische Grundlage für die Theorie der Musik.* 1st ed. Braunschweig, 1863. (On the Sensations of Tone as a Physiological Basis for the Theory of Music.)

[22]H. von Helmholtz. Über das Muskelgeräusch. *Arch. Anat. Physiol. Med.,* 766–768 (1864). (On Muscle Sounds.)

the skin covering the muscle. It can best be heard in a quiet place, preferably at night, if one stuffs one's ears with wax or wet paper and then contracts the muscles in the head in a violent manner (for example the masticatory muscles). As long as the muscle remains at equal tension, one hears a hollow, roaring sound, whose key-note is not greatly changed by an increase in tension, the roaring sound becoming however louder and higher.

Helmholtz then went on to examine the tone he could hear from a contracting muscle. He reported as follows (in translation):

> . . . I repeated these observations, this time however contracting my muscles not by voluntary action, but rather by use of an induction apparatus with an oscillating spring which could produce up to 130 oscillations of the spring. The induction apparatus was situated in a room separated from me by two closed doors, so that none of its noise could be heard. When I placed the electrodes on my masticatory muscles, and made them contract, I heard the tone of the spring of the induction apparatus. If the oscillation was changed (by my assistant), then I heard the change.
>
> The tone was produced from the contracted muscle, and not from direct effect of the current of the ear. I was able to determine this, since the tone only became audible when the current strength was high enough to produce contraction of the muscle.

It is perhaps of interest that he failed to detect tones from frog muscles.

Helmholtz died in 1894 at the age of 72, before his friend Du Bois-Reymond, who wrote a fine eulogy of him. In this long (50-page) article, Du Bois-Reymond outlined the many major contributions made by the great scientist, starting with his formation of the law of the Conservation of Energy (using, as did Helmholtz, the word "Kraft," though it is "Energy" and not "Force" which is the core of the work). Helmholtz's other contributions were meticulously described as were the colleagues and friends he made. Quoting Du Bois-Reymond's own words (in translation): "He is no more. Nothing is left to us but the questionable comfort of the poet: he was ours. We will never see his like again: yet it is a question whether a figure such as his can ever appear again."[23]

Emil Heinrich Du Bois-Reymond (1818–1896)

The principal attack on Carlo Matteucci came from the scientist who was to dominate the field of electricity of the nervous system in the 19th century—Du Bois-Reymond.

Du Bois-Reymond was the son of a Swiss family that had moved to Berlin; and there, except for a brief period at Bonn, he spent the rest of his long life, almost to the end of the century. Although he took his degrees at the University of Berlin and eventually rose to be one of its most distinguished professors, the short period at the University of Bonn charted his whole career, for there he came under the influence of the great physiologist Johannes Müller. Müller was born in 1801 and died young in 1858, but during his short life his influence on physiology was great. In his youth he was drawn to the *Natürphilosophie* of Schelling, which was the last potent movement of vitalism.

Toward the half-century, a marked swing away from the metaphysics of *Natürphilosophie* characterized neurophysiology. Du Bois-Reymond considered himself (with some right) to be the champion of the movement that strove to explain all physiology on chemical and physical grounds. And, in fact, it was the physicists of this period who were contributing most of the new experiments and concepts of muscle and peripheral nerve action. Before this, neurophysiologists had reached a stage in their work in which progress was hampered by the lack of sufficiently sensitive instruments. The physicists came to their help

---

[23]E. Du Bois-Reymond. Gedächtnisrede auf Hermann von Helmholtz. Akademie der Wissenschaften, Berlin, 1896.

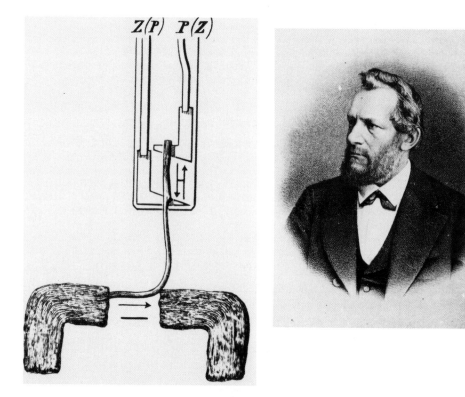

FIG. 38.   Left: The apparatus designed by Du Bois-Reymond to demonstrate the change in flow of current when the nerve is stimulated. The nerve current flows between two damp pads (*below*), and when the nerve is stimulated by electrodes (*above*), the change in current can be observed. (From: *Untersuchungen über thierische Elektricität.* Reimer, Stuttgart, 1838, 1848). **Right:** Emil Du Bois-Reymond (1818–1896) at the height of his experimental period.

and indeed were themselves intrigued by the types of physical phenomena that biological preparations provided.

In 1841 the young Du Bois-Reymond went to Müller who gave him a copy of Matteucci's short essay[24] on animal electricity. The path for his career as a research worker was set (as Müller described in his famous *Handbook*[25]). By November of that year, he had already completed a preliminary note,[26] but his major work, the *Untersuchungen über thierische Elektricität,*[27] did not appear until 1848. (This work remained the master text on the subject until the publication of Biedermann's *Elektrophysiologie* in 1895 from the University of Jena.[28]) The first part of these long and detailed volumes, unlike its later sections, shows little originality in scientific ideas, the author with a chip on his shoulder being carried along in the wake of Matteucci of whose publications he was outspokenly critical.

[24]C. Matteucci. *Essai sur les Phénomènes électriques des Animaux.* Carilian, Goeury et Dalmot, Paris, 1840.

[25]Johannes Müller. *Handbuch der Physiologie des Menschen für Vorlesungen.* 2 vols. Hölscher, Coblenz, 1834–1840. (Handbook of Human Physiology for Lecturers.)

[26]E. Du Bois-Reymond. Vorläufiger Abriss einer Untersuchung über die elektromotorischen Fische. *Ann. Physik. Chem.,* 58:1 (1843). (Preliminary description of an investigation of electromotive fish.)

[27]E. Du Bois-Reymond. *Untersuchungen über thierische Elektricität Reimer.* Berlin, 1848–1884. (Investigation into animal electricity.)

[28]Wilhelm Biedermann (1852–1923). *Elektrophysiologie.* Jena, 1895.

FIG. 39. Two giants in the field of electricity who have left their names on units used. **Left:** Michael Faraday (1791–1867). **Right:** André Marie Ampére (1775–1836). (Said to be a self-portrait.)

However, where Du Bois-Reymond shines, and what makes his work classic, is his skill in instrumentation (Fig. 38), far surpassing that of Matteucci, so that he was able to extend and improve on these earlier observations.

Moreover, not being hampered (as was Matteucci) by residual traces of a belief in "nerve force," he brought clearer inductive reasoning to the interpretation of his observations.

Du Bois-Reymond confirmed Matteucci's demonstration that not only nerve-muscle preparations but muscles themselves could produce electricity.[29] With some acerbity, he claimed priority for naming this the "muscular current" ("Muskelstrom"). Both Matteucci and Du Bois-Reymond distinguished muscle current from the "frog current" ("la correnta propria della rana") so named by Nobili to describe the current flow between the feet of the prepared frog and any other part of the animal. Neither Nobili[30] nor Matteucci nor even Du Bois-Reymond at this time recognized that the so-called frog current was an injury current consequent to their having spinally transected their frogs. Nobili had thought it was a thermoelectric effect due to differential cooling times of nerve and muscle.

Du Bois-Reymond, using faradic stimulation (Fig. 39), also confirmed Matteucci's finding that the muscle current was reduced during tetanic stimulation, and he named this the "negative variation." It is what is now called the action current of muscle. Du Bois-Reymond went on to demonstrate the same negative variation in nerve activity and thus

[29]E. Du Bois-Reymond. Beobachtungen und Versuche an lebend nach Berlin gelangten Zitterwelsen (Malopterurus electricus). Chap. XXVIII. *Gesammelte Abhandlungen zur Allgemeinen Muskel- und Nervenphysik.* Vol. 2. Veit, Leipzig, 1877, pp. 601–647. (Observations and Experiments on Malopterurus Electricus, Brought Live to Berlin.)

[30]L. C. Nobili. Analyse expérimentale et théorique des phénomènes physiologiques produits par l'électricité sur le grenouille; avec un appendice sur la nature du tetanos et de la paralysie, et sur les moyens de traiter ces deux maladies par l'électricité. *Ann. Chem. Phys.,* 44:60–94 (1830).

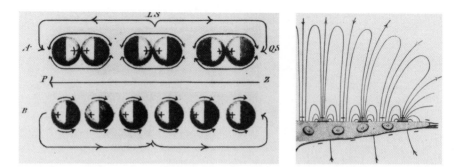

FIG. 40. Sketches designed by Du Bois-Reymond to illustrate his concepts of current flow in nerve. **Left:** Schema denoting change of orientation of molecules during polarization. This schema was to receive criticism from Bernstein. (From: *Untersuchungen über thierische Elektricität.* Vol. 2. Stuttgart, 1849.) **Right:** Flow of currents during activity of nerve. (From: *Gesammelte Abhandlungen zur Allgemeinen Muskel-und Nervenphysik.* Vol. 1. Veit, Leipzig, 1875.)

discovered the action current of nerve which Matteucci had failed to find with his less sensitive instruments. Impressed by the number of workers drawn into the field, Du Bois-Reymond wrote: ''It may be said that wherever frogs were to be found, and where two kinds of metal could be procured everybody was anxious to see the mangled limbs of frogs brought to life in this wonderful way.''

In the light of future discoveries, it is of interest that Du Bois-Reymond thought of looking in the brain for currents similar to those he found in the nerve. He attempted this with frogs' brains, placing one electrode on a cut surface and one on the cortex. It is now known that these are currents of injury, or demarcation potentials, though they are indeed evidence of electricity intrinsic to neuronal tissue. He was clearly unaware that the electric currents of the brain had been discovered in 1875 by the Englishman Richard Caton.[31]

But of those whose writings he read, Du Bois-Reymond was merciless in his criticism, especially of the publications of Matteucci. As late as 1850, a collection of Matteucci's writings was published in the English language by Bence Jones in London.[32] This was immediately attacked by Du Bois-Reymond.[33]

As related in the first volume[34] of the present work, the electric fish had claimed the attention of almost all of the great writers of the past, including Plato, Aristotle, and Pliny—and continues to fascinate biologists to the present day.[35] Du Bois-Reymond was no exception.[36] He had marine torpedoes imported to Berlin, and there he made a long series of experiments (Fig. 40), many of which paralleled Matteucci's, which he had derided.

[31]R. Caton (1842–1926). The electric currents of the brain. *Brit. Med. J.*, 1:278 (1875).

[32]C. Matteucci. Electro-physiological researches. On induced contraction—Ninth Series. *Phil. Trans. Roy. Soc.*, 140:645–649 (1847–1850).

[33]On Signor Carlo Matteucci's Letter to H. Bence Jones, M.D. ERS, etc. Editor of an Abstract of Dr. Du Bois-Reymond's Research in Animal Electricity. Churchill, London, 1858.

[34]See the first volume in this study: M. A. B. Brazier. *A History of Neurophysiology in the 17th and 18th Centuries: From Concept to Experiment.* Raven Press, New York, 1984.

[35]Modern research has revealed that the chemistry of the electric organ of the fish resembles the acetylcholine process of the familiar end-plate of muscle, but, having no output, the current passes in a series into the cells of the organ reaching considerable current of only moderate voltage.

[36]E. Du Bois-Reymond. Beobachtungen und Versuche an lebend nach Berlin gelangten Zitterwelsen (Malopterurus electricus). *Gesammelte Abhandlungen zur Allgemeinen Muskel und Nerven-physik*, Vol. 2. Leipzig, 1877, pp. 601–647. (Observations and experiments on malopterurus electricus brought live to Berlin.)

FIG. 41. Left: Model made by Du Bois-Reymond to illustrate his theory of muscle. He envisaged this as innumerable fine cylinders, segments of which are divided once again into innumerable particles. **Right:** Each of these particles was held to be positive at its equator and negative at the poles. (From: *Untersuchungen über thierische Elektricität.* Reimer, Berlin, 1838–1848.)

Du Bois-Reymond designed some very intricate experiments for measuring the shock but found he had to simplify them because the fish (he reported) failed to lie still. The experiments were then simplified to the one illustrated in Figure 41. In translation his description reads:

> I therefore contented myself with the less rational but much simpler arrangement diagrammatically represented in my Figure 13, which was sufficiently satisfactory. A glass vessel 30 cm. wide and 10 cm. deep was used. On the bottom of it rested a circular zinc plate of about the same width as the fish forming a ventral shield v°v, a portion of which vv′ was bent and hung hook-like over the side of the vessel. One end of the circuit was brought into contact with this hook. There are two handles in the circuit, to one of which Hv, the wire v′H is connected. A circular piece of flannel ff′ was laid on the ventral shield, soaked with sea-water, to prevent the edge of the dorsal shield d°d from touching the ventral. The fish rests on the flannel and is represented in the figure as if in cross-section through the organ. The dorsal shield is an arched zinc plate with the edge turned up, the upper surface of which is lackered, having a wooden knob in the middle through which the lead-off wire d′Hd is conducted insulated to the second handle. I had two shields for Torpedoes of different sizes, the one 22 and the other 18 cm. in diameter. The sea-water in the vessel was in sufficient quantity for its surface just to touch the back of the fish.[37]

The great work for which Du Bois-Reymond became so widely known at a young age, *Untersuchungen über thierische Elektricität,* written while he was in his 30s, has not received a full translation into the' English language; even London's famous physiologist Burdon-Sanderson,[38] who had hired translators of some of Du Bois-Reymond's' articles,

[37]E. Du Bois-Reymond. Beobachtungen und Versuche an lebend nach Berlin gelangten Zitterwelsen (Malopterurus electricus). *Gesammelte Abhandlungen zur allgemeinen Muskel und Nerven-physik,* Vol. 2. Leipzig, 1877, pp. 601–647. (Observations and experiments on malopterurus electricus brought live to Berlin.)

[38]J. Burdon-Sanderson (1828–1905). *Translations of Foreign Biological Memoirs.* Clarendon Press, Oxford, 1887.

did not tackle the big two-volume book. Some parts of the work were translated and incorporated in a book by an American, Charles Edward Morgan,[39] who had visited the laboratory in Berlin. Morgan died in 1868, so his book, published posthumously, covers only the first quarter-century of Du Bois-Reymond's work.

Morgan, a physician with a degree from Columbia, although a minor figure himself, stands out as one of the first scientists from the United States to show interest in the centuries-old field of biological electricity. Apparently, he had visited Du Bois-Reymond and published some studies of his own in the German language,[40] but on returning to his own country, he did not pursue this interest and instead turned to dermatology. Somewhat oddly, therefore, he needs to be recognized as the first from the United States to publish in neurophysiology, as distinct from clinical neurology.

Morgan's book is largely a review of the opinions of Du Bois-Reymond and is of interest in that he quotes (in his own translation) the attacks on Du Bois-Reymond's ideas by others, including Pflüger, as to what quality of the electrical current actually was the stimulus to the nerve or muscle. In Morgan's words:

> The fundamental law of electrical stimulation, according to Du Bois-Reymond, is as follows: "The motor nerve is not stimulated by the absolute value of the current-density at any given moment, but by its variations from one instant to another, and the effect produced by these rapid changes increases with their rapidity and their greatness in a given time.

Also, Morgan noted that Du Bois-Reymond believed that the nerve is not stimulated when the current flows parallel with the axis of the nerve.

Morgan went on to interpret Du Bois-Reymond's views as follows:

> 1. The stimulation of the nerve and induction both obtain only at the instant the circuit is opened or closed, that is whilst the current force rises from and falls back suddenly to zero, and not whilst the circuit is kept closed. 2. There is moreover, no direct and essential relation between the quantity of electricity discharged through the nerve in a given space of time.

Morgan tackled a difficulty raised by du Bois-Reymond's rule, namely:

> . . . Again, it has long been known that sensitive nerves act not only on the circuit being closed or opened but during the entire continuance of the current. This has been explained by saying that though sensitive nerves obey the same general laws as the motor nerves, they have in addition a special action during the continuance of the current. But this explanation becomes unsatisfactory when it is found that motor nerves likewise react under the same circumstances. However, Du Bois-Reymond explains this by saying that exceedingly strong constant currents excite motor nerves by electrifying them; and Eckhard,[41] without giving any direct proof assumes that in such cases the seemingly constant current is really inconstant, owing to the existence of polarization. Nevertheless, Pflüger,[42] using an arrangement by which polarization was avoided and the current maintained absolutely constant, found that the very feeblest currents able to excite contraction (i.e., the closing contraction) do not tetanize; that, however, on increasing the current-strength to about equal that of the muscular current, tetanus soon arises, but quickly ceases on increasing the current-strength beyond a certain point, though it again appears on returning to the feebler

[39]C. E. Morgan (1832–1868). *Electro-physiology and Therapeutics.* Wood, New York, 1868. (Morgan's book is rare, but a copy can be found in the John Crerar Library, Chicago.)

[40]C. E. Morgan. Einige Versuche mit dem Strom des ruhenden Nerven. *Arch. Anat. Physiol. Wiss. Med.,* 338–344 (1863). (Experiments with the Current in Resting Nerve.)

[41]Conrad Eckhard (1822–1915). *Beiträge zur Anatomie u. Physiologie.* Vol 1. Gissen, 1855. p. 41. (Contributions to Anatomy and Physiology.)

[42]E. F. W. Pflüger (1820–1910). Ueber der tentanisirende Wirkung des constanter Ströme der allgemeine Gesetz der Nerven-erregung. *Virchow's Arch.,* 13:437. (The General Law of Nerve Arousal.)

current. Pflüger[43] explains this, and the facts embraced Du Bois-Reymond's law, by the following general law: "The nerve is excited the instant any extraneous force whatever changes with a certain rapidity the molecular condition of the nerve, whereas a stationary condition of the latter is never connected with excitement."

Throughout his long life, Du Bois-Reymond preached what to him was almost a crusade, namely, to convince the world that our bodies work according to physical-mechanical laws. He felt that he himself had established this for the nerves. He declared in print: "If I do not greatly deceive myself, I have succeeded in realizing in full actuality (albeit under a slightly different aspect) the hundred years' dream of physicists and physiologists, to wit, the identity of the nervous principle with electricity."

Du Bois-Reymond had the advantage over Galvani in that, during the intervening years, the nerve cell had been seen and the old concept by which every nerve had to have its direct connection with the brain (in order that nervous fluid might flow down it from the ventricles) had been swept away. This, the hoariest of false hypotheses, had reigned for 1,700 years—such was the strength of Galen's authority. Near the end of the century, when Du Bois-Reymond was to take up the controversy, he wrote:

1. Animals have an electricity peculiar to themselves, which is called animal electricity.
2. The organs to which this animal electricity has the greatest affinity, and in which it is distributed, are the nerves, and the most important organ of its secretion is the brain.
3. The inner substance of the nerve is specially adapted for conducting electricity, while the outer oily layer prevents its dispersal, and permits its accumulation.
4. The receivers of the animal electricity are the muscles, and they are like a Leyden jar, negative on the outside and positive on the inside.
5. The mechanism of motion consists in the discharge of the electric fluid from the inside of the muscle via the nerve to the outside, and this discharge of the muscular Leyden jar furnishes an electrical stimulus to the irritable muscle fibers, which therefore contract.

Du Bois-Reymond's conviction led to his making outspoken attacks on any scientist suspected of some trace of vitalism. In 1877 and 1887, Ernst von Brücke, one of Du Bois-Reymond's favorite colleagues, collected and published two volumes entitled *Reden;*[44] the first volume consists of a collection of Du Bois-Reymond's writings on literature, philosophy, and the history of the times. The second volume covers his biography, his scientific work and his lectures. It is this volume that reproduces the charming picture Du Bois-Reymond had an artist make of Galvani hanging frogs' legs on the balcony of his home.[45] Long after Galvani's death, Du Bois-Reymond had made a pilgrimage to Bologna. He sought out the house where Galvani had lived and made a rough sketch of it, which he later had perfected by an artist. It shows the great man hanging his frogs' legs for stimulation by atmospheric electricity. Galvani and his wife lived on the third story of the house, at the back of which was an addition with a wing built onto it. This had a flat roof with railings surrounding an area. At the bottom of this area was the garden that Galvani described in his *Commentary.* It was on these iron railings that Galvani hung his frogs' legs, and there he noticed their twitchings during a thunderstorm. He did not recognize that the electricity came from the copper hooks swinging against the iron rails during the storm.

It is in this work, *Reden,* that we find Du Bois-Reymond's pungent attack on vitalism.

---

[43]E. F. W. Pflüger. *Untersuchungen über die Physiologie des Elektrotonus.* Hirschwald, Berlin, 1859. (Studies on the Physiology of Electrotonus.)

[44]E. Du Bois-Reymond. *Reden.* Collected by E. von Brücke. 2 vols. Veit, Leipzig, 1886–1887.

[45]See the first volume in this study: M. A. B. Brazier. *A History of Neurophysiology in the 17th and 18th Centuries: From Concept to Experiment.* Raven Press, New York, 1984 (Figure 95).

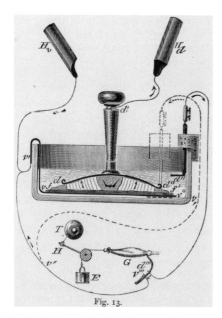

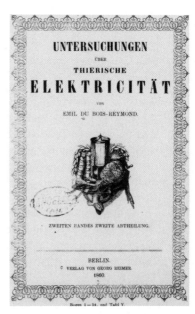

FIG. 42. Left: One of Du Bois-Reymond's experiments on the electricity evoked from an electric fish. **Right:** The frontispiece of the first volume of his classic text *Untersuchungen über thierische Elektricität*. (Note the marine torpedo.)

Opening rather strangely with a German translation of a quotation from Shakespeare,[46] Du Bois-Reymond's attack reads:

> This is perniciousness. For this vitalism is the genial resting place where, according to Kant's statement:[47] "Reason is put to rest upon the bed of obscure qualities." It is the not-to-be-leapt-over wide ditch of which the runner at the race has falsely heard of and which he now imagines to lie behind every hedge, thus becoming morally paralyzed. This ghost must be banished at last. And I don't believe that it will be so difficult to confront those who do not have perception for the drastic effect of such a picture as the above one, with an analysis whose binding power they can hardly escape.
>
> A deficiency in the conception of vitalism is first of all very much on the surface. We have seen above that all motion, that is also the forces which they are supposed to cause, are at the final end divisible into straightlined movements and forces between the presumed particles of matter. This has not been taken into consideration at all with that idea. If, for example, the cut-off leg sprouts again in the salamander, the theory in question plainly sees in this the work of vitalism. It does not consider that the structure which is discussed at this point refers to the movement and suitable order of innumerable cells. . . . Therefore, not ONE vitality must be assumed, if there are to be vitalities, but there ought to be at least innumerable ones.
>
> In one word, the so-called vitalism, of the type as it is currently presumed to be present in all points of the living body is nonsense.

---

[46]*Henry IV*. Act I. Francis Percy is speaking and says:

> And now I will unclasp a secret book
> And to your quick conceiving discontent
> I'll read you matter deep and dangerous,
> As full of peril and adventurous spirit
> As to o'er-walk a current roaring loud
> On the unsteadfast footing of a spear.

[47]Immanuel Kant (1724–1804). *Kritik der reinen Vernunft. 1781 (The Critique of Pure Reason.)*

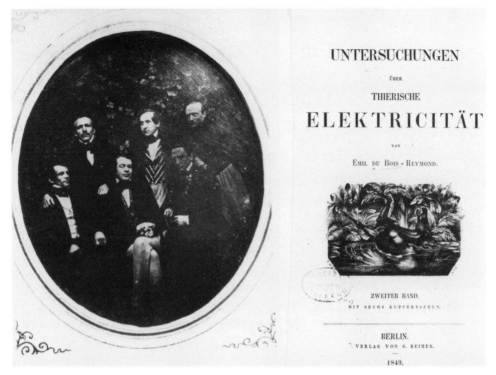

FIG. 43. Left: The group from Du Bois-Reymond's laboratory in 1852. *At the rear:* Du Bois-Reymond, H. Knoblauch, E. Brücke; *in front:* W. Heitz, G. Karsten, W. Beetz (Original photograph is the gift of Brücke's grandson, the late Professor Ernst von Brücke.) **Right:** Title page of the second volume of Du Bois-Reymond's famous book.

Du Bois-Reymond's research was aided by his brilliant development of electrical apparatus. To begin with, he extended Oersted's electromagnetic design of a magnet surrounded by a single coil of wire to a coil of over 2,080 windings. Within a few years, he had built another—this with nearly 4,500 windings. Also, in his search to understand muscle currents, he designed and made models of his concept of current flow in the nerve and muscle, models that later were to come under criticism from his contemporary Julius Bernstein (Figs. 41, 42).

Du Bois-Reymond was a prolific writer, and his style of wording was also prolific, the reason perhaps why so few of his writings have been translated into other languages. He received recognition from his peers (Fig. 43), including election as secretary of the Akademie der Wissenschaften in Berlin, and he was the founder of the Berlin Physical Society.

He lived to a great age (Fig. 44), working persistently through the turbulence of his country in a century that saw the Revolution of 1848 (that brought so many immigrants to the New World) and the Franco-Prussian War in 1870, with the seizure of Alsace. This brought out in Du Bois-Reymond a hatred of the French for what he called ''the rape of Alsace'' and caused him to be sensitive about his French name. He tried to deny his lineage, claiming to be ''a pure Celt.'' Speaking as a Prussian, he gave a long speech at the University of Breslau in 1870 that was published later under the title ''Der deutsche Krieg.''[48] In 1871 came the unification of Germany with the installment of Wilhelm the First as Emperor. Du Bois-Reymond just missed reaching the 20th century, dying in 1896—a year in which his country lost another great German, Johannes Brahms.

[48]E. Du Bois-Reymond. The German war. *Reden.* Vol. 1. Veit, Leipzig, 1886, pp. 65–94.

 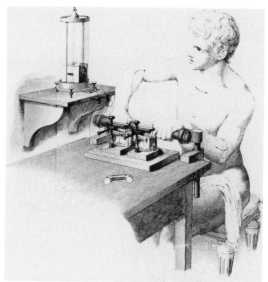

FIG. 44. Left: Emil Du Bois-Reymond in older age. **Right:** One of Du Bois-Reymond's experiments to record muscle contraction in man—what would now be called an electromyogram. (From: K. Rothschuh. *Von Boerhaave bis Berger.* Fischer, Stuttgart, 1964.)

Ernst von Brücke (1819–1892)

One of Du Bois-Reymond's favorite pupils was the young Berliner Ernst von Brücke. Although educated in Berlin and Heidelberg and having briefly served as Professor of Physiology and General Pathology at Königsberg, his name is most often associated with Vienna. In Vienna, to which he moved in 1849, his interest turned away from the type of experiments he had been pursuing in Königsberg; he left nerves and muscles for the physical chemistry of the digestive system. But neurophysiologists remember him for the pioneer studies he made with Müller on the branching of the nerve as it entered the muscle, for the histology of the neuromuscular endings was still not known. They did this work in the eye of the pike.[49]

Under the influence of Du Bois-Reymond, Brücke wrote two studies in electrophysiology, one on electrical stimulation of motor nerves[50] and another on direct stimulation of denervated muscle.[51]

Brücke's position was that all the operations of our bodies were in fact physicochemical actions. In this, of course, he was echoing his teacher, Du Bois-Reymond, and the views of Carl Ludwig, who joined him in Vienna in 1855 (Fig. 45). Du Bois-Reymond commented on this viewpoint as follows: ''Brücke and I, we have both sworn to expose the truth, namely that there are no other forces operating in the organism except those physicochemical ones.''

[49]E. von Brücke and J. Müller. Über die physiologische Bedeutung der stabförmigen Körper und Zwillingszapfen in den Augen der Wirbeltiere. *Handbuch der Physiol.*, 1:414–524 (1884). (On the physiological significance of the rhabdoidal bodies and twin cones in the eyes of vertebrates.)

[50]F. von Brücke. Über die Reizung der Bewegungsnerven durch elektrische Ströme. *Akad. Wiss. Wien.* 58:451–466 (1868). (On the stimulation of motor nerves by electrical current.)

[51]E. von Brücke. Über das Verhalten entnervter Muskeln gegen discontinuierliche elektrische Ströme. *Sb. Abt.* 58:125–128 (1868). (On the behavior of denervated muscles to discontinuous electrical currents.)

FIG. 45. Two of Du Bois-Raymond's favorite students. **Left:** Ernst von Brücke (1819–1892). (Portrait by gift of his grandson, the late Professor E. T. Brücke.) **Right:** Carl Ludwig (1816–1895). From a portrait when young by Krais.

Brücke's interests were extremely wide, and his publications stretched from the single cell to the power of speech in man. An example of the first is a treatise he wrote in 1862 on molecular movements in animal cells. He presented this treatise to a session of the Mathematisch-Natürwissenschaftlichen division of the Academy of Science in Vienna.[52] He also, five years later, read a report on his work on the red cells of the blood.[53]

In contrast to these studies in lower animals, he published in 1856[54] a long treatise on the anatomy and physiology of the mechanism of speech. This was profusely illustrated. After specifying the multiple parts of the mouth, the larynx, and the pharynx, he proceeded to identify how each structure played its part in, for example, vowel sounds. He wrote (in translation):

> I first illustrated the 3 main vowels, a, i, u, and also ü, to show positions between i and u. I have omitted the consonants of the first double row, since they can easily be observed. But I have illustrated the four modifications of the stops of the second row, and two modifications of the stops of the third row. In the same position a small opening will result in

[52]E. von Brücke. Über die sogennante Molekularbewegung in thierischen Zellen, insonderheit in den Speichel-körperchen. *Akad. Wiss. Wien,* 45:629–643 (1862). (On the so-called molecular movement in animal cells, especially in the salivary corpuscles.)

[53]E. von Brücke. Über den Bau der roten Blutkörperchen. *Akad. Wiss. Wien,* 56:79–89 (1867).

[54]E. von Brücke. *Grundzüge der Physiologie und Systematik der Sprachlaute für Linguisten und Taubstummen-lehrer.* Gerold, Vienna, 1856. (Fundamentals of physiology and system of language sounds for linguists and teachers of deaf mutes.)

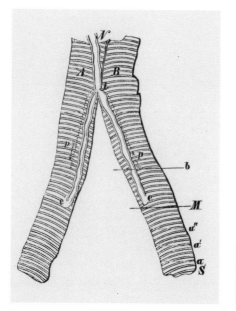 

FIG. 46. **Left:** Kühne's diagram designed to demonstrate two-way stimulation in a nerve-muscle prepa-
ration. (From: Untersuchungen über Bewegungen und Veränderungen der contraktilen Substanzen.
Arch. Anat. Physiol., 564–643, 1859.) (Investigations on movements and changes of contractile tissues.)
Right: Willy Kühne (1837–1900).

the corresponding fricative. I did not illustrate the L-sounds, since they differ only in side
openings from the stops of the second row. I also omitted the voiced sounds, since one
cannot illustrate their main distinguishing feature, vibration. One resonant, (n) is illustrated
to show how it is only distinguished from the corresponding stop by a hanging soft palate.
Finally, one figure is devoted to German ''sch,'' and its two articulation points.''

The last publication of his life was entitled *Beauty and Flaws of the Human Figure.*[55]
This book was (he said) ''written for the artist and for friends of art.'' The book has illustrations, including Botticelli's *Primavera.*

Brücke's distinguished long life is recounted in the biography[56] written by his (also distinguished) grandson, the late Professor von Brücke, holder of the Chair of Biochemistry
at the University of Vienna.

During the height of interest in the spread of excitation in stimulated muscle, one of
Brücke's favorite pupils came into the field. This was Willy Kühne. He had been trained
by Claude Bernard in Paris and with Brücke and Ludwig in Vienna, where he set out to
identify the motor plate in muscle. In an ingenious experiment with two muscle fibers (Fig.
46) served by branches of a nerve (N), he showed that stimulation of one of the fibers
below the entry of the nerve produced contraction of that fiber (B) only, whereas with stimulation of the point where the nerve entered the fiber (i.e., the motor plate), the stimulus
passed up, over, and into the second fiber (A). He was careful to use a chemical stimulus,
not an electric one, which could cause current spread. These experiments led to his identification of myosin.

He published this work in 1859,[57] and in 1868 he moved to Amsterdam, having received

[55]E. von Brücke. *Schönheit und Fehler der menschlichen Gestalt.* Braumüller, Vienna, 1893.

[56]E. T. Brücke. *Ernst Brücke.* Verlag von Julius Springer. Vienna, 1928.

[57]Willy R. Kühne (1837–1900). Untersuchungen über Bewegungen und Veränderungen der contraktilen Substanzen. *Arch. Anat. Physiol.*, 564–643 (1859). (Investigations on movements and changes of contractile tissues.)

the appointment of Professor of Physiology, and later to Heidelberg to follow Helmholtz. This led to his changing his field of research to the visual system that had interested Helmholtz so deeply. In all his work, Kühne explored the chemistry of the system he examined. He died in 1900.

BIBLIOGRAPHY

Hermann von Helmholtz (1821–1894)

Selected Writings

De fabrica systematis nervosi evertebratorum (thesis). Berlin, 1842. (Structure of the nervous system of invertebrates.)

Über den Stoffverbrauch bei der Muskelaction. *Arch. Anat. Physiol. Wiss. Med.*, 375–744 (1845). (On the consummation of matter during muscle movement.)

Über die Erlangung der Kraft. Berlin, 1847. (On the conservation of energy.)

Ueber die Wärmeentwickelung bei Muskelaction. *Arch. Anat. Physiol. Wiss. Med.*, 144–164 (1848). (On the development of heat during muscle action.)

Über die Fortpflanzungsgeschwindigkeit der Nervenreizung. *Arch. Anat. Physiol. Wiss. Med.*, 1–3 (1850). (On the speed of propagation of nerve currents.)

Über die Fortpflanzungsgeschwindigkeit der Nervenreizung. *Verh. Press. Akad. Wiss. Berlin*, 14–15 (1850). (On the speed of propagation of nerve stimulation.)

Vorläufiger Bericht über die Fortpflanzungsgeschwindigkeit der Nervenreizung. *Arch. Anat. Physiol. Wiss. Med.*, 71–73 (1850). (Preliminary report on the speed of propagation of nerve stimulation.)

Messungen über den zeitlichen Verlauf der Zuckung animalischer Muskeln und die Fortpflanzungsgeschwindigkeit der Reizung in den Nerven. *Arch. Anat. Physiol. Wiss. Med.*, 276–364 (1850). (Measurements of the chronological progress of spasms in animal muscles and the speed of conduction of nerve stimulation.)

Bericht über die zur Bekanntmachung geeigneten Verhandlungen der Königl. Pruss Akademie der Wissenschaften zu Berlin aus dem Jahre 1850). (Report on those proceedings of the Royal Prussian Academy of Science in Berlin selected for publication; reprinted in Founders of Experimental Physiology, edited by J. F. Lehmans. Verlag, Munich, 1971.)

Note sur la vitesse de la propagation de l'agent nerveux dans les nerfs rachidiens. *C. R. Acad. Sci.*, 30:204–206 (1850).

Deuxième note sur la vitesse de la propagation de l'agent nerveux. *C. R. Acad. Sci.*, 33:262–265 (1851).

Über die Methoden, kleinste Zeittheile zu messen, und ihre Anwendung für physiologische Zwecke. *Königsberger Naturwissenschaftl. Untersuchung*, 2:169–189 (1851). (On the methods of measuring the smallest units of time, and their use for physiological ends.)

Beschreibung eines Augenspiegels zur Untersuchung der Netzhaut im lebenden Auge. *Arch. Anat. Physiol. Wiss. Med.*, 229–260 (1851). (Description of an ophthalmoscope for the investigation of the retina of the living eye.)

Über den Verlauf und die Dauer der durch Stromesschwankungen inducirten elektrischen Ströme. *Arch. Anat. Physiol. Wiss. Med.*, 429–462; 554–557 (1851). (On the progress and duration of electric currents induced by current variation.)

Messungen über Fortpflanzungsgeschwindigkeit der Reizung in den Nerven, Zweite Reihe. *Arch. Anat. Physiol. Wiss. Med.*, 199–246; 844–861 (1852). (Measurements of the speed of conduction of nerve stimulation.)

Die Resultate der neueren Forschungen über thierische Elektricität. *Arch. Anat. Physiol. Wiss. Med.*, 886–923 (1852). (The results of the latest investigations into animal electricity.)

Ein Theorem über die Vertheilung elektrischer Ströme in körperlichen Leitern. *Arch. Anat. Physiol. Wiss. Med.*, 562–564 (1852). (A theory of the distribution of electrical currents in bodily conductors.)

Über eine neue einfachste Form des Augenspiegels. *Arch. Anat. Physiol. Wiss. Med.*, 261–279 (1852). (On a new simple form of the ophthalmoscope.)

Über die Theorie der zusammengesetzten Farben. *Arch. Anat. Physiol. Wiss. Med.*, 461–482 (1852). (On the theory of composite colors.)

Über die Geschwindigkeit einiger Vorgänge in Muskeln und Nerven. *Arch. Anat. Physiol. Wiss. Med.*, 881–885 (1854). (On the speed of certain processes in muscle and nerve.)

Über die Accommodation des Auges. *Arch. Anat. Physiol. Wiss. Med.*, 283–345 (1855). (On the accommodation of the eye.)

Handbuch der physiologischen Optik. Part II. Leipzig, 1860. (Handbook of Physiological Opticism, Part II.)

The Application of the Law of the Conservation of Force to Organic Nature. *Proc. R. Inst.*, 3:347–357 (1861).

Über das Muskelgeräusch. *Arch. Anat. Physiol. Wiss. Med.*, 25–43 (1864). (On muscle sound.)

Lectures on the Conservation of Energy, delivered at the Royal Institution. *Med. Times Gaz.*, 1:385–388, 415–418, 443–446, 471–474, 499–501, 527–530 (1864).

Die Lehre von den Tonempfindungen als physiologische Grundlage für die Theorie der Musik. 1st ed. Braunschweig, 1863; 2nd ed. 1877. (The theory of perception of sound as physiological basis for a theory of music.)

Über den Muskelton. *Arch. Anat. Physiol. Wiss. Med.*, 2:928–931 (1866). (On muscle tone.)

Versuche über Fortpflanzungsgeschwindigkeit der Reizung in den motorischen Nerven des Menschen (zusammen mit N. Baxt). *Monatsber. König. Press Akad. Wiss. Berlin*, 228–235 (1867). (Experiments into the speed of transmission of stimulus in human motor nerves [together with N. Baxt].)

Über die Zeit, welche nöthig ist, damit ein Gesichtseindruck zum Bewusstsein kommt. *Monatsber. Königl. Press. Akad. Wiss. Berlin*, 333–337 (1871). (On the time which is necessary for a visual impression to enter consciousness.)

On the later views of the connection of electricity and magnetism. Annual Report of the Smithsonian Institution for 1873. Washington, D.C., 1874, pp. 247–253.

Review of Lord Rayleigh's theory of sound. *Nature* (London), 17:237–239; 19:117–118, 1878.

The modern development of Faraday's conception of electricity. *Arch. Anat. Physiol. Wiss. Med.*, 52–87 (1881); 2:249–291, 407–410 (1881).

Note on stereoscopic vision. *Phil. Mag.* 5th series. 11:407–408 (1881).

Bestimmung magnetischer Momente mit der Waage. Akademie der Wissenschaften. Berlin, 1883. (The determination of magnetic moments with a weighing scale.)

Zur Thermodynamik chemischer Vorgänge Folgerungen die galvanische Polarisation betreffend. Akademie der Wissenschaften. Berlin, 1883. (On the thermodynamics of chemical processes. Conclusions concerning galvanic polarization.)

Vorträge und Reden. 2 vols. Braunschweig, 1892. (Lectures and Speeches.)

Goethe's Vorahnungen kommender naturwissenschaftlicher Ideen. *Vortrage und Reden*, Vol. 2. 1892, pp. 335–361. (Goethe's anticipation of subsequent scientific ideas.)

Addresse an Hrn. E. Du Bois-Reymond bei Gelegenheit seines 50 Jährigen Doctorjubiläums, verfasst im Auftrage der Königl. Akademie der Wissenschaften zu Berlin. Berliner Sitzungsberichte. February 16, 1893, pp. 93–97. (Addresses given to Emil Du Bois-Reymond, on the occasion of his 50th anniversary as Doctor, given on behalf of the Royal Academy of Science in Berlin.)

Handbuch der physiologischen Optik, 2nd ed. Section 8. Hamburg Nachtrag zu dem Aufsatze: Über das Princip der kleinsten Wirkung in der Elektrodynamik. *Wiss. Abhand.*, 3:596–603 (1894). (Handbook of Physiological Optics. Addendum to the Essay. On the Principle of the Smallest Effect in Electrodynamics.)

Suggested Readings

Davis, J. The conduction velocity of nerve. In: *Founders of Experimental Physiology*. (English translation by J. W. Boylan.) Lehmanns, Munich, 1971, pp. 159–168.

Geschichte der Physiologie. Springer-Verlag, Berlin, Göttingen, Heidelberg, 1953.

Henatsch, H-D. Allgemeine Elektrophysiologie der erregbaren Structuren. In: *Lehrbuch der Physiologies des Menschen*. Landois-Roemann, Munich, Berlin, 1962.

Kahl, R. *Selected Writings of Hermann von Helmholtz*. Wesleyan University Press, Middletown, Conn., 1971.

Kline, M. *Popular Scientific Lectures by Hermann von Helmholtz*. Dover Publications, New York, 1962.

Königsberger, Leo. *Hermann v. Helmholtz*. Braunschweig, 1902–1903.

Königsberger, L. *Hermann von Helmholtz*. English translation by F. A. Welby with preface by Lord Kelvin. Dover Publications, New York, 1965.

McKendrick, J. G. *Hermann Ludwig Ferdinand von Helmholtz*. Longman, London, 1899.

Pouillet, C. S. M. *Comp. Rend. Acad. Sci.*, 19:1384–1389 (1844).

Rothschuh, K. E. *Entwicklungsgeschichte physiologischer Probleme in Tabellenform*. Munich, Berlin, 1952.

Rothschuh, K. E. *History of Physiology*. Kriger, New York, 1973.

Emil Du Bois-Reymond (1818–1896)

Selected Writings

Quae apud veteres de piscibus electricis exstant argumenta. Rypis Nietackinis, Berlin, 1843.

Vorläufiger Abriss einer Untersuchung über den sogenannten Froschstrom und über die elektromotorischen Fische. *Ann. Phys. Chem.*, 58:1–30 (1843). (Preliminary draft of an investigation into the so-called frog current and into the electric fish.)

Untersuchungen über thierische Elektricität, Vol. 1. Reimer, Berlin, 1848; Vol. 2, Part 1. Reimer, Berlin, 1849; Vol. 2, Part 2. Reimer, Berlin, 1860–1884. (Investigations into animal electricity.)

Sur la loi courant musculaire, et sur la modification qu'eprouve cette lois par l'effet de la contraction. *Ann. Chim. Phys.*, 3rd ser., 30:119–127 (1850).

On Signor Carlo Matteucci's letter to H. Bence Jones, editor of an abstract of Dr. Du Bois-Reymond's researches in animal electricity. Churchill, London, 1853.

De fibrae muscularis reactione ut chemicis visa est acida. Reimer, Berlin, 1859.

Über das Gesetz des Muskelstromes, mit besonderer Berücksichtigung des n. Gastroknemius des Frosches. Hofbuchdrucker, Berlin, 1863. (On the law of muscle current, with especial consideration of the gastrocnemius muscle of the frog.)

Ueber den Einfluss körperlicher Nebenleitungen auf den Strom des M. gastrokneimus des Frosches. *Arch. Anat. Physiol. Wiss. Med.*, 561–607 (1871). (On the influence of body secondary connections on the current of the M. gastrocnemius of the frog.)

Anleitung zum Gebrauch des runden Compensators. *Arch. Anat. Physiol. Wiss. Med.*, 608–618 (1871). (Instructions for use of the round compensator.)

Fortgesetzte Beschreibung neuer Vorrichtungen für Zwecke der allgemeinen Nerven- und Muskelphysik. Chapter 11. In: *Gesammelte Abhandlungen zur Allgemeinen Muskel- und Nervenphysik*, Vol. 1. Veit, Leipzig, 1875. (Continued description of new apparatus for uses in general nerve and muscle physics. In: *Collected Papers on General Muscle and Nerve Physics.*)

Beobachtungen und Versuche an lebend nach Berlin gelangten Zitterwelsen (Malopterurus electricus). Chapter 28. Gesammelte Abhandlungen zur Allgemeinen Muskel- und Nervenphysik, Vol. 1. Veit, Leipzig, 1875. (Observations and experiments on electric catfish brought live to Berlin. In: *Collected Papers on General Muscle and Nerve Physics.*)

Ueber die negative Schwankung des Muskelstroms bei der Zusammenziehung. *Arch. Anat. Physiol. Wiss. Med.*, 342–381 (1876). (On the negative variation of muscle current during contraction.)

Beobachtungen und Versuche an lebend nach Berlin gelangten Zitterwelsen (Malopterurus electricus). Chapt. 28. In: *Gesammelte Abhandlungen zur Allgemeinen Muskel- und Nervenphysik, Vol. 2.* Veit, Leipzig, 1877. (Observations and experiments on electric catfish brought live to Berlin. Chapter 28, [Malopterurus Electricus] In: *Collected Papers on General Muscle and Nerve Physics.*)

Experimentalkritik der Entladungshypothese über die Wirkung von Nerv auf Muskel. Chapter 31. In: *Gesammelte Abhandlungen zur Allgemeinen Muskel- und Nervenphysik, Vol. 2.* Veit, Leipzig, 1877. (Experimental critique of the discharge hypothesis of the effect of nerve on muscle. In: *Collected Papers on General Muscle and Nerve Physics.*)

Der physiologische Unterricht sonst und Jetzt. Rede bei Eröffnung des neuen physiologischen Instituts der Königl. Friedrich-Wilhelms Universität zu Berlin am 6. November 1877. Hirschwald, Berlin, 1878. (Physiological education then and now. Speech at the opening of the new physiological institute of the Royal Friedrich-Wilhelm University in Berlin on 6 November 1877.)

Über secondär-elektromotorische Erscheinungen an Muskeln, Nerven und elektrischen Organ. *Sitzungsberichte der Königlichpreussischen Academie der Wissenschaften zu Berlin,* Vol. 16. 1883. (On secondary electromotive phenomena in muscles, nerves, and electric organs.)

Living Torpedos in Berlin. *Sitzungsberichte der Königlich Preussisschen Akademie der Wissenschaften.* Berlin, March 13, 1883.

Ueber secundär-elektromotorische Erscheinungen an Muskeln, Nerven und elektrischen Organen. *Arch. Anat. Physiol. Wiss. Med.*, 1–63 (1884). (On secondary electromotor phenomena in muscles, nerves, and electric organs.)

Reden von Emil Du Bois-Reymond. Literatur, Philosophie, Zeitsgeschichte. Collected by Ernst von Brücke. Veit, Leipzig, 1886, 1887. (Speeches of Emil Du Bois-Reymond. Literature, Philosophy, History.)

Eine Wirkung galvanischer Ströme auf Organismen. *Pflüger's Arch. f. d. ges. Physiol.*, 37:457–460 (1885). (An effect of galvanic currents on organisms.)

Ergebnisse einiger in Dissertationen veröffentlichter Untersuchungen. *Pflüger's Arch. f. d. ges. Physiol.*, 37:460–478 (1885). (Results of experiments published in several dissertations.)

Ueber die Ursache des Electrotonous. *Pflüger's Arch. f. d. ges. Physiol.*, 38:153–181 (1886). On the cause of electrotonus.)

Über indirekte Muskelreizung durch Kondensator-entladungen. *Pflüger's Arch. f. d. ges. Physiol.*, 111:537–566 (1906). (On indirect muscle stimulation by condenser discharges.)

Suggested Readings

Bence Jones, H. (Ed.). *On Animal Electricity: Being an Abstract of the Discoveries of Emil Du Bois-Reymond.* Churchill, London, 1852.

Borruttan, Heinrich Johannes. *Emil Du Bois-Reymond.* Springer, Vienna, 1922.

Du Bois-Reymond, Estelle (Ed.). *Zwei Grosse Naturforscher des 19 Jahrhunderts. Ein Briefwechsel zwischen Emil du Bois-Reymond und Karl Ludwig.* Johann Ambrosius Barth, Leipzig, 1927.

Two Great Scientists of the Nineteenth Century: Correspondence of Emil Du Bois-Reymond and Carl Ludwig. Johns Hopkins University Press, Baltimore, 1982.

Rothschuh, K. E. (Ed.). Emil Du Bois-Reymond (1818–1896) und die Elektrophysiologie der Nerven. In: *Von Boerhaave bis Berger.* Gustav Fischer, Stuttgart, 1964.

Jugendbriefe, von Emil Du Bois-Reymond an Eduard Hallmann; zu seinem hundertsten Geburtstag den 7 November 1918, herausgegeben von Estelle Du Bois-Reymond. Reimer, Berlin, 1918. (Letters from the Young Emil Du Bois-Reymond to Eduard Hallmann; on his Hundredth Birthday 7 November 1918, edited by Estelle Du Bois-Reymond.)

Ernst von Brücke (1819–1892)

Selected Writings

Dissertation: "De diffusione humorum per septa mortua et viva" (Als Selbstreferat unter dem Titel "Beiträge zur Lehre von Diffusion tropfbarer flüssiger Körper durch poröse Scheidewände" *Poggendorfer Ann.,* 58:77–94 (1843). (Dissertation, published as report. Contributions to the theory of the diffusion of liquid bodies of dripping consistency through porous diaphragms.)

Anatomische Untersuchungen über die sogenannten leuchtenden Augen bei den Wirbeltieren. *Arch. Anat. Physiol. Wiss. Med.,* 387–406 (1845). (Anatomical investigations of the so-called luminous eyes in vertebrates.)

Anatomische Beschreibung des menschlichen Augapfels. Reimer, Berlin, 1847. (Dedicated to Johannes Müller.) (Anatomical Description of the Human Eyeball.)

Grundzüge der Physiologie und Systematik der Sprachlaute für Linguisten und Taubstummenlehrer. Gerold, Vienna, 1856. (Fundamentals of Physiology and System of Language Sounds for Linguists and Teachers of Deaf Mutes.)

Über die sogenannte Molekularbewegung in thierischen Zellen, insonderheit in den Speichlkörperchen. *Akad. Wiss. Wien.,* 45:629–643 (1862). (On the so-called molecular movement in animal cells, especially in the salivary corpuscles.)

Über den Bau der roten Blutkörperchen. *Akad. Wiss. Wien,* 56:79–89 (1867). (On the structure of red blood corpuscles.)

Über die Reizung der Bewegungsnerven durch elektrische Ströme. *Akad. Wiss. Wien,* 58:451–466 (1868). (On the stimulation of nerves of movement by electric current.)

Über das Verhalten entnervter Muskeln gegen discontinuierliche elektrische Ströme. *Sb. Abt.,* 58:125–128 (1868). (On the behavior of muscles having had their nerves removed to discontinuous electrical currents.)

Über eine neue Methode, Dextrin und Glykogen aus thierischen Flüssigkeiten und Geweben abzuschieden und über einige damit erlangte Resultate. *Sb. Abt.,* 63:214, 222 (1871). (On a new method of extracting dextrin and glycogen from animal liquids and tissues, and some of the results.)

Grundzüge der Physiologie und Systematik der Sprachlaute für Linguisten und Taubstummenlehrer. 2nd. ed. Gerold, Vienna, 1876. (Fundamentals of Physiology and a System of Language Sounds for Linguists and Teachers of Deaf Mutes.)

Schönheit und Fehler der menschlichen Gestalt. Braumüller, Vienna, 1893. (Beauty and Flaws of the Human Form.)

CHAPTER VII

The Triumph of Electrophysiology

Julius Bernstein (1839–1917)

A younger scientist followed Helmholtz and Du Bois-Reymond, those giants of neuro-physiological electricity; this was Julius Bernstein, a German physicist in the same tradition, whose work was to stretch over to the 20th century.

A pupil of Du Bois-Reymond and of Helmholtz, he was a Berliner, born in 1839, and initially educated under Heidenhain[1] at Breslau. At that time Breslau was the capital of lower Silesia but was transferred by treaty in 1945 to Poland, resuming its original name of Wroclaw. In 1860 Bernstein joined the active group at the University of Berlin and, after working there for four years, followed Helmholtz to Heidelberg. On receiving in 1872 an invitation to succeed Goltz[2] as Professor at Halle, he accepted this position and remained for the rest of his working life at this university (now named for Martin Luther).

Bernstein, like the workers before him (Galvani, Matteucci, Du Bois-Reymond), was drawn to the electric fish for his first studies on animal electricity, and we find again the familiar illustration of the marine torpedo but he was soon to move to the frog. Bernstein's ingenuity with instrumentation led to his invention of the rheotome, a device enabling the superposition in time of many nerve signals—signals which alone were difficult to record with the slow galvanometers of his day. The instrument consisted of a commutating switch which controlled the interval between the electrical signal of the stimulator and the onset of the galvanometer recording. Many responses could thus be superposed. In the following decade, and from France, was to come the invention of the capillary electrometer—the design of Marey.[3] Bernstein's expertise with instrumentation led to the publication in 1871 of

---

[1]Rudolf Heidenhain (1834–1897).

[2]Friedrich Leopold Goltz (1834–1902).

[3]Étienne Jules Marey (1830–1904). Les variations électriques des muscles et du coeur en particulier etudiées au moyen de l'électromètre de M. Lippmann. *C. R. Acad. Sci.*, 82:975–977 (1876).

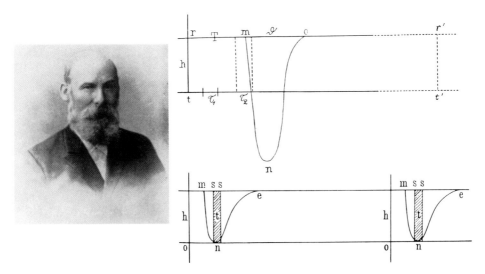

FIG. 47. Left: Julius Bernstein (1839–1917). **Right:** The overshoot of the response in nerve to an electric stimulus (*top*). Bernstein's failure to find an overshoot in muscle (*bottom*). (From: J. Bernstein. *Untersuchungen über den Erregungsvorgang im Nerven und Muskelsystem.* Winter, Heidelberg, 1871.)

a small book[4] (dedicated to Helmholtz and to his [then] friend Ludimar Hermann) describing the improvements in instrumentation he had made while at Heidelberg. Among Bernstein's researches of most interest to neurophysiologists were those on the nerve and muscle. He confirmed Helmholtz's figure for the speed of conduction in nerve and went on to establish the speed of conduction in the muscle fiber. His knowledge of biophysics led him to formulate a membrane theory that is essentially the foundation of the modern ionic theory, which is based on the difference in ionic content across the nerve membrane. Bernstein held that the leading potential of nerve was maintained by the difference in concentration of potassium ions on either side of a membrane selectively impermeable to anions. According to such an hypothesis, the membrane potential would be proportional to the logarithm of the ratio of the internal to the external concentration of potassium as expressed by the Nernst equation.[5]

It was in his work on the frog nerve that Bernstein discovered the overshoot, namely, that after responding to a stimulus with an action spike, the potential does not immediately return to the baseline but overshoots it (Fig. 47). Bernstein's own statement[6] reads (in translation):

> These experiments give therefore sufficient evidence that the growth of the negative variation does not reach its limit at the point at which the current becomes zero. On the contrary, during stronger stimuli the strength of the negative variation can become larger than that of the nerve current, and even exceed it by many times.

[4]J. Bernstein. *Untersuchungen über den Erregungsvorgang im Nerven und Muskelsystem.* Winter, Heidelberg, 1871. (Investigations into the Stimulation Process in Nerve and Muscle.)

[5]The Nernst equation is: $E = 0.058 \log \dfrac{K_2}{K_1}$ V, where E = potential difference, K_1 = potassium concentration inside the fiber, K_2 = potassium concentration of surrounding medium, and 0.058 = the factor of proportionality at 18°C (frog). Although many scientists have evidence suggesting a linear relationship between E and $\log K_2$ K_1, the full calculated value for E has not been found experimentally for the nerve membrane.

[6]J. Bernstein. *Über den zeitlichen Verlauf der negativen Schwankung des Nervenströmes. Pflüger's Arch. f. d. ges. Physiol.,* 1:180–207 (1868). (On the chronological progress of negative variation of the nerve current.)

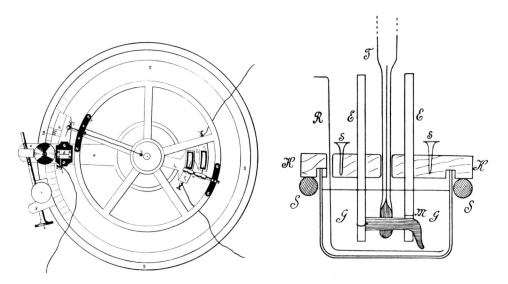

FIG. 48. Two examples of apparatus used by Bernstein. **Left:** The differential rheotome. The electrical stimulus was delivered by one pair of contacts; the second pair was connected to a galvanometer each time the rheotome turned; the time interval between stimulus and response was set by the investigator. (From: *Untersuchungen über den Erregungsvorgang im Nerven und Muskelsystem.* Winter, Hieldelberg, 1871.) **Right:** Apparatus for testing the effect of temperature on the response of muscle (Untersuchungen der Thermodynamik der bioelektrischen Ströme. *Pflüger's Arch. f. d. ges. Physiol.,* 92:521–562, 1902.)

Ludimar Hermann, another worker in Du Bois-Reymond's group, had failed to find the overshoot in nerve. His publication of this negative report caused so much dissension that he was forced to leave Berlin. Later, however, when working at Zurich, Hermann[7] was able to record the overshoot in nerve, and he immediately published this positive result. Bernstein failed to find this phenomenon in muscle (see Fig. 47). This difference between the nerve fiber and the muscle fiber was disturbing to his ionic hypothesis. Some years later he commented in a footnote as follows:[8]

> Apparently contradicting this, the negative variation in nerves can be stronger than the nerve current. However, for the muscles it is certain that the current can only be neutralized by the negative variation, and, since this cannot be a coincidence, I would like to postpone my observation of the nerve until my calculation of the electrotonus, on which, however, further knowledge is necessary.

Bernstein's research is remembered best in his famous book *Elektrophysiologie,*[9] published early in the 20th century. The book became the standard text for students, especially those of the German schools. But nerve and muscle had not been Bernstein's only interest; he published a small book in English, entitled *The Five Senses of Man,*[10] which was a profusely illustrated text with woodcuts, including his design for a rheotome (Fig. 48). Having

[7]H. Hermann. Untersuchungen über die Actionsströme des Nerven II. *Pflüger's Arch. f. d. ges. Physiol.,* 18:574–586 (1879). (Studies on the action currents of nerve.)

[8]J. Bernstein. Über den Elektrotonus und die innere Mechanik des Nerven. *Pflüger's Arch. f. d. ges. Physiol.,* 8:40–60 (1874). (On electrotonus and the inner mechanics of the nerves.)

[9]J. Bernstein. *Elektrophysiologie.* Brunschweig, 1912.

[10]King & Co., London, 1876.

outlived the several great men with whom he had worked, Bernstein died in 1917 during a war that crippled research in his field and temporarily silenced Pflüger's *Archiv*. Fine obituaries of Bernstein can be found in Pflüger's *Archiv*[11] and in *Medizinischer Klinik*.[12]

Ludimar Hermann (1838–1914)

Another Berliner and a friend (at first) of Bernstein was Ludimar Hermann, his elder by a year. Unlike Bernstein, Hermann remained in Berlin in the laboratory of Müller until the latter's retirement, and then he stayed on with Du Bois-Reymond who succeeded Müller in the Chair. With this background, it is not surprising that Hermann wrote his thesis on the tonus of skeletal muscle. A grave disagreement, however, led to Hermann's leaving Du Bois and moving in 1868 to Zürich from whence most of his publications came until 1884, when he moved to Königsberg (whose university was made so famous by Helmholtz). Like the rest of the pupils from Du Bois-Reymond's school, he continued to work on nerve and muscle (Fig. 49), but he also began research on acoustics, no doubt following the leadership of Helmholtz. His scientific work suffered a short break when Prussia attacked Denmark and seized Schleswig Holstein. Hermann at this time served as an army doctor.

It is clear that Hermann felt so strongly about experimental work that he quarrelled bitterly with those with whom he did not agree. This included Bernstein on the theory of electrotonus, as well as Fleischl von Marxow[13] and Du Bois-Reymond for his views on after-potentials. Hermann wrote a long article[14] attempting to destroy Du Bois-Reymond's theories. This was published in Pflüger's *Archiv* after Hermann had left Berlin. Du Bois-Reymond's views of the origin of injury currents of resting muscles were that in resting muscles currents are flowing at all times. Injury merely augments them. His concept was that there were present, in both muscle and nerve, electromotive molecules with clearly defined positive and negative surfaces, the middle of the nerve or muscle being positive to the two ends. He held that this inner "resting" potential was the source of the flow of current when the fiber was injured. This he called the negative potential, or current of injury.

This was the root of the long polemic written by Hermann in 1884.[15] In this he maintained that the resting nerve or muscle was isoelectric and that Du Bois-Reymond's results were due to inadequate care in preparing his experiments to prevent minor injury,[16] for Hermann held that the so-called demarcation current came only from damaged tissue. He made

[11]A. V. Tschermak. Julius Bernstein's Lebensarbeit. Zugleich ein Beitrag zur Geschichte der neueren Biophysik. *Pflüger's Arch. f. d. ges. Physiol.*, 1–89 (1919). (Julius Bernstein's life work. Being at the same time a contribution to the history of modern biophysics.)

[12]E. Abderhalden. Dem Andenken von Julius Bernstein gewidmet. *Med. Klinik*, 9:(1917). (In memory of Julius Bernstein.)

[13]L. Hermann. Ueber E. v. Fleischl's zweite vermeintliche Widerlegung meiner Theorie des Electrotonus. (Aus dem physiologischen Laboratorium in Zürich.) *Pflüger's Arch. f. d. ges. Physiol.*, 20:388–394 (1879). (On E. v. Fleischl's second supposed refutation of my theory of electrotonus. [From the physiological laboratory in Zürich.])

[14]L. Hermann. Über sogenannte secundär-electromotorische Erscheinungen an Muskeln und Nerven. *Pflüger's Arch. f. d. ges. Physiol.*, 33:103–168 (1884). (The so-called secondary electromotive phenomena of muscle and nerve.)

[15]L. Hermann. Physiologie der Bewegungasapparate. I. Allgemeine Muskelphysik. In: *Handbuch der Physiologie*, Vol. 1, 1870, pp. 1–260. (Physiology of the organs of movement.)

[16]L. Hermann. Über sogenannte secundär-electromotorische Erscheinungen an Muskeln und Nerven. *Pflüger's Arch. f. d. ges. Physiol.*, 33:103–162 (1884). (On so-called secondary electromotor phenomena in muscles and nerves.)

FIG. 49. Above: Ludimar Hermann (1838–1914) at the age of 70. A pupil of Du Bois-Reymond and Professor at the University of Königsberg. (Photograph the gift of Professor H. Lullies, Physiologische Institut der Universität, Kiel.) **Below:** His sketch of his concept of current flow in excited nerve. (From: L. Hermann. Algemeine Nervenphysiologie. *Handbuch der Physiologie,* Vol. 2. Erster Theil, Leipzig, 1879.)

exactly timed experiments from which he claimed that the current did not flow immediately the nerve was cut but developed only as the tissue disintegrated.

The most intensive attack that Hermann made was in 1874 on Bernstein. This led to a long altercation. Hermann had easy access to Pflüger's *Archiv* and used this channel to launch his attacks.[17] The argument was over electrotonus and comparative roles of catelec-

[17]L. Hermann. Experimentelles und Kritisches über Electrotonus. *Pflüger's Arch. f. d. ges. Physiol.,* 8:258–275 (1874). (Experimental and critical matters concerning electrotonus.)

trotonus and anelectrotonus.[18] He was particularly critical of results published by Bernstein[19] earlier in the same year. Hermann wrote:

> The results of [Bernstein's] experiments are as follows: the maximum arousal contraction which can be achieved by a strong stimulus is increased by anelectrotonus at the point of stimulation and is decreased by catelectrotonus.
>
> This experiment, which does not seem to be compatible with my ideas on the influence of electrotonus on the arousal results, seemed to me to be very suspicious. As I have already pointed out (Pflüger's, 7, p. 326), and as Bernstein himself mentions (p. 46), the nerve response can, during increased stimulation, climb far above the value necessary to evoke a maximal muscle contraction. It therefore seemed very unlikely that the contractions produced by maximal nerve stimulation can be increased by changes in the point of stimulation; even weakening seemed very unlikely. I therefore proceeded to repeat these experiments with weak currents. The usual phenomena of electrotonus appeared, but with strong currents the polarizing current had absolutely no effect on the height of the contraction. If one moves from weak to strong stimuli, then one sees the effect of catelectrotonus disappear. With somewhat stronger stimuli the effects on anelectrotonus also disappear. However, there is never a reversal of effects. Even when I started with strong stimuli and proceeded to weaker ones, I was never able to obtain the results reported by Bernstein. . . . I must therefore throw doubt on Bernstein's theory.[20]

Bernstein did not, of course, leave this attack unanswered, and in the same journal in the same year he published a rebuttal.[21] Referring to his paper published earlier in the same year, he commented:[22]

> In an article published in this journal (8, 1874, p. 40), I remarked in an appendix that I could not agree with Hermann's theory on the negative variation of the nerve in electrotonic condition. This remark, which was totally justified, since Hermann had formulated his theory on the basis of my own experiments, has been the cause of a polemic against me by Hermann. However, his polemic does not deal mainly with my remark, but immediately goes into the offensive, questioning the veracity of an observation made during my work.
>
> This is my reply.
>
> The experiment I set up led to the result that in anelectrotonic condition of the nerve the maximum arousal released by sufficiently strong stimuli rises, while in the catelectrotonic condition it sinks. Hermann claims I was led by theoretical argument to this experiment. This is not the case, I was guided by facts alone. I proceeded from the fact that the negative variation produced by sufficiently strong stimuli is increased in the anelectrotonic condition, and decreased in the catelectrotonic condition. Without thinking out any theory, I then asked myself how the muscle contraction would occur if the nerve were aroused in the

[18]The effect of switching on a nonphasic current (a rectangular pulse) has the effect of exciting at the cathode. This causes a depolarization of the membrane, i. e., a decrease in negativity inside the membrane at the cathode and the region surrounding it (catelectrotonus), and a simultaneous increase in the internal positivity at the anodal region (anelectrotonus). On the break of the current, this positive charge at the anode collapses, and the membrane potential at the region returns to its resting level. This local reversal of charge is equivalent to a depolarization and accounts for the excitation at the anode on the break of the current. Especially in the central nervous system and in experiments *in vivo,* care is taken to avoid the duration of depolarization being long enough to cause injury.

[19]J. Bernstein. Über den Elektrotonus und die innere Mechanik des Nerven. *Pflüger's Arch. f. d. ges. Physiol.,* 8:40–60 (1874). (On electrotonus and the inner mechanics of the nerve.)

[20]Francis Gotch. In: *Translations of Foreign Biological Memoirs,* edited by J. Burdon-Sanderson and Henry Frowde. Oxford University Press, 1877.

The arguments of Hermann and of Bernstein are reported here at some length because they were available only in the German language.

[21]J. Bernstein. Über den Elektrotonus und die innere Mechanik des Nerven. *Pflüger's Arch. f. d. ges. Physiol.,* 8:40–60 (1874). (On the electrotonus and the inner mechanics of the nerve.)

[22]J. Bernstein. Über Electrotonus. *Pflüger's Arch. f. d. ges. Physiol.,* 8:498–505 (1874). (On electrotonus.)

anelectrotonic and catelectrotonic conditions by sufficiently strong stimuli. I arrived at the above result, in which I saw a new correlation between the behavior of the negative variation and the contraction releasing procedure. My experiment is therefore not at all ad hoc, in order to confirm a previously formulated theory—something which Hermann has done in his recent work.

According to Hermann the results observed by me are improbable since ''the contractions caused by a maximum nerve stimulus, which are certainly the maximum contractions possible, cannot be strengthened by changes in the point of stimulation.

Is it possible, I ask, that a physiologist, ''who,'' as he himself says ''has all the necessary equipment for the experiments in question'' should have completely failed to notice that during tetanic stimulation a muscle is capable of much stronger contraction than is produced by the strongest single induction surge? This happens, as we learned from Helmholtz, through the summation of maximum arousals, by which two arousals following closely one on the other will produce a greater contraction than a single one.

Needless to say, as argumentative a person as Hermann was, he did not let the issue lie there. In the next volume of Pflüger's *Archiv,*[23] he responded with vigor as follows:

The readers of this archive will have read with amazement the article full of personal attacks by J. Bernstein, in which he answers my calm and purely scientific reply to his latest work. Herr Bernstein would have done better to have repeated his experiments while avoiding the mistakes which I pointed out. He would then have used weaker polarising currents, and have observed the fundamental rule that every polarisation experiment should take place immediately between two directly comparable experiments, so that differences in height lying more or less within the margin of error gain some significance. He has achieved nothing with his scream of indignation that I would think *him* guilty of having current loops.

Hermann was incensed that Bernstein had suggested shortening of the current might explain his (Hermann's) failure to obtain the same results.

Hermann's two later attacks on Bernstein in which he denied the phenomenon of the overshoot were published in 1879[24] and in his *Handbuch*[25] in the same year. He did, however, continue to experiment on the issue, and by 1881[26] he had succeeded in finding the overshoot and the controversy died.

In 1884[27] Hermann launched a polemic against Du Bois-Reymond. Hermann did not mince his words in his attack on Du Bois-Reymond's interpretation of his experiments on the electricity of nerve-muscle preparations. He wrote that his own investigations (in translation):[28]

. . . accomplished the task of explaining the phenomena of intrapolar after-currents as given by Du Bois-Reymond; it has introduced new facts; and further, it has widened our knowledge of the extrapolar after-currents, has extended these to muscle, and fully explained them. . . . In respect to nerve, Du Bois-Reymond's own statements are dubious, but in respect to muscle they are rather decisive, although, strangely enough, the most con-

[23]L. Hermann. Zur Aufklärung und Abwehr. *Pflüger's Arch. f. d. ges. Physiol.,* 9:28–34 (1874). (For enlightenment and defense.)

[24]L. Hermann. Untersuchungen über die Actionsströme des Nerven. *Pflüger's Arch. f. d. ges. Physiol.,* 18:574–586 (1879). (Investigation into the action currents of nerve.)

[25]L. Hermann. Allgemeine Nervenphysiologie. *Handbuch der Physiologie,* Vol. 2, edited by L. Hermann. Erster Theil, Leipzig, 1879, pp. 1–96. (General Neurophysiology.)

[26]L. Hermann. Untersuchung über die Actionsströme des Nerven II. *Pflüger's Arch. f. d. ges. Physiol.,* 24:246–294 (1881). (Investigation into the action currents of nerve, II.)

[27]L. Hermann. Über sogenannte secundär-electromotorische Erscheinungen an Muskeln und Nerven. *Pflüger's Arch. f. d. ges. Physiol.,* 33:103–162 (1884). (On so-called secondary electromotor phenomena in muscle and nerves.)

[28]L. Hermann. *Pflüger's Arch. f. d. ges. Physiol.,* 33:103–162 (1884).

clusive method of experiment has not been employed, viz. that of leading through the entire muscle and leading-off the after-currents from two intrapolar regions in the two muscle-halves at the same time. But, before acquiescing in the far-reaching conclusions which Du Bois-Reymond draws from these phenomena, it might be well to inquire what share the above enunciated simple laws, which have entirely escaped Du Bois-Reymond's notice, may have in the production of the phenomena. . . . Du Bois-Reymond had, at the extremities of both muscles, fibers of which the ends were for the most part injured, as his intimations with regard to the currents of rest show. This circumstance which has so often proved a source of danger, must, in accordance with the above disclosed law, more or less impair the development of the positive after-current, in the case of a terminal (from dead to living) direction of current, so that it would appear to be favoured in the case of a terminal direction of current, i.e., a direction like that of the excitation wave. There is, therefore, no subject in the whole range of electrophysiology in which the connection of the phenomena is more transparent than that of the ''secondary electromotive phenomena,'' which now extend to the extrapolar region. To quote Du Bois-Reymond's own terms: ''curiosity'' as to what ''auxillary hypothesis'' I shall advance to dispose of his polarization-currents, will now be fully satisfied. The question, how far it is true that ''all that I have discovered concerning electrotonus'' has ''been exploded,'' that ''the elucidation of electrotonus must be approached anew,'' and finally, that, I have been ''compelled'' to ''import some changes into the scheme,'' as to this everyone can now judge for himself. Happily none of these suggestions is true. The confusion which the molecular theory caused, in that it overlooked, whilst expatiating, what was most obvious, is luckily done away with now . . . and everything has so turned out as to furnish fresh support to my explanation of the nature of the phenomena of animal electricity. . . . The molecular theory owes its production to the error of fact that uninjured muscles show as strong a muscle-current at their natural cross-section, as injured muscles do at their artificial cross-sections. . . . The molecular theory was especially unhappy in its application—electrotonus. It was incapable of explaining either the galvanic or the excitatory phenomena, although it pretended to explain the former and held out the prospect of explaining the latter, together with the function both of nerve and of muscle. . . . On the one hand the theory claims to have discovered a factual basis for the supposed intrapolar arrangement of molecules (of course, only by disregarding obvious explanations), while on the other it abandons its explanations of extrapolar currents, notwithstanding that this intrapolar rotation was invented (it need scarcely be said without any basis of fact, or discussible theory) for the very purpose of explaining the extrapolar currents. . . . I maintain, however, that this theory which was sufficiently dangerous to lead its author (the creator of this rich field, the discoverer of its methods, the man who has helped to teach his contemporaries in medicine to think in accordance with physics) from error to error, that this theory must now at length be given up.

Hermann was not the only scientist to dispute the conclusions Du Bois-Reymond drew from his experiments. Ewald Hering, whose major interest was the physiology of vision, wrote a paper in 1883 attacking Du Bois-Reymond's views. Hering had held the Chair in Physiology at the Military Medical-Surgical Academy at the Josephenum in Vienna. When the war in 1870 closed this famous center, Hering moved to Prague to take the position vacated by Purkyně. In 1895 he went to Leipzig to follow Carl Ludwig in the Chair. But it was from Prague that he released his attack on Du Bois-Reymond. He wrote (in translation):[29]

> In Du Bois-Reymond's treatise on ''Secondary Electromotive Phenomena in Muscle, Nerve and in Electrical Organs,''[30] he arrives at deductions directly contradicted by facts

[29]Ewald Hering (1884–1918). Sitzungenberichte der Kaiserlichen Academiė der Wissenschaften. *Acad. Wiss. Wien,* 88:445 (1883). (Proceedings of the Imperial Academy of Sciences.) [Translated in: R. J. Herrstein and E. L. Boring (Eds). *A Source Book for the History of Psychology.* Harvard Univ. Press, Boston, 1965.]

[30]E. Du Bois-Reymond (1819–1896). Über secondär-elektromotorische Erscheinungen an Muskeln, Nerven und elektrischen Organen. *Sitzungsberichte der Königlichpreussischen Academie der Wissenschaften zu Berlin.* Vol. 16, 1883, pp. 343–404. (On secondary electro-motor phenomena in muscles and electrical organs.)

which I have established. Du Bois-Reymond concludes from his research that when an electrical current flows through a muscle, it polarizes the whole of the region through which it flows. If this is so, every part of the intrapolar region lying between the galvanometer electrodes should, on breaking the circuit, give an after- or polarization-current, which according to the density and duration of the primary current, now opposes now reinforces it. Equal lengths of the intrapolar region should give equal polarization-currents, provided the sectional area be the same throughout, so that it would be of no importance from what portion of the intrapolar region of the polarization-current was led off, so long as the length of that portion remained the same.

Provided the resistance in the galvanometer-circuit is sufficient, the current of polarization should be correspondingly stronger if the leading off electrodes include a longer tract than if they include a shorter track, on account of the greater number of electromotive parts between the galvanometer electrodes, and therefore strongest when the whole intrapolar tract is interposed. In this respect the negative or positive polarization-currents would be quite analogous to the negative polarization-currents of a charcoal cylinder steeped in weak sulphuric acid, on which account, moreover, Du Bois-Reymond assumes an "internal polarization" for the whole of the intrapolar track.

In contradiction to this, I have arrived at the conclusion that the polarization-currents in a muscle consequent on breaking the primary or stimulating current, result solely from altercations which the latter undergoes at any point of the contractile substance of the muscle-fibers at which it enters or leaves that substance; in short, at the anodes and cathodes. So long as the current flows in a direction which is mathematically parallel with the muscle-fibers, in all probability it does not polarize them at all: at least there is not a single recorded observation which would lead one to such a conclusion. If polarization of the muscle-substance occurred in the intrapolar tract (in the strict sense of the term), in any case it would be so trifling that it might be provisionally quite neglected in comparison with the polarization-currents which accompany changes at the anode and cathode. Without prejudging in any way the essential nature of these changes, they may be regarded as a polarization which has an analogy with external polarization, but certainly not with internal.

In 1877 Hermann published a volume called (in translation) *An Outline of Human Physiology*,[31] in which he covered his views on secretions, respiration, metabolism, and circulation. For neurophysiologists, the most interesting part of his book is on the nervous system, for in this we find the work that he himself contributed (Fig. 50). This book essentially covers his teaching courses as Professor at the University of Zürich. Although published seven years after Fritsch and Hitzig's[32] discovery of cortical location by electrical stimulation, Hermann does not mention either their classic finding or his own attack[33] on their interpretations of their results.

Hermann was a very profuse writer, using his friend Pflüger's *Archiv* in which to publish almost every experiment, mostly designed to refute Du Bois-Reymond's notion that an electrical charge lay in the completely rested muscle or nerve. He lived into the present century, dying in 1914. He earned a long obituary in the *Lehrbuch der Königlichen Akademie der Wissenschaften* in Munich.[34] He was remembered not only for his work on the muscle and nerve but also for his theories of voice production and his interest in the chemistry of the digestive system. In his long life, he had also studied the physiology of the blood and respiration.

---

[31]L. Hermann. *Grundriss der Physiologie des Menschen*. Hirschwald, Berlin, 1877. (An Outline of Human Physiology.)

[32]G. T. Fritsch and E. H. Hitzig. Über die elektrische Erregbarkeit des Grosshirns. *Arch. Anat. Physiol. Wiss. Med. Leipzig*, 300–332 (1870). (On electrical sensitivity of the cerebrum.)

[33]L. Hermann. Über electrische Reizversuche an der Grosshirnrinde. *Pflüger's Arch. f. d. ges. Physiol.*, 10:77–85 (1875). (On electrical stimulus experiments on the cerebral cortex.)

[34]*Jahrbuch der Königlichen Akademie der Wissenschaften*. Munich, 1914, pp. 105–114.

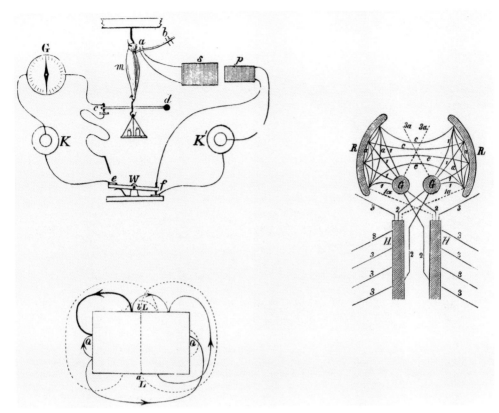

FIG. 50. Three of Hermann's sketches of alternative concepts of the circuits involved in stimulation of muscle. **Left:** *Above:* In this experiment contraction of the muscle is prevented by weights. The muscle cannot contract until its energy overcomes the weight. These can be added to until the muscle cannot contract. This gives a measure of its energy. Contact to the galvanometer (G) is made at C. *Below:* The rectangle represents a muscle. The heavy line indicates the current path of strong stimuli, the light line that of weak stimuli. The current flow too weak to cause contraction is indicated by the broken line. (From: L. Hermann. *Grundriss der Physiologie des Menschen.* Hirschwald, Berlin, 1877.) **Right:** Hermann's schema of the pathways of peripheral impulses travelling to the brain.

Edouard Friedrich Wilhelm Pflüger (1820–1910)

Edouard Pflüger, a student of Du Bois-Reymond, after preliminary training in Marburg, was inevitably drawn to the field of the master. Electrotonus in muscle and nerve proved to be his special interest. Without a laboratory of his own, he persevered with these experiments on frogs in his rented rooms. By 1859 he was able to publish his results.[35]

One of his special neurophysiological interests was in the phenomenon of tetanization in the nerve. In this he disagreed with Du Bois-Reymond who had originally held that contraction followed only make-and-break of the current. Pflüger's principle was that

> when a nerve is traversed into two zones by an electrical current, the intrapolar piece is divided by a neutral point situated between the two poles; in the one next to the cathode the irritability is increased, while in that next to the anode it is diminished, and, as this last zone decreases in extent as the current-strength sinks, when the latter is exceedingly feeble,

[35]E. F. W. Pflüger. *Untersuchungen über die Physiologie des Elektrotonus.* Hirschwald, Berlin, 1859. (Studies on the Physiology of Electrotonus.)

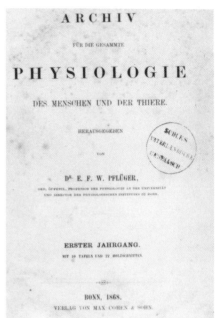

FIG. 51. Edouard Friedrich Wilhelm Pflüger (1820–1910). Scientist and experimenter in his own right but remembered most for his *Archiv für die gesammte Physiologie,* which became the major outlet for the work of physiologists. Founded in 1868 and published with only one lapse caused by the Franco-Prussian War of 1870–1871.

almost the entire intrapolar length may be in a state of increased irritability, just as in the contrary case it may be in one of lower irritability.

The success of his first book led to Pflüger's appointment to Professor of Physiology in Bonn, a Chair he held until his death. At Bonn Pflüger's interests widened; he left the nervous system for studies of the oxygen composition of the blood—work which led to another book.[36] Teaching these views he remained in the Chair for the rest of his life. His field of experimentation broadened as also did his philosophy. He became an ardent proponent of the role of teleology in physiology. He published many articles in this vein, using the position he gained in 1868 as editor of the *Archiv für die gesammte Physiologie* (Fig. 51). One of the scientists to publish most profusely in its pages was Pflüger himself. But apart from his writings on the nerve and the inhibitory actions of the splanic nerves of the gut, he is remembered as being the last eminent physiologist to retain a trace of vitalism—perhaps first seeded in his days as a young man in Müller's laboratory. In the dismay following research on spinal reflexes in decapitated animals, which shook the belief that the brain was the seat of the soul,[37] some strove to locate it in the spinal cord. Pflüger, as late as 1853, was one of the last to take this stand,[38] insisting that even in headless animals

---

[36]*Die sensorischen Functionen des Rückenmarks der Wirbelthiere nebst einer neuen Lehre über die Leitungsgesetze der Reflexionen.* Hirschwald, Berlin, 1853. (The Sensory Functions of the Spinal Cord of Vertebrates, together with a New Theory on the Transmission Laws of the Reflexes.)

[37]See the first volume of this study: M. A. B. Brazier. *A History of Neurophysiology in the 17th and 18th Centuries: From Concept to Experiment.* Raven Press, New York, 1984.

[38]Über die physiologische Verbrennung in den lebendigen Organismen. *Pflüger's Arch. f. d. ges. Physiol.,* 10:461–644 (1875). (On the physiological combustion in living organisms.)

there must be a conscious purpose as well as sensibility. Pflüger lived into the next century, dying in 1910.

Rudolf Peter Heinrich Heidenhain (1834–1897)

Heidenhain was educated at Königsberg[39] and at Halle,[40] completing his medical training under Du Bois-Reymond in Berlin. Under Du Bois-Reymond's direction, he wrote his dissertation, entitled "Disquisitiones des nervis organisque centralibus cordis cordiunque ranae lymphaticorun experimentis illustrate," which he successfully defended in 1854, and in which he tested Schiff's[41] theory that the vagus nerves caused the heart to beat. He showed that these nerves only regulate activity.

Heidenhain stayed 18 months more in Berlin, then returned to Halle. He worked for a short time as an assistant to the pathologist Karl Julius Vogel (1814–1880) before gaining a post in the Physiological Institute. In 1859 he moved to Breslau, where he became Director of the Physiological Institute, staying there for the rest of his life. His major contributions were to glandular physiology rather than to neurophysiology.

His own research in neurophysiology centered on the nerve supply to blood vessels. In its way, this was a continuation of the work he had done under Du Bois-Reymond on the role of the vagus nerve in cardiac activity. In contrast to Du Bois-Reymond, Heidenhain supported the hypothesis (of Schiff) that the vagus was a regulator of the heartbeat. Heidenhain is, however, better known for his work on the fatigue of muscles, though later in life he transferred his interest to the glandular system and the digestive system. For neurophysiologists, there is his interest in hypnosis that led him to experiment on inhibition evoked by electrical stimulation of the motor cortex of dogs under narcosis, work done with a visiting Russian N. Bubnoff.[42] Of their results he wrote:

> . . . It is well known that we are able not only to innervate muscles voluntarily, but also voluntarily to put muscles out of action. The question, however, whether the voluntary interruption of muscular activity is due simply to the cessation of impulses from the motor centers or to positive antagonistic effects which inhibit the action of these motor centers, has hardly ever been seriously considered and much less subjected to experiments. The observations reported here show that it is indeed possible to evoke from the periphery antagonistic effects putting motor centers out of action. These peripheral stimuli were surprisingly small; in fact, much smaller than those which elicited activity of the centers from the same receptor apparatus. The foregoing observations necessarily lead to the question whether slight direct stimulation of the motor centers might not act in a way similar to slight peripheral stimuli and terminate an excitatory state.
>
> Experimentation gave a positive answer. If either reflexly, or by strong electrical stimulation of the cortex, a continuous muscular contraction was induced, it could be released by a much weaker stimulation of the self-same cortical point. This occurred either completely after a single stimulus or in steps after repeated stimuli. . . .

They were writing 11 years after Fritsch and Hitzig's classic work on stimulation of the motor cortex and 15 years after Sechenov's *Reflexes of the Brain*. Heidenhain and his student reached the following conclusion:

---

[39]Now Kaliningrad in the Soviet Union.

[40]Now Martin Luther University in East Germany.

[41]Moritz Schiff (1823–1896).

[42]N. Bubnoff and R. Heidenhain. Ueber Erregungs und Hemmungsvorgänge innerhalb der motorischen Hirncentren. *Pflüger's Arch. f. d. ges. Physiol.*, 26, 137–200 (1881). Translated by G. von Bonin and W. S. McCullough as "On excitatory and inhibitory processes within the motor centers of the brain." *Illinois Monographs in the Medical Sciences*, 4:173–210 (1944).

FIG. 52. **Left:** Rudolf Peter Heinrich Heidenhain (1834–1897), who researched for inhibitory influences in the brain and invented the "Heidenhain pouch." **Right:** The gastric pouch as used in one of Pavlov's experiments.

> The assumption of inhibitory processes accompanying excitatory processes in the motor centers of the brain appears also to make intelligible the differences in the effect which a stimulation of the cortex and a stimulation of the subjacent white matter induce. . . . Other conditions being equal, the reaction time is longer, the contraction of the musculature generally smaller, and the muscular curve drawn out longer in the latter than in the former case. These differences can be understood by the assumption that direct cortical stimulation induces not only processes of excitation but also processes of inhibition.

At Breslau Heidenhain had many visiting scientists who came to spend a period of training in his laboratory. In 1875 one of them was the 26-year-old Ivan Petrovich Pavlov. The work Heidenhain was then doing was research on secretions, and he had developed an ingenious sac attached to an animal's stomach to collect the secretions. This "Heidenhain pouch" was used later by Pavlov (Fig. 52) in his experiments of conditioned reflex secretion.

Heidenhain died in 1897, one year after the death of Du Bois-Reymond. He had received many honors, especially from his old contact with Halle, where he was made a member of the Academia Leopoldina.

Carl Friedrich Wilhelm Ludwig (1816–1895)

Among the group of Du Bois-Reymond's friends who closely shared his materialistic approach to physiology was Carl Ludwig, although only a small portion of his voluminous scientific output was targeted on the nervous system. For the neurophysiologist his most important work in this field was on the inhibitory action of certain nerves—a startling revelation in a scientific world focused at that time on the exciting effects of nerve stimulation. His work on the control of the heart by the vagus nerve was of importance to him in his studies on circulation, for the physiology of the circulatory system was the major interest of his scientific life (Fig. 53). In all his work it was the physicochemical relationships that interested him.

FIG. 53.   Left: Carl Ludwig (1816–1895), in whose laboratory many distinguished experimenters were trained. **Right:** Ludwig's first design of a kymograph for recording blood flow in the aorta of the dog. (Courtesy of Dr. Dietmar Biesold, Leipzig.)

Ludwig was born in a small town in Hesse and there received his early education. In 1834 he was admitted for medical studies at Marburg where, on qualifying in 1841, he was appointed as an assistant in the Anatomical Institute at the university. In 1842 he produced his first paper[43] (on the secretion of urine), a subject that heralded his lifelong interest in physical chemistry. Ludwig remained at Marburg for nine years, becoming Professor of Comparative Anatomy.

Needing a method for recording changes in blood pressure, Ludwig invented a kymo-graph[44] (an original instrument which he had given to his pupil Mosso[45, 46] can be seen in the museum in Turin). With this instrument he was able to make some of the first recordings of blood pressure. He published a description of the apparatus in 1847. This included the smoked drum so familiar to later generations of students. In Ludwig's apparatus the drum was made to revolve by a falling weight (Fig. 54), the rise and fall of mercury level with the pulse being recorded on the smoked drum.

[43]C. Ludwig. *De viribus physics secretion urinae adjuvantibus.* Habilitationschrift, Marburg, 1842.

[44]C. Ludwig. Beiträge zur Kenntniss des Einflusses der Respirationsbewegungen auf den Blutlauf. *Arch. Anat. Physiol. Wiss. Med.,* 242–302 (1847).

[45]Angelo Mosso (1846–1910). Professor of Physiology at Turin, he had as a student spent two years in Ludwig's laboratory just at the period when the master was concentrating on improving graphic methods for recording physiological events.

[46]A. Mosso. Action physiologique de la cocaine et critique expérimentale des travaux publiés sur son mécanisme d'action. (*Arch. Ital. Biol.* XIV, 3, p. 247). *Zbl. Physiol.,* 5:332–333 (1891).

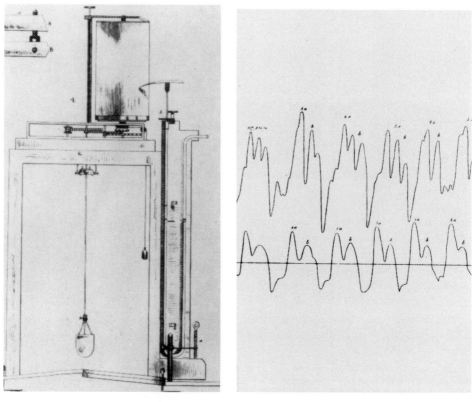

FIG. 54. Left: Kymograph designed by Carl Ludwig. **Right:** One of the first recordings of Ludwig's by kymograph. They represent blood pressure curves in the external carotid artery of a dog (*above*) and the intraplural pressure (*below*). (Courtesy of Dr. Dietmar Biesold, Leipzig.)

In 1849 Ludwig moved to Zürich to become Professor of Anatomy and Physiology. There his interests moved further away from neurophysiology and concentrated almost entirely on such subjects as diffusion and endosmosis in saliva and body fluids. However, his interest returned to the nervous system by his discovery of the secretory nerves in the salivary gland. Using the experience from his early days with Du Bois-Reymond, he did make experiments with electrical stimulation that demonstrated increase of flow from the lymph glands. He designed an experiment in which by tying a manometer to the outlet of the submaxillary gland, he could watch the effect of electrical stimulation of the chorda tympani. When the current flowed, the pressure, recorded on the kymograph, was seen to rise. The secretion of the gland was thus shown to be an active process under nervous control and not merely a filtrate from the blood.

Ludwig achieved prominence by his famous book *Textbook of Human Physiology*,[47] the opening lines of which read (in translation): "Scientific physiology has the task of determining the activities of the living body and deriving them as following necessarily from its elementary conditions." This book is dedicated to Brücke, Du Bois-Reymond, and Helmholtz, and it reflects strongly the goal he shared with them to oust vitalism from physiology. Ludwig left Zurich to become Professor of Anatomy and Physiology at the famous Josephe-

---

[47]C. Ludwig. *Lehrbuch der Physiologie des Menschen.* 2 vols. Veit, Heidelberg, 1852–1856.

num in Vienna. The Medical-Surgical Institute there had been founded by the Emperor Joseph II in 1786, for whom it is named. Specifically endowed to serve the military forces, the Emperor had declared:

> This is to express our esteem for those of our nation who risk their lives in much danger for the defense of our common fatherland, for the rights of Our throne and for the security of its citizens. In order to ease in our capital their honorable but heavy duty, we have erected our own complete Military Medical-Surgical Academy.

At the Institute Ludwig was joining his friend Brücke who had come to Vienna in 1849; Brücke also had left the intensive research on the nervous system that he had learned with Du Bois-Reymond to study the physical chemistry of the digestive system.

One article of marked importance that came from Ludwig's work at the Josephenum was a paper published with Thiry,[48] which revealed Ludwig's recognition that the brain and spinal cord had direct influence on the blood pressure. Many experiments in animals after section of the spinal cord convinced Ludwig to continue this work, and by 1871 he had located the vasomotor center in the medulla.

In his textbook Ludwig made an attack on Marshall Hall's theory of reflexes.[49] This was long before there was any understanding of nerve networks and synaptic connections. What Ludwig was also unwilling to accept was the proposal of Wagner that there were direct connections between the afferent and efferent nerves serving a spinal pathway. Neither antagonist had the basis to prove his point. Neurophysiology had to wait for the demonstration of nerve fibers' endings on nerve cells. The argument took place before the recognition of the neuron theory and the idea that direct connections existed between all elements (known as the "reticular theory"). Doubts began to come in from the microscopists, for example, from the findings by Held[50] of bushy endings (controller of afferent fibers). This information was coming to physiologists at the very end of the century under review and before the neuron theory had been generally accepted—a theory which destroyed the concept of direct connection between neurons within the brain. In the opening of the next century, Sherrington was to name this the synapse. But the well-known teacher Sharpey-Schäfer warned readers of the coming revelations in the famous textbook he masterminded.[51] "Every cell," he wrote, "forms a structural element anatomically isolated. . . ."

Ludwig continued his interest in the neuronal basis of spinal reflexes when he moved later to Leipzig and gave this subject for research to several of his students. The discovery of the synapse still lay in the future and was not to be revealed (by Sherrington) before the end of Ludwig's life. At Leipzig a school grew up around him. On the retirement of Ernst Weber (the discoverer of vagus inhibition of the heart), Ludwig moved into the Chair of Physiology, a position he held until his death. The Institute that Ludwig himself had planned had originally been financed in 1869 by Johann, King of Saxony, but was destroyed in the Second World War. A new building was erected after the war and named the Carl Ludwig Institute in the newly named Karl Marx University. He had collected many scientists around him—some who later made their mark in other countries—for example,

---

[48]C. Ludwig and L. Thiry. Über den Einfluss des Halsmarks auf den Blutstrom. *Sitz. Math. Nat. Wiss. Wien II. Abt.*, 421 (1864). (On the influence of the medulla oblongata on the bloodstream.)

[49]Marshall Hall (1790–1857). On the reflex functions of the medulla oblongata and the medulla spinalus. *Phil. Trans. R. Soc.*, 123:625–665 (1833).

[50]H. Held. Beiträge zur Struktur der Nervenzellen und ihrer Fortsätze. *Arch. Anat. Physiol. Leipzig* (Suppl. 27), 204 (1897).

[51]Ernest Sharpey-Schäfer (1850–1925). *Textbook of Physiology*. 2 vols. Pentland, Edinburgh, 1898–1900.

Sechenov who was to startle the world with his *Reflexes of the Brain* and Elie Cyon with whom Ludwig published frequently. Investigators came from afar to work with him, including two from the United States (for example, Henry Bowditch). Another American to visit Ludwig was W. P. Lombard, who was to make his career later in Michigan. Ludwig gave him a problem of the spinal cord, and he published one paper[52] (written for him by Ludwig since his own German was inadequate). But he did not carry this interest in neurophysiological research back home with him, though it inspired in him an interest in muscle physiology. Yet the 19th century was to close without a contribution to neurophysiology originating from the New World.

Neurophysiologists remember Ludwig for his discoveries of nervous control of glandular secretion, this having been his principal love in research. He died just before the turn of the century, the same decade that carried away Helmholtz, Du Bois-Reymond, and Brücke. The great friendships with Brücke in Vienna and Du Bois-Reymond and Ludwig in Germany happily survived the outbreak in 1866 of war between Austria and Germany. But, clearly, there was a break, for in a collection[53] of correspondence of Ludwig with Du Bois-Reymond, the last letter between them is dated May 1865. Du Bois-Reymond outlived Ludwig by one year and wrote a penetrating obituary of his friend.[54]

Ludwig stands out as one of the great teachers of physiology in the 19th century. Younger scientists came to him at all stages of his career—to Vienna, to Zurich, and to Leipzig. Among them from Russia were Cyon and Pavlov; from the Ukraine, Danilevsky and Betz; and from Italy, Luigi Luciani.

While under Ludwig's tutelage, Pavlov began work on his dissertation which was on the nervous control of the heart. Sechenov worked on blood gases, and Danilevsky on the vagus nerve; on his return to the Ukraine, Danilevsky worked on the electrical activity of the brain (an activity not attempted by Ludwig). One who worked most closely with Ludwig and published with him was Cyon, for they shared an interest in questions of circulatory regulation.

Among the many obituary notices of Ludwig's death was one published from the University of Kazan written by N. A. Mislavsky, a pupil of Ludwig's.[55]

BIBLIOGRAPHY

Julius Bernstein (1839–1917)

Selected Writings

Untersuchungen über den Mechanismus des regulatorischen Herznervensystems. *Archiv Anat. Physiol. Wiss. Med.*, 633–666 (1864). (Investigations into the mechanism of the regulatory heart nerve system.)

Über den zeitlichen Verlauf der negativen Schwankung des Nervenstromes. *Pflüger's Arch. f. d. ges. Physiol.*, 1:173–207 (1868). (On the chronological progress of negative variation of the nerve current.) (English translation by S. Schutze: The differentiation of rheotome and its application to the study of nervous activity. In: *Electrical Engineering 573*. December 1, 1972.)

[52]Warren Plimpton Lombard (1855–1892). Die räumliche und zeitliche Aufeinanderfolge reflectorish contrahirter Muskeln. *Arch. Anat. Physiol.*, 408–489 (1885).

[53]Estelle Du Bois-Reymond. *Zwei Grosse Naturforscher des 19 Jahrhunderts.* Leipzig-Verlag, 1927. (Two Great Scientists of the 19th Century.)

[54]Emil Du Bois-Reymond. C. Ludwig. *Wien. Klin. Woch.*, 19:1–11 (1895).

[55]N. A. Mislavsky. *Neurol. Vestnik. Kazan.*, 4:1 (1896).

Untersuchungen über den Erregungsvorgang im Nerven und Muskelsysteme. Winter, Heidelberg, 1871. (Investigations into the stimulation process in the nerve and muscle system.)

Über das myophysische Gesetz des Herrn Preyer. *Pflüger's Arch. f. d. ges. Physiol.*, 6:403–412 (1872). (On the myophysical law of Mr. Preyer.)

Über den Electrotonus und die innere Mechanik des Nerven. *Pflüger's Arch. f. d. ges. Physiol.*, 8:40–60 (1874). (On the electrotonus and the inner mechanics of the nerve.)

Über Electrotonus. *Pflüger's Arch. f. d. ges. Physiol.*, 8:498–505 (1874). (On electrotonus.)

Über die Höhe des Muskeltones bei elektrischer und chemischer Reizung. *Pflüger's Arch. f. d. ges. Physiol.*, 11:191–196 (1875). (On the extent of muscle tone during electrical and chemical stimulation.)

Die Fuenf Sinne des Menschen. Leipzig, 1875. (The Five Senses of Man.)

The Five Senses of Man. Henry S. King, London, 1876.

Über die Ermüdung und Erholung der Nerven. *Pflüger's Arch. f. d. ges. Physiol.*, 15:289–327 (1877). (On the fatigue and recovery in nerves.)

Weiteres über die Entstehung der Aspiration der Thorax nach der Geburt. *Pflüger's Arch. f. d. ges. Physiol.*, 21–37 (1884). (Further report on the origin of aspiration of the thorax following birth.)

Ueber das angebliche Hören labyrinthloser Tauben. *Pflüger's Arch. f. d. ges. Physiol.*, 61:113–122 (1895). (On the supposed hearing ability of doves without labyrinths.)

Das Beugungspektrum des quergestreifen Muskels bei der Contraktion. *Pflüger's Arch. f. d. ges. Physiol.*, 61:285–291 (1895). (The flexion spectrum of the striated muscle during contraction.)

Lehrbuch der Physiologie des thierischen Organismus, im Speziellen des Menschen. 2nd ed. Stuttgart, 1900. (Textbook of the physiology of the animal organism, especially that of humans.)

Untersuchungen zur Thermodynamik der bioelektrischen Ströme. *Pflüger's Arch. f. d. ges. Physiol.*, 92:521–562 (1902).

Elektrobiologie. Braunschweig, Vierneg, 1912. (Electrobiology.)

Suggested Readings

Geddes, L. A. A Short History of the Electrical Stimulation of Excitable Tissue. Suppl. to the *Physiologist,* Vol. 27, No. 1, 1984.

Grundfest, H. Jules Bernstein, Ludimar Hermann and the discovery of the action spike. *Arch. Ital. Biol.*, 103:483–490 (1965).

Hoff, H. E. and Geddes, L. A. The rheotome and its prehistory: A study in the historical interrelation of electrophysiology and electromechanics. *Bull. Hist. Med.*, 31:212–234, 327–347 (1957).

Rothschuh, K. E. *History of Physiology.* Krieger, New York, 1973.

Rudolf, G. Jules Bernstein. In: *Founders in Experimental Physiology.* Lehmanns, Munich, 1971, pp. 249–299.

Schutze, W. The differential rheotome and its application to the study of nervous activity. *Elect. Eng.*, 1972, p. 573. (English translation of Bernstein's paper: Über den zeitlichen Verlauf der negativen Schwankung das Nervenstroms. *Pflüger's Arch. f. d. ges. Physiol.*, 1:173–207, 1868.)

Tschermak, A. V. Julius Bernstein's Lebensarbeit. *Pflüger's Arch. f. d. ges. Physiol.*, 171:1–89 (1919).

Tschermak, A. V. Julius Bernstein's Lebensarbeit. Zugleich ein Beitrag zur Geschichte der neueren Biophysik. *Pflüger's Arch. Physiol.*, 174: 1–89 (1919). (Julius Bernstein's life work. Together with a contribution to the history of modern biophysics.)

Ludimar Hermann (1838–1914)

Selected Writings

Untersuchungen über den Stoffwechsel der Muskeln. Hirschwald, Berlin, 1867. (Investigations of the metabolism of muscle.)

Ueber die Krämpfe bei Circulationsströrungen im Gehirn. (With Escher.) *Pflüger's Arch. f. d. ges. Physiol.*, 3:3–8 (1870). (On the convulsions during circulation disturbances in the brain.)

Eine Erscheinung simultanen Contrastes. *Pflüger's Arch. f. d. ges. Physiol.*, 3: 13–15 (1870). (A phenomenon of simultaneous contrast.)

Weitere Untersuchungen über die Ursache der electromotorischen Erscheinungen an Muskeln und Nerven. I. Ueber das Fehlen des Stromes in unversehrten ruhenden Muskeln. *Pflüger's Arch. f. d. ges. Physiol.*, 3:15–39 (1870). (Further investigations on the cause of electromotor phenomena in muscles and nerves.)

Versuche über den Verlauf der Stromentwicklung beim Absterben. *Pflüger's Arch. f. d. ges. Physiol.*, 3:39–46 (1870). (Experiments on the process of current development during death.)

Über die Abnahme der Muskelkraft während der Contraction. *Pflüger's Arch. f. d. ges. Physiol.*, 5: 223–275 (1872). (On the reduction of muscle during contraction.)

Über eine Wirkung galvanischer Ströme auf Muskeln und Nerven. *Pflüger's Arch. f. d. ges. Physiol.*, 5:280–282 (1872). (On an effect of galvanic current on muscles and nerves.)

Ueber eine Wirkung galvanischer Ströme auf Muskeln und Nerven. *Pflüger's Arch. f. d. ges. Physiol.*, 5:317–360 (1872). (On an effect of galvanic currents on muscles and nerves.)

Das galvanische Verhalten einer durchflossenen Nervenstrecke während der Erregung. *Pflüger's Arch. f. d. ges. Physiol.*, 5:560–567 (1872). (The galvanic behavior of an electrified section of nerve during arousal.)

Ueber eine Wirkung galvanischer Ströme auf Muskeln und Nerven. *Pflüger's Arch. f. d. ges. Physiol.*, 6:312–360 (1872). (On an effect of galvanic currents on muscles and nerves.)

Das galvanische Verhalten einer durchflossenen Nervenstrecke während der Erregung. *Pflüger's Arch. f. d. ges. Physiol.*, 6:560–574 (1872). (The galvanic behavior of a nerve fiber during electrical stimulation.)

Weitere Untersuchungen über den Electrotonus, insbesondere über die Erstreckung desselben auf die intramusculären Nervenenden. *Pflüger's Arch. f. d. ges. Physiol.*, 7:301–322 (1873). (Further investigations into electrotonus, especially into its extension into the intermuscular nerve endings.)

Untersuchungen über das Gesetz der Erregungsleitung im polarisirten Nerven. *Pflüger's Arch. f. d. ges. Physiol.*, 7:323–364 (1873). (Investigations into the law of transmission of stimulus in the polarized nerve.)

Ein Versuch über die sog, Sehnenverkürzung. *Pflüger's Arch. f. d. ges. Physiol.*, 7:417–420 (1873). (An experiment into the so-called tendon contraction.)

Berichtigender Zusatz zu den Untersuchungen über die Erregungsleitung im polarisirten Nerven. *Pflüger's Arch. f. d. ges. Physiol.*, 7:497–498 (1873). (Additional report on the investigations into the transmission of stimulus in the polarized nerve.)

Experimentelles und Kritisches über Electrotonus. *Pflüger's Arch. f. d. ges. Physiol.*, 8:258–275 (1874). (Experimental and critical matters concerning electrotonus.)

Fortgesetzte Untersuchungen über die Beziehungen zwischen Polarisation und Erregung in Nerven. *Pflüger's Arch. f. d. ges. Physiol.*, 10:215–239 (1875). (Continued investigations into the connections between polarization and stimulus in the nerve.)

Die Fortpflanzungsgeschwindigkeit der Reizung in der quergestreiften Muskelfaser. *Pflüger's Arch. f. d. ges. Physiol.*, 10:465–467 (1875). (The speed of transmission of stimulus in transversely striated muscle fiber.)

Über electrische Reizversuche an der Grosshirnrinde. *Pflüger's Arch. f. d. ges. Physiol.*, 10:77–85 (1875). (On electrical stimulus experiments on the cerebral cortex).

Der Querwiderstand des Nerven während der Erregung. *Pflüger's Arch. f. d. ges. Physiol.*, 12:151–156 (1876). (The transverse resistance of the nerve during stimulation.)

Notizen zur Muskelphysiologie. *Pflüger's Arch. f. d. ges. Physiol.*, 13:369–372 (1876). (Notes on muscle physiology.)

Untersuchungen über die Entwicklung des Muskelstroms. *Pflüger's Arch. f. d. ges. Physiol.*, 15:191–232 (1877). (Research into the development of the muscle current.)

Versuche mit dem Fall-Rheotom über die Erregungsschwankung des Muskels. *Pflüger's Arch. f. d. ges. Physiol.*, 15:233–245 (1877). (Experiments with the fall-rheotom on the stimulation variation of the muscle.)

Grundriss der Physiologie des Menschen, Hirschwald, Berlin, 1877. (Fundamentals of Human Physiology.)

Untersuchungen über die Actionsströme des Muskels. *Pflüger's Arch. f. d. ges. Physiol.*, 16:191–262 (1878). (Investigations into the action currents of the muscle.)

Zur Lehre vom Einflusse der Reizstelle und Reizstromrichtung im Nerven. *Pflüger's Arch. f. d. ges. Physiol.*, 16:262–263 (1878). (On the theory of the influence of the point of stimulation and the direction of the stimulus current in the nerve.)

Ueber Brechung bei schiefer Incidenz, mit besonderer Berücksichtigung des Auges. Theil I. *Pflüger's Arch. f. d. ges. Physiol.*, 18:443–455 (1878). (On refraction at oblique angle, with especial attention to the eye.)

Ein Beitrag zur Theorie der Muskelcontraction. *Pflüger's Arch. f. d. ges. Physiol.*, 18:455–457 (1878). (A contribution on the theory of muscle contraction.)

Notizen über einige Gifte der Curaregruppe. *Pflüger's Arch. f. d. ges. Physiol.*, 18:458–460 (1878). (Notes on some poisons of the curare group.)

Luchsinger, B. (co-author). Ueber Secretionsströme an der Zunge des Frosches, nebst Bemerkungen über einige andere Secretionsströme. *Pflüger's Arch. f. d. ges. Physiol.*, 18:460–472 (1878). (On secretion streams on the tongue of the frog, with comments on some other secretion streams.)

Untersuchungen über die Actionsströme des Nerven. I. *Pflüger's Arch. f. d. ges. Physiol.*, 18:574–586 (1878). (Investigations of the action currents of the nerves.)

Ueber den atelectatischen Zustand der Lungen und dessen Aufhören bei der Geburt. Nach Versuchen des Herrn stud. med. Otto Keller. (Aus dem physiologischen Laboratorium in Zurich.) *Pflüger's Arch. f. d. ges. Physiol.*, 20:365–369 (1879). (On the atelectatic condition of the lungs, and its ceasing to be at birth. Following experiments by Otto Keller.)

Ueber Brechung bei schiefer Incidenz mit besonderer Berücksichtigung des Auges. Theil II. Mit 2 Holzschnitten. (Aus dem physiologischen Laboratorium in Zurich.) *Pflüger's Arch. f. d. ges. Physiol.*, 20:370–387 (1879). (On refraction at oblique angle, with especial consideration of the eye. Part II. With two woodcuts. [From the physiological laboratory in Zurich.])

Ueber E. v. Fleischl's zweite vermeintliche Widerlegung meiner Theorie des Electrotonus. (Aus dem physiologischen Laboratorium in Zurich.) *Pflüger's Arch. f. d. ges. Physiol.*, 20:388–394 (1879). (On E. v. Fleischl's second supposed refutation of my theory of electrotonus. [From the physiological laboratory in Zurich.])

Allgemeine Nervenphysiologie. *Handbuch der Physiologie*, Zweiter Band, Edited by L. Hermann. Erster Theil. Leipzig, 1879, pp. 1–96. (General Neurophysiology.)

Bemerkungen über das galvanische Verhalten einer durchflossenen Nervenstrecke. *Pflüger's Arch. f. d. ges. Physiol.*, 19:416–417 (1879). (Remarks on the galvanic behavior of an electrified piece of nerve.)

Untersuchung über die Actionsströme des Nerven. II. *Pflüger's Arch. f. d. ges. Physiol.*, 24:246–294 (1881). (Investigations into the action currents of nerve. II.)

Neue Untersuchungen über Hautströme. *Pflüger's Arch. f. d. ges. Physiol.*, 27:280–288 (1882). (New investigations of skin currents.)

Notiz über eine Verbesserung am repetirenden Rheotom. *Pflüger's Arch. f. d. ges. Physiol.*, 27:289–290 (1882). (Note on an improvement to the repeating rheotom.)

Untersuchungen zur Lehre von der electrischen Muskel- und Nervenreizung. *Pflüger's Arch. f. d. ges. Physiol.*, 30:1–17 (1883). (Investigations on the theory of electrical muscle and nerve stimulation.)

Das Verhalten des kindlichen Brustkastens bei der Geburt. *Pflüger's Arch. f. d. ges. Physiol.*, 30:276–288 (1883). (The behaviour of the infant thorax at birth.)

Zur electrophysiologischen Literaturgeschichte. *Pflüger's Arch. f. d. ges. Physiol.*, 30:620 (1883). (On the history of writings on electrophysiology.)

Untersuchungen zur Lehre von der electrischen Muskel- und Nervenreizung. *Pflüger's Arch. f. d. ges. Physiol.*, 31:99–118 (1883). (Investigations on the theory of electrical muscle and nerve stimulation.)

Untersuchungen zur Lehre von der electrischen Muskel -und Nervenreizung. *Pflüger's Arch. f. d. ges. Physiol.*, 31:308–378 (1883). (Investigations on the theory of electrical muscle and nerve stimulation.)

Eine modificirte Construction des Differential-Rheotoms. *Pflüger's Arch. f. d. ges. Physiol.*, 31:600–608 (1883). (A modified construction of the differential rheotom.)

Zur Bestimmung der Umlaufszeit des Blutes. *Pflüger's Arch. f. d. ges. Physiol.*, 33:169–173 (1884). (To determine the circulation rate of the blood.)

Über sogenannte secundär-electromotorische Erscheinungen an Muskeln und Nerven. *Pflüger's Arch. f. d. ges. Physiol.*, 33:103–162 (1884). (On so-called secondary electromotor phenomena in muscle and nerves.) (English translation in: J. Burdon-Sanderson, *Translation of Foreign Biological Memoirs*, Frowde, London, 1887.)

Untersuchung zur Lehre von der electrischen Muskel- und Nervenreizung. *Arch. Physiol.*, 35:1–25 (1885). (Investigation on the theory of the electrical muscle and nerve current.)

Ewald Hering (1884–1918)

Suggested Writings

Die sogenannte Raddrehung des Auges in ihrer Bedeutung für das Sehen bei ruhendem Blicke. *Arch. Anat. Physiol. Wiss. Med.*, 278–285 (1864). (The so-called rotation of the eye, in its significance for sight during gaze and rest.)

Das Gesetz der identischen Sehrichtungen. *Arch. Anat. Physiol. Wiss. Med.*, 27 (1864). (The law of identical sight directions.)

Über den Einfluss der Athmung auf den Kreislauf. I. Über Athembewegungen des Gefässsystems. *Sb. Abt.*, 60, 829–856 (1869). (On the influence of respiration on circulation. I. On respiratory movements on the vessel system.)

Über das Gedächtnis als eine allgemeine Funktion der organischen Materie. *Ann. Alm.* 8:253–278 (1870). (On the memory as a general function of organic material.)

On Du Bois-Reymond's Researches on Secondary Electromotive Phenomena of Muscle, *Sitzungenberichte der Kaiserlichen Academie der Wissenschaften*, 88:(Part 3) 445 (1883). (Proceedings of the Imperial Academy of Sciences) (English translation in: J. Burdon-Sanderson. *Translations of Foreign Biological Memoirs*, Oxford University Press, 1887.)

The Positive Variation of the Nerve Current after Electrical Stimulation. *Sitzungsberichte der Kaiserlichen Akademie der Wissenschaften*, 89: (Part 3) 137 (1884). (Proceedings of the Imperial Academy of Sciences.) (English translation in: J. Burdon-Sanderson. *Translations of Foreign Biological Memoirs*, Oxford University Press, 1887.)

Edouard Friedrich Wilhelm Pflüger (1820–1910)

Selected Writings

Die sensorischen Functionen des Rückenmarks der Wirbelthiere nebst einer neuen Lehre über die Leitungsgesetze der Reflexionen. Hirschwald, Berlin, 1853. (The Sensory Functions of Spinal Cord Vertebrates, Together with a New Theory of the Laws of Transmission of Reflexes.)

Untersuchungen über die Physiologie des Elektrotonus. Hirschwald, Berlin, 1859. (Studies on the Physiology of Electrotonus.)

Über die Abhängigkeit der Leber von dem Nervensystem. *Pflüger's Arch. f. d. ges. Physiol.*, 2:459–491 (1869). (On the dependence of the liver on the nervous system.)

Zum Nachweis der Nervendungen in den Acinösen Drüsen und der Leber. *Pflüger's Arch. f. d. ges. Physiol.*, 4:50–53 (1871). (On evidence of nerve endings in the acinose glands and the liver.)

Theorie des Schlafes. *Pflüger's Arch. f. d. ges. Physiol.*, 10:468–478 (1875). (Theory of sleep.)

Über die physiologische Verbrennung in den lebendigen Organismen. *Pflüger's Arch. f. d. ges. Physiol.*, 10:461–464 (1875). (On physiological combustion in living organisms.)

Beiträge zur Lehre von der Respiration. Über die physiologische Verbrennung in den lebendigen Organismen. *Pflüger's Arch. f. d. ges. Physiol.*, 10:251–367 (1875). (Contributions to a theory of respiration. On physiological combustion in living organisms.)

Über den Einfluss des Auges auf den thierischen Stoffwechsel. *Pflüger's Arch. f. d. ges. Physiol.*, 11:263–272 (1875). (On the influence of the eye on animal metabolism.)

Über den Einfluss der Athemmechanik auf den Stoffwechsel. *Pflüger's Arch. f. d. ges. Physiol.*, 14:1–37 (1877). (On the influence of the breathing mechanism on metabolism.)

Über den Einfluss der Temperatur auf die Respiration der Kaltblüter. *Pflüger's Arch. f. d. ges. Physiol.*, 14:73–77 (1877). (On the influence of temperature on the respiration of cold blooded animals.)

Die teleologische Mechanik der lebendigen Natur. *Pflüger's Arch. f. d. ges. Physiol.*, 15:57–103 (1877). (The teleological mechanics of living nature.)

Bemerkungen zur Physiologie des centralen Nervensystems. *Pflüger's Arch. f. d. ges. Physiol.*, 15:150–152 (1877). (Observations of the physiology of the central nervous system.)

Die Physiologie und ihre Zukunft. *Pflüger's Arch. f. d. ges. Physiol.*, 15:361–365 (1877). (Physiology and its future.)

Ueber Wärme und Oxydation der lebendigen Materie. *Pflüger's Arch. f. d. ges. Physiol.*, 18:247–380 (1878). (On heat and oxidation of living material.)

Zur Kenntniss der Gase der Organe. *Pflüger's Arch. f. d. ges. Physiol.*, 18:381–387 (1878). (On the knowledge of organ gases.)

Über die Einwirkung der Schwerkraft und anderer Bedingungen auf die Richtung der Zelltheilung. *Pflüger's Arch. f. d. ges. Physiol.*, 34:607–616 (1884). (On the effect of gravity and other conditions on the direction of cell division.)

K. Bohland (co-author). Ueber eine Methode das Stickstoffgehalt des menschlichen Harnes schnell annährungsweise zu bestimmen. *Pflüger's Arch. f. d. ges. Physiol.*, 38:575–624 (1886). (On a method of quickly obtaining an approximate estimate of nitrogen content of human urine.)

Nachschrift zu dem vorhergehenden Aufsatze betreffend ein neues Grundgesetz der Ernährung und die Quelle der Muskelkraft. *Pflüger's Arch. f. d. ges. Physiol.*, 51:317–320 (1892). (Postscript to the previous article, concerning a new basic law of nourishment and the source of muscle power.)

Untersuchungen über die quantitative Analyse des Traubenzuckers. *Pflüger's Arch. f. d. ges. Physiol.*, 69:399–471 (1898). (Investigations on quantitative analysis of glucose.)

Rudolf Peter Heinrich Heidenhain (1834–1897)

Selected Writings

Über A. Fick's experimentellen Beweis für die Gültigkeit des Gesetzes von der Erhaltung der Kraft bei der Muskelzusammenziehung. Nach Versuchen der Herren Studirenden Leopold Landau und Carl Pacully mitgetheilt von R. Heidenhain in Breslau. *Pflüger's Arch. f. d. ges. Physiol.*, 2:423–432 (1869). (On A. Fick's experimental evidence for the validity of the law of the retention of power during muscle contraction. Following experiments by Leopold Landau and Carl Pacully.)

Über bisher unbeachtete Einwirkungen des Nerven-systems auf die Körpertemperatur und den Kreislauf. *Pflüger's Arch. f. d. ges. Physiol.*, 3:504–565 (1870). (On effects of the nerve system on body temperature and circulation, previously not considered.)

Über arhytmische Herzthätigkeit. *Pflüger's Arch. f. d. ges. Physiol.*, 4:143–153 (1872). (On arrhythmical heart activity.)

Erneute Beobachtungen über den Einfluss des vasomotorischen Nervensystems auf den Kreislauf und die Körpertemperatur. *Pflüger's Arch. f. d. ges. Physiol.*, 5:77–113 (1872). (New observations into the influence of the vasomotor nerve system on the circulation and on body temperature.)

Ueber die Wirkung einiger Gifte auf die Nerven der glandula submaxillaris. *Pflüger's Arch. f. d. ges. Physiol.*, 5:209–318 (1872). (On the effect of some poisons on the nerves of the glandula submaxillaris.)

Bemerkungen zu Herrn Dr. Franz Riegel's Aufsatz: "Ueber die Beziehung der Gefässnerven zur Körpertemperatur. *Pflüger's Arch. f. d. ges. Physiol.*, 6:20–22 (1872). (Remarks with reference to Dr. Franz Riegel's article: "On the relationship of the vessel nerves to body temperature.")

Versuche über den Vorgang der Harnabsonderung. *Pflüger's Arch. f. d. ges. Physiol.*, 9:1–27 (1874). (Experiments on the process of urine secretion.)

Die Einwirkung sensibler Reizung auf den Blutdruck. *Pflüger's Arch. f. d. ges. Physiol.*, 9:250–262 (1874). (The effect of sensory stimulation on blood pressure.)

P. Grutzner (co-author). Beiträge zur Kenntniss der Gefässinnervation. Über die Innervation der Muskelgefässe. *Pflüger's Arch. f. d. ges. Physiol.*, 16:159 (1878). (Contributions to knowledge of vessel innervation.)

Über secretorische und tropische Drüsennerven. *Pflüger's Arch. f. d. ges. Physiol.*, 17:1–67 (1878). (On secretory and trophic glandular nerves.)

Ueber die Pepsinbildung in den Pylorusdrüsen. *Pflüger's Arch. f. d. ges. Physiol.*, 18:169–171 (1878). (On the formation of pepsin in the pylorus glands.)

N. Bubnoff (co-author). Ueber Erregungs- und Hemmungsvorgänge innerhalb der motorischen Hirncentren. *Pflüger's Arch. f. d. ges. Physiol.*, 26:137–200 (1881). (On arousal and inhibition processes within the brain motor centers.)

Untersuchungen über den Einfluss des nv. vagus auf die Herztthätigkeit. *Pflüger's Arch. f. d. ges. Physiol.*, 27:368–411 (1882). (Investigations on the influence of the nv. vagus on heart activity.)

Karl Friedrich Wilhelm Ludwig (1816–1895)

Selected Writings

De viribus physics secretionem urinae adjuvantibus. Habilitationschrift. Marburg, 1842.

Beiträge zur Lehre vom Mechanismus der Harnsecretion. N. G. Elwert, Marburg, 1843. (Contributions to the Theory of Mechanism of Urine Secretion.)

Beiträge zur Kenntniss des Einflusses der Respirationsbewegungen auf den Blutlauf im Aortensystem. *Arch. Anat. Physiol. Wiss. Med.*, 242–302 (1847). (Contributions to the knowledge of the influence of respiratory movements on the blood flow in the aortic system.)

Über die Herznerven des Frosches. *Arch. Physiol. Wiss. Med.*, 139 (1848). (On the cardiac nerves of frogs.)

Lehrbuch der Physiologie des Menschen. 2 vols. Heidelberg, 1852. (Handbook of Human Physiology.)

Zur Ablehnung der Anmuthungen des Herrn R. Wagner. in Göttingen. *Z. Med. N. F.*, 5:269 (1854). (A rejection of the suggestions of R. Wagner in Göttingen.)

Über die Kräfte des Nervenprimitivrohres. *Wien. Med. Woch.* 729:129 (1861). (On the powers of the primitive nerve tube.)

L. Thiry (co-author). Über den Einfluss des Halsmarks auf den Blutstrom. *Sitzungsber. Math. Nat. K. Akad. Wiss. Wien*, 2:421 (1864). (On the influence of the medulla oblongata on the bloodstream.)

Zusammenstellung der Untersuchungen über Blutgase, welche aus der physiologischen Anstalt der Josefs-Akademie hervorgegangen sind. *Med. Jb.,* 21:145–166 (1865). (Summary of investigations on blood gases carried out in the physiological institute of the Josephenum.)

Die physiologischen Leistungen des Blutdrucks. Hirzel, Leipzig, 1865. (The Physiological Properties of Blood Pressure.)

Physiol. Anst. Leipzig, 1:128–149 (1867). (On vasomotor reflexes.)

Suggested Readings

Bauereisen, E. Carl Ludwig as the founder of modern physiology. *Physiologist.* Vol. 4, No. 4, November 1962.

Hoff, H. E., and Geddes, L. A. Graphic registration before Ludwig: the antecedents of the cymograph. *Isis,* 50:1–21 (1959).

Hoff, H. E., and Geddes, L. A. Graphic recording before Carl Ludwig, an historical summary. *Arch. Int. Hist. Sci.,* 12:3–25 (1959).

Morton, H. F., and Weiss, H. J. The "Introduction" to Carl Ludwig's textbook of human physiology. *Med. Hist.,* 10:76–86 (Jan. 1, 1966).

Mosso, M. A. Charles Ludwig. *Rev. Sci.* Paris, 1896.

Richet, Charles. Carl Ludwig. *Dictionnaire de Physiologie.* Vol. 10. Librarie Félix Alcan, Paris, 1928, pp. 267–276.

Rosen, G. Carl Ludwig and his American students. *Bull. Hist. Med.,* 4:609–650 (1936).

Rothschuh, K. E. von. Carl Ludwig (1816–1895). Gestalt und Bildnis des grossen deutschen Physiologen. *Z. Kreislaufforschung,* 49:2 (1960).

Schröder, H. Carl Ludwig—Begründer der messenden Experimental-physiologie, 1816–1895. Wissenschaftliche Verlagsgesellschaft M. B. H., Stuttgart, 1967.

Stasch, P. *Ludwig's Kymograph and Its First Uses.* Leipzig, 1936.

CHAPTER VIII

The Question of Localization in the Brain

Only in the 17th century did the brain displace the heart as the controller of our actions. Such a role for the heart still lingers on in our language: we describe ourselves as kind-hearted, downhearted, and heartless. By the 18th century, recognition had come to the brain, but the dominating question was whether or not this was the site of the soul. To the neurophysiologist, it may seem strange that the evidence used for action by the soul was the movement of muscles. Using this as the criterion of its actions, many loci in the brain were chosen as its home. Down the ages, the soul has been said to be located in the heart,[1] the ventricles,[2] the pineal,[3] the stomach,[4] the blood,[5] the corpus callosum,[6] and, even as late as the 19th century, in the medulla spinalis and even in the spinal fluid.[7]

When the soul appeared to be established in the brain, a new disturbance arose, owing to the discoveries of motor movements in decapitated animals (for example, the work of Alexander Stuart[8] and of Marshall Hall[9]). The elicitation of muscular contractions was still

[1]Aristotle (384 B.C.–322 B.C.). *De Anima.* Translated by W. S. Hett. Loeb Classical Library. Heinemann, London, 1935, pp. 29, 49, 79, 121.

[2]Galen (130–200 A.D.). *De usu partium. Opera omina 3 and 4.* Translated by C. G. Kuhn. Leipzig, 1822.

[3]René Descartes (1596–1650). *De Homine.* Moyardus & Leffen, Leyden, 1662.

[4]Johannes Batista van Helmont. *De Magnetica.* Paris, 1621.

[5]William Harvey (1578–1657). *Exercitiones de generatione animalium.* Pulleyen, London, 1651.

[6]Thomas Willis (1621–1675). *Cerebri anatome cui accessit nervorum descriptio et usus.* Martyn and Allestry. London, 1664. (English translation by Samuel Portage, London, 1684.)

[7]S. T. Soemmering (1775–1830). *Über das Organ der Seele.* Mainz, 1796. (On the Organ of the Soul.)

[8]Alexander Stuart (1673–1742). Croonian Lectures. *Proc. Trans. R. Soc.,* 40:36–48 (1739).

[9]Marshall Hall (1790–1857). Synopsis of the Diastolic Nervous System. Croonian Lectures. Mallett, London, 1850.

considered to be in control of the soul; consequently, the activity of headless animals was a challenge. Robert Whytt,[10] the brilliant Scottish physician (and firm believer) felt forced to acknowledge that the soul must also have some extension in the spinal cord. He was not alone in this belief, which remained in the literature until the 19th century.[11] The movements of animals would not have worried Descartes, for he denied their having a rational soul.

The first, and most important recognition of localization of function in the brain, was the discovery in the early 18th century that nerve pathways coming to the periphery crossed in the pyramids, thus relating one side of the body to the contralateral brain. We owe the first recognition of this important fact to a young Pisan (who later became Professor) named Domenico Misticelli. He was puzzled by finding lesions on one side of the brain of a patient whose clinical signs were on the opposite side. In a tract[12] that he wrote on apoplexy, he stated (in translation):

> . . . what I have recently observed, that is, that the medulla oblongata externally is interwoven with fibers that have the closest resemblance to a woman's plaited tresses. Whence it occurs that many nerves that spread out on one side have their roots on the other; so for example, those that extend to the right arm, through such plaiting, can readily have their roots in the left fibers of the meninges. The same may be understood of those on the left proceeding from the right; and so one may go on describing, if not all the other nerves, that have origin immediately from the spinal cord.''

Misticelli's observations were only of the efferent meningeal fibers in the medulla, but a confirmation and more extensive conclusion soon came from a surgeon in the army of Louis Quatorze in the war of the Spanish Succession, fought in Flanders fields. His name was Pourfour du Petit,[13] and his observations were made on soldiers with unilateral head wounds but with obvious motor loss on the contralateral side of their bodies.

In all the early attempts to localize brain functions, it is the clinical neurologists (about whom there are many distinguished histories) who made the first progress.[14] In other words, progress was first made not by neurophysiological experiment but by clinical observation of neurological signs that were followed up, where possible, by postmortem examination of the brain. Outstanding in those years was the study of the loss of speech and the localization of its representation in the brain.

In the middle of the 19th century, those in search of localization in the brain were startled by the rise of what came to be known as phrenology. Adherents to this strange concept held that many of the brain's faculties caused localized bulging of discrete areas of the skull. What is perhaps most surprising in this strage hypothesis is that the man who postulated it, Franz Gall, was himself an excellent anatomist.

---

[10]Robert Whytt (1714–1766). *Observations on the Nature, Causes and Cure of those Disorders which are commonly called Nervous, Hypochondriac, or Hysteria, to which are prefixed some remarks on the sympathy of the nerves*. Becket, Hondt and Balfoor, Edinburgh, 1765.

[11]Edouard Friedrich Wilhelm Pflüger (1820–1910). *Die sensorischen Functionen des Rückenmarks der Wirbelthiere nebst einer Lehre über die Leitungsgesetze der Reflexionen*. Hirschwald, Berlin, 1853. (The Sensory Function of the Spinal Cord in Vertebrates together with a New Theory of the Laws of Transmission of Reflexes.)

[12]D. Misticelli (1675–1715). *Trattato del Apoplesia*. Rossi, Rome, 1709.

[13]F. Pourfour du Petit (1677–1714). *Trois lettres d'un Médecin des Hôpitaux du Roi*. Albert, Namur, 1710.

[14]A more detailed account of these early workers can be found in the first volume of this study: M. A. B. Brazier. *A History of Neurophysiology in the 17th and 18th Centuries: From Concept to Experiment*. Raven Press, New York, 1984.

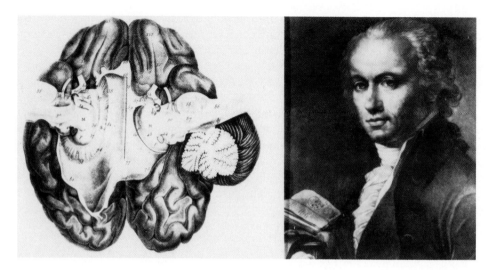

FIG. 55. **Left:** One of Gall's dissections of the interior of the human brain. **Right:** Franz Josef Gall (1758–1828) at the age when he was working in Vienna.

Franz Josef Gall (1758–1828)

Born of Italian stock in 1758 in a village in the Duchy of Baden, Gall received his medical education in Strasbourg, then in French Alsace, the university where, later, a totally different approach to the search for localization of function in the brain was to be taken by Goltz. In Strasbourg Gall concentrated on anatomy, moving later to Vienna for his doctorate in medicine. Vienna at this time was a magnet for those in medicine, for in 1784 the famous Allgemeine Krankenhaus opened—the first general hospital in Europe. Having graduated in 1785, Gall chose medical practice for his career, but by 1791 he had published a book[15] which gives the first clue to his developing concepts in psychology and philosophy.

The great contribution that Gall made to the anatomy of the brain was due to his technique (Fig. 55). Instead of slicing with a sharp knife, he manipulated with a dull instrument, following the nerve tracts from cortex to thalamus by gently separating them from surrounding tissue. His anatomy was also based on material from autopsies, and hence by relating his findings to the mental faculties of patients he had known, the seed for phrenology was sown. On reading through Gall's works one does not find the word "phrenology." He used instead "cranioscopy" and "organology." He thought that the "skull was molded on the brain."

The interest in localization of brain function was augmented by the lectures of Gall in Vienna. Suggestions of brain locations, even for sensory and motor functions, were not readily accepted at the time, but a major reason for antagonism toward Gall was his claim for localization of mental functions. Although one may not agree with the sites he chose or their categories, one must admit that today's localization of memory in the temporal lobe and of affect in the frontal lobes makes his general concept less ludicrous. It was when he

---

[15]Franz Josef Gall. *Philosophische medizinische Untersuchungen über Natur und Kunst im kranken und gesunden Zustand des Menschen.* 1791. (Philosophical Medical Investigations of Nature and Art in Man in Sickness and in Health.)

began to claim the ability to localize these features by bumps on the skull that he brought ridicule on himself.

Yet, surprisingly enough, there were many prominent men who gave him their support. One of these was Herbert Spencer, whose name is certainly well known to psychologists. Herbert Spencer[16] was a self-trained psychologist-philosopher who acquired a large following in the 1850s. To him, Gall's claim for localizing mental faculties by feeling the skull was attractive. Spencer went to a phrenologist to have his bumps read and came away with a report that opened with the sentence: "Such a head ought to be in the church." Seeking the basis for this statement in the itemized score for Spencer's bumps, we find both Firmness and Self-esteem reported as "very large," Language "rather full," and Wit and Amativeness only "moderate." Another distinguished subject to have his bumps read (by George Combe[17]) was the future King Edward VII when Queen Victoria despaired of her young son not taking life seriously enough.

Gall's lectures in Vienna, however, raised condemnation from both the church and the court, causing Gall to leave Vienna and eventually to settle in Paris where he lived for the rest of his life. One of his students, Johann Caspar Spurzheim, attracted by Gall's lectures in Vienna, followed him. Spurzheim became Gall's closest colleague, sharing authorship with him in the first two of the famous series of volumes of phrenology.[18] Apart from its bizarre approach, this series spurred a growing interest in the localization of function in the brain (Fig. 56). But the outstanding quality of Gall's tenets was his attack on the mind-brain problem and the claim to localize mental functions.

Gall believed that striking behavior of a given kind implied enlargement of a certain cortical region. This concept survived into this century when Oscar Vogt at the previously named Kaiser Wilhelm Institute in Berlin-Buch was asked to search for prominence of some kind in the brain of Lenin. Other prominent men whose brains were examined postmortem for enlargements of special regions of the brain (though not the skull) were the psychologists Wundt and·Tichener.

Gall's ideas came to grief because he believed that the enlargement of a cortical center caused the skull above it to protrude. It is surprising that so fine an anatomist as Gall should have devoted his career to phrenology, but it must be granted that his work emphasized that the brain is the organ of the mind—a concept not generally accepted at that time. It was this clearly materialistic concept that led to his condemnation in Vienna by both church and state. Gall and Spurzheim left Vienna at an extremely critical year for Austria and for Europe generally; 1805 was the year of Napoleon's victory at Austerlitz (for England it was the year of Trafalgar). By its defeat Austria lost the Tyrol and Venice, and the French occupied Vienna. The next year saw the end of the Holy Roman Empire.

Gall left Vienna in 1805, forbidden to teach there. One of the places he visited was Halle and there in 1805 he met Goethe. Goethe wrote afterwards: "He joked about us all, and maintained that, to judge by my forehead, I would not be able to open my mouth without pronouncing a trope, at which he was able to catch me out, admittedly, at every moment. . . ." Goethe saw in Gall the great pioneer of brain anatomy. "For the brain remains his main concern . . ..Gall's lecture can be seen as the pinnacle of comparative anatomy," words that remain true today. Gall was to wander over Europe lecturing widely for

---

[16]Herbert Spencer (1820–1903). *Principles of Psychology.* Longman, Green, London, 1855.

[17]George Combe (1788–1858). *Essays on Phrenology,* 1822.

[18]F. J. Gall and J. C. Spurzheim (1776–1832). *Recherches sur le Système Nerveux en générale et celui du Cerveux en particulier.* 4 vols. Schoell & Nicolle, Paris, 1809.

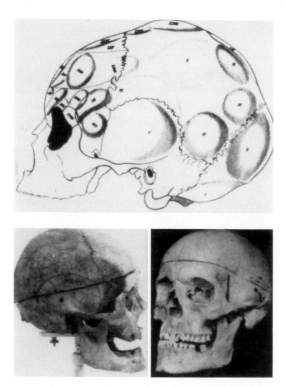

FIG. 56. **Above:** Gall and Spurzheim's map of a skull with certain areas marked for correspondence with different mental faculties. **Below:** For comparison: Gall's skull on the *left,* that of Spurzheim on the *right.* (The skull of Gall is in the Musée de l'Homme in Paris and is reproduced here by the kindness of Dr. Ardvège; that of Spurzheim is in the Warren Museum at the Harvard Medical School and was photographed by permission of the late Dr. P. I. Yakovlev.)

two years before settling in Paris. This was the Paris of Raspail,[19] of Pinel,[20] and of Cuvier,[21] and among them they blocked his acceptance by the Institute. This setback did not, however, dissuade him from taking out French nationality and writing his book in French. He ran a medical practice and continued his lectures, receiving acclaim mainly from his nonmedical audiences. Spurzheim stayed with him until 1813, and then he left to settle in England and later in America where he continued to publish, frequently introducing the name "phrenology."

In his publication of 1825,[22] Gall summed his philosophical concept of brain function as follows:

> In all my researches, my object has been to find out the laws of organization, and the functions of the nervous system in general, and of the brain in particular. The exposition of the nervous systems of the soul and the brain, and give the history of the discovery of each primitive moral and intellectual power, its natural history in a state of health and of disease, and numerous observations in support of the seat of its organ.

[19]François Vincent Raspail (1794–1878).

[20]Philippe Pinel (1745–1826).

[21]Georges Cuvier (1769–1832).

[22]F. J. Gall. *Sur les Fonctions du Cerveau et sur Celles de chacune de ses Parties.* Baillière, Paris, 1825. (Translation by Winslow Lewis, *Gall's Works,* Boston, 1835.)

An examination of the forms of heads of different nations, a demonstration of the futility of physiognomy, a theory of natural language, or pathognomy, added new weight to preceding truths.

The thorough development of the physiology of the brain, has unveiled the defects of the theories of philosophers, on the moral and intellectual powers of man, and has given rise to a philosophy of man founded on his organization, and consequently, the only one in harmony with nature.

. . . .

The physiology of the brain is entirely founded on observations, experiments, and research for the thousandth time repeated, on man and brute animals. Here, reasoning has had nothing more to do with it, than to seize the results, and deduce the principles that flow from the facts; and therefore it is, that the numerous propositions, though often subversive of commonly received notions, are never opposed to or inconsistent with one another. All is connected and harmonious; every thing is mutually illustrated and confirmed. The explanation of the most abstruse phenomena of the moral and intellectual life of man and brutes, is no longer the sport of baseless theories; the most secret causes of the difference in the character of species, nations, sexes, and ages, from birth to decrepitude, are unfold; mental derangement is no longer connected with a spiritualism that nothing can reach; man, finally, that inextricable being, is made known; organology composes and decomposes, piece by piece, his propensities and talents; it has fixed our ideas of his destiny, and the sphere of activity; and it has become a fruitful source of the most important applications to medicine, philosophy, jurisprudence, education, history, and surely, these are so many guarantees of the truth of the physiology of the brain—so many titles of gratitude to HIM, who has made them known to me!

Gall died in 1828, his pupil Spurzheim, four years later, but Gall's influence continued for many years. Unfortunately, his reputation as a phrenologist overshadowed his more important work on the fiber tracts of the white matter of the brain, work that clarified the previously contradictory ideas as to the anatomy of the commissures and of the pyramidal decussation. But, while his contemporaries concerned themselves with sites for motor functions, Gall proposed localization of mental faculties, and he may be regarded as a pioneer in emphasizing the importance of the gray matter for intellectual processes. Needless to say, in view of the insistence on the meaning of the cranial surface, their own skulls evoked interest. That of Gall is preserved in the Musée de l'Homme[23] in Paris (together with many of his papers). The skull of Spurzheim is in the Warren Museum of Harvard Medical School. Of the many attacks on his views that Gall had received in his lifetime, the most damaging at the time was that of the neurologist Flourens.

Pierre Jean Marie Flourens (1794–1867)

Pierre Flourens, who held the Chair of Comparative Anatomy at the University of Paris, was determined to destroy Gall's claims. He not only lectured against his "cranioscopic"[24] theories but also ran several experiments to refute them. Many of these were made on pigeons—a strange choice for refuting theories of cortical localization in view of the different structure of the avian brain. Flourens recognized three major functional regions of the brain—the cerebral hemispheres, the medulla, and the cerebellum—but within these entities he envisaged their action as global and their roles as being sensory, vital, and motor, respectively.

[23]This collection has been catalogued by E. H. Ackerknecht.

[24]P. Flourens. *Recherches Expérimentales sur les Propriétés et les Fonctions du Système Nerveux, dans les Animaux Vertébrés.* Crevot, Paris, 1824.

FIG. 57.   Left: Pierre Jean Marie Flourens (1794–1867). **Right:** Pigeon deprived of its cerebral hemispheres in position described by Flourens. (From Luigi Luciani (1840–1919). *Fisiologia del homo.* 3 vols. Le Monnier, Firenze, 1901–1911; *Human Physiology,* English edition. Macmillan, London, 1915.)

Concerning the cerebral hemispheres, he said that animals that survive their removal "lose perception, judgment, memory and will . . . therefore the cerebral hemispheres are the sole site of perception and all intellectual abilities." He did not hesitate to infer subjective qualities and faculties. In one of the more renowned of his experiments, he had kept a pigeon alive after removal of its cerebral hemispheres (Fig. 57). The bird was "blind" and "deaf" and appeared to be asleep, although it stirred when poked. Flourens went so far as to say that the bird lost its volition and "even the faculty of dreaming." He noted that it retained the sense of equilibrium and that its pupils reacted to light. Others repeating Flourens' experiments were unconvinced, for their decerebrate pigeons could be startled by a loud noise and could avoid obstacles. Since sudden death followed section of the medulla, Flourens concluded that there lay the essential mechanism for respiration and the maintenance of life. In this conclusion he had, of course, been anticipated by LeGallois.[25] Much of Flourens' fame as an experimentalist derived from his observation that expiration of the cerebellum (in birds and mammals) caused loss of coordinated movement. In this context Flourens quotes the multiple experiments of François Lauret[26] that aimed at relating the size of the cerebellum to sexual physiology. His experiments contrasted the weight of the cerebellum in stallions as compared with geldings. The weight was found greater in the geldings, a finding contrary to the expectations of Gall.

The global form of localization was promoted by Flourens, but he was adamant in his opposition to any such analysis of mental function for which he insisted the whole brain was essential. He was equally opposed to Gall's claims for special locations of aspects of the human character (such as pride and generosity). Flourens' concentration on lower cen-

[25]Julien Jean Cesar LeGallois (1770–1814). *Expériences sur le Principe de la Vie, Notamment sur celui des Mouvements du Coeur, et sur le Siège de ce Principe.* D'Hautel, Paris, 1812.

[26]François Lauret. *Anatomie comparé du Système nerveux omideré dans ses Raports avec l'intelligence.* Baillière, Paris, 1839.

ters ignored the cortex, though one must remember that at that time, even to him, an anatomist, the connecting pathways to the cortical mantle were unknown. Flourens believed in
the unity of the mind and its independence of anatomical connections. In contrast to Gall's
repudiation by the Académie des Sciences, Flourens was welcomed as a member at the age
of 35 and eventually inherited from Cuvier the post of permanent secretary to the Académie. Flourens' experimental technique was essentially ablation followed by extreme care
in keeping his animals alive for a long period of observation. A typical experiment in which
he had removed both cerebral lobes of a bird was reported four months later as follows:[27]

> I let this hen starve several times for as long as three days. Then I brought nourishment
> under her nose, I put her beak into grain, I put grain into her beak, I plunged her beak into
> water, I placed her on a shock of corn. She did not smell, she did not eat anything, she
> did not drink anything, she remained immobile on the shock of corn, and she would cer-
> tainly have died of hunger if I had not returned to the old process of making her eat myself.
>   Twenty times, in lieu of grain, I put sand into her beak; and she ate this as she would
> have eaten grain.
>   Finally, when this hen encounters an obstacle in her path, she throws herself against it,
> and this collision stops her and disturbs her, but to collide with an object is not the same
> as to touch it. . . .

Flourens was aware that some work already existed that contradicted his views on the
lack of influence by the cortex, and among those were the experiments of Rolando.[28] In
experiments that antedated the epoch-making discovery of Fritsch and Hitzig[29] and that produced movements of the body by electrical stimulation of the cortex, Flourens rejected this
as being due to spread of current to lower centers, saying: "My experiments establish that
the hemispheres of the brain do not produce any movement." These words essentially sum
up Flourens' position in the puzzle of brain physiology.

Gall, whose views he had done so much to destroy, had died before the publication of
Flourens' most important book (somewhat oddly dedicated to Descartes). Flourens was a
man of very strong opinions, and his authority made him the most important opponent of
Gall. In the acrimonious dispute over priority for defining the sensory and motor roots of
the spinal cord, he took the side of Bell.[30] Flourens made his position in this controversy
quite clear and would not change it in spite of a spirited defense by Magendie. This altercation took place at a meeting of the Académie des Sciences in 1847.[31]

In the future, long after all protagonists had died, an extremely important discovery
brought new light to the problem. This was the 20th-century demonstration by Leksell[32] of
sensory innervation of muscle spindles by fine fibers in the ventral root. This demonstration

[27]P. Flourens. *Recherches Expérimentales sur les Propriétés et les Fonctions du Système Nerveux dans les Animaux Vertébrés*. Crevot, Paris, 1824. (Translation by G. von Bonin in *The Cerebral Cortex*. Thomas, Spring-field, Ill., 1960.)

[28]Luigi Rolando (1773–1831). *Inductions physiologiques et pathologiques sur les différentes Especes d'Exci-tabilité et d'Excitement, sur l'Irritation, et sur les Puissances excitantes, débilitantes et irritantes*. Caille et Ra-vier, Paris, 1822. (Physiological and Pathological Inductions on the Different Kinds of Excitability and Excite-ment, on the Irritation and on the Exciting Powers, Debilitating and Irritating.)

[29]G. T. Fritsch (1838–1891) and E. Hitzig (1838–1927) Über die elektrische Erregbarkeit des Grosshirns. *Arch. Phys. Wiss. Med.*, 37: 300–314 (1870).

[30]Charles Bell (1774–1842). *Idea of a New Anatomy of the Brain*. Submitted for observation of his friends. Privately printed, 1811. (Reproduced in: J. F. Fulton, *Selected Reading in the History of Physiology*. Thomas, Springfield, Ill., 1930, p. 251.)

[31]P. Flourens. *C. R. Acad. Sci.*, 24: 253–320 (1847).

[32]L. Leksell. The action potential and excutatory effects of the small ventral root fibers to skeletal muscle. *Acad. Sci. Physiol. Scand.* (Suppl. 31), 10 (1945).

destroyed the concept of a pure dichotomy of function of the roots so dear to the 19th century.

The arguments in the Académie took place after Bell's death, with Flourens arguing on his side. But what emerges far more clearly than a scientific interest in nerve roots is Flourens' antagonism to Magendie. This antagonism was deep set and long lasting; Flourens had opposed him for membership in the Académie and had persisted in ridiculing his character, even in an obituary[33] about Magendie after his death—circumstances when *de mortuis nihil nisi bonum* is usually observed. Flourens added an appendix especially referring to the spinal roots. To the modern neurophysiologist, Flourens' method of serial slicing through the cerebrum to reveal unity of function is ludicrous. His work on the cerebellum was more valid.

It is tragic that Flourens, whose interest lay so deeply in the elucidation of the control of voluntary movement, was himself to suffer paralysis for a long period before his death in 1867.

Friedrich Leopold Goltz (1834–1902)

In the 20th century, so much has been revealed about localization of function in the brain, and even in the human brain, that the initial researches in the 19th century seem crude indeed. The startling results of cerebral ablations in submammalian species (by Alexander Stuart, Marshall Hall, Sechenov, and others) were a challenge to those interested in the mammalian brain, and among these was Friedrich Leopold Goltz.

Goltz led a life in countries plagued by war and changing nationality. He was born in 1834 in Posen, brought up in Danzig, and educated at Königsberg—all then in East Prussia but now no longer German. His most important contribution to neurophysiology was the work he did at the University of Halle and later in Strasbourg, at the university founded after the Franco-Prussian War. After his medical training at Königsberg, he began to devote himself to the laboratory where he made intensive studies of the reactions of decerebrate frogs, greatly expanding the work of Marshall Hall. His findings were published (Fig. 58) in a small book[34] illustrated with line drawings of the condition of his frogs and their movements after surgical interference with their nervous systems.[35]

While at Königsberg, although Goltz had not yet found his major field, he instigated some experiments on shock, on reflex croaking in frogs, and on the physiology of the semicircular canals. The major work for which he is remembered was done at Strasbourg. There he moved to the mammalian brain, using dogs as his subjects, a choice that may seem strange in view of the variability in size and shape of the skull in the species. To the modern neurophysiologist, the surgical ablations he made appear so gross that one wonders not only what conclusions could be made but also how the animals could survive.

He was, however, successful in keeping many of his decorticate dogs alive and some survived for many months. He recognized that the animal's vision depended on his sparing its locale in the brain, but apart from this sense he related the amount of deficit to the volume of tissue removed. Goltz was careful to have the brains he removed from his dogs photographed postmortem. Several of these he took to a professional photographer. All were not published, and in 1940, when a later generation was clearing drawers at the Uni-

[33]P. Flourens. Mémoire de Magendie. *C. R. Acad. Sci.* (1858).

[34]F. L. Goltz. *Beiträge zur Lehre von den Functionen der Nervenzentren des Frosches*. Hirschwald, Berlin, 1869. (Contributions to the Theory of the Function of the Nerve Centers of the Frog.)

[35]F. L. Goltz. Über die Verrichtungen des Grosshirns. *Arch. Physiol.*, 13: 1–44 (1876). (On the functions of the brain.)

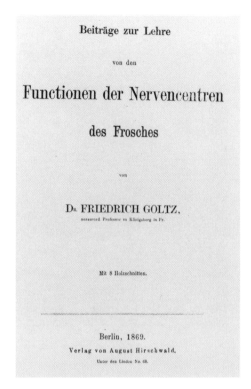

FIG. 58. Left: Goltz's first book describing his ablation experiments on frogs. **Right:** Friedrich Leopold Goltz (1834–1902).

versity of Strasbourg at the threat of invasion, more photographs were found.[36] Goltz took the dog pictured in Fig. 59 to the International Medical Congress in London in 1881; understandably, he tangled with Ferrier who was describing precise location of function in the cortex of the monkey.[37, 38] Goltz sacrificed the dog he had shown at the Congress and gave the material to Langley, Professor of Physiology at the University of Cambridge. Langley became very prominent and powerful in the field of physiology in that he was from 1884 the editor of the *Journal of Physiology,* the major outlet for this field.[39–44] He died in 1925 with a long list of publications in the 20th century as well as those listed here.

[36]When Alsace was invaded in 1940, these original photographs of Goltz's dogs were rescued by the late Professor Paul Dell and, with great kindness, given to this author.

[37]David Ferrier (1843–1928). The localization of function in the brain. *Proc. Trans. R. Soc.,* 22: 229–252 (1873–1874).

[38]E. A. Schäfer. Report on the lesions, primary and secondary, in the brain and spinal cord of the macacque monkey exhibited by Professors Ferrier and Yeo. *J. Physiol.,* 4: 316–326 (1883).

[39]E. Klein, J. N. Langley, and E. A. Schäfer. On the cortical areas removed from the brain of a dog and the brain of a monkey. *J. Physiol.,* 4: 231–247 (1883).

[40]J. N. Langley. The structure of the dog's brain. *J. Physiol.,* 4: 248–309 (1883).

[41]E. Klein. Report on the parts destroyed on the left side of the brain of the dog operated on by Prof. Goltz. *J. Physiol.,* 5: 286–332 (1884).

[42]J. N. Langley. *The Automatic Nervous System.* Cambridge University Press, 1821.

[43]J. N. Langley and C. S. Sherrington. Secondary degeneration of nerve tracts following removal of the cortex on the cerebrum in the dog. *J. Physiol.,* 5: 49–65 (1884).

[44]J. N. Langley and C. S. Sherrington. On the degeneration resulting from removal of the cerebral cortex and corpora striata in the dog. [*J. Physiol.,* 11 (Suppl.), p. 606] *Cblt. Physiol.,* 2: 57–58 (1884).

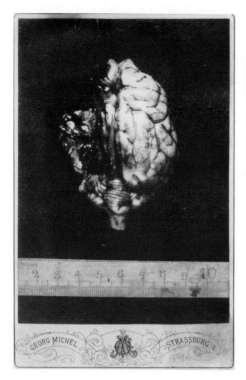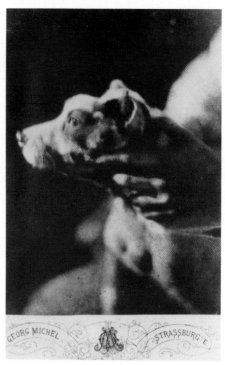

FIG. 59.   Left: Brain removed from the dog **(right)** that was demonstrated alive by Goltz at the International Medical Congress in 1881. Photographs from a professional studio. (Gift of the late Paul Dell).

John Newport Langley was born in 1852 and educated at Cambridge, where he spent most of his life—first as a student of Michael Foster and eventually as Professor of Physiology. His interests were in the autonomic nervous system. Most of his important contributions came in the 20th century, as did the outstanding contributions of some of his students, one of whom was to become one of the most outstanding physiologists of the next century—Charles Sherrington. He and another young man in Langley's laboratory, A. S. Grunbaum, were drawn (as was Langley himself) into the current interest in localization of function in the brain. Langley made known the anatomical findings and published them with the student who had helped in the dissections. The student's name was Charles Sherrington. Langley's technique was to look for degeneration of nerve tracts in the thalamus and spinal cord caused by removal of cortical regions. He made an extensive study of what he called "the fissures and convolutions" of the dog's brain (Fig. 60). Langley reported as follows:

> The whole of the right anterior pyramid was markedly sclerosed, the mesial portion being least, the ventral portion most sclerosed.
> The sub-olivary tract was normal; it cannot then be in direct continuation with the dorsal part of the crusta; in the decussation its fibers become placed lateral to the pyramid fibers; the two sets cross over together throughout, or nearly throughout the whole region of the decussation. The transverse area of the sclerosed region diminishes considerably in passing through the pons, and in the decussation of the pyramids. The sclerosis can be traced in the lateral pyramidal tract on each side of the cord as far as the upper part of the lumbar regions; throughout the sclerosed area are many normal nerve-fibres; for the form of the

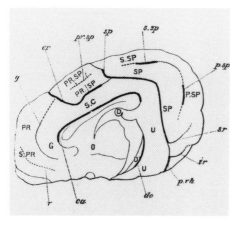 

FIG. 60. **Left:** Langley's map of the cortical regions of the dog's brain (*J. Physiol.,* 4 248–309, 1883). **Right:** John Newport Langley (1852–1925).

area in different parts of the cord. . . . The patch of greatest sclerosis is on the left side of the cord, and is placed more ventrally than the patch of greatest sclerosis on the right side. Not improbably, the dorsal part of the lateral pyramid tract is connected with the cortex posterior to the sigmoid gyrus.

Presumably because of the size of the lesions he made, Goltz concluded that claims (and especially those of Ferrier) that localization of function exists in the cortex were false. He expressed himself quite forcibly about the conclusions he drew from his surgical experiments. Of the behavior of the dogs after removal of one hemisphere, he wrote:[45]

We have seen that after a thorough destruction of one half of the brain, that in the initial stages following the operation very great disturbances are present in the opposite side of the body, perhaps as badly as to cause blindness, paralysis and numbness. However, very soon the animal improves, and the only permanent damage may be a dulling of sensation, a disturbance in sight, and a disturbance in certain movements.

Goltz, of course, had to meet the challenge evoked by the findings of Fritsch and Hitzig[46] that electrical stimulation of given cortical areas produced specific muscle movements. In summarizing his stubborn resistance to cortical localization, he wrote:

1. If one destroys a part of a dog's cortex, then one can by observation conclude that such an animal still possesses a conscious will, especially evident in its search for food. *It is not possible, by damaging the cortex in any way, to produce a long-lasting paralysis of any muscle.*
2. Such an animal still possesses conscious sensation. *It is not possible to produce complete numbness in any part of the body by destroying any part of the cortex.* The serges of sight, hearing, smell and taste can also not be permanently destroyed by destroying any part of the cortex. However, all the senses are deadened.

[45]E. Klein. Report on the parts destroyed on the left side of the brain of the dog operated on by Prof. Goltz. *J. Physiol.,* 4: 310–335 (1883).

[46]G. T. Fritsch and E. Hitzig. Über die elektrische Erregbarkeit des Grosshirns. *Arch Anat. Physiol. Wiss. Med.,* 37: 300–382 (1870). (On the electrical excitability of the cerebrum.)

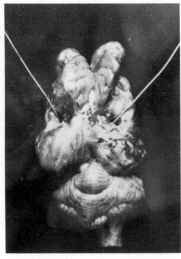

FIG. 61.   Left: Friedrich Goltz toward the end of his life, still defending his theory of the lack of a motor cortex. **Right:** One of his photographs of an almost total cortical ablation. (Gift of the late Paul Dell.)

3. Such an animal shows a long-term impairment of intelligence, which increases according to the amount of cortex destroyed.
4. It has not been established whether each part of the cortex is of the same importance. Animals with destroyed parietal lobes displayed a greater deadening of feeling; those with destroyed napex lobes showed greater disturbance of sight.
    I conclude that each part of the cortex takes part in the process of will, sensation, imagination and thought. Each part is independent of the rest, connected to all voluntary muscles and sensitive nerves.

Goltz's reply to the objection of Edouard Hitzig, the psychiatrist, was that the latter mistook the temporary phenomenon of inhibition following surgery to be a loss that would be permanent. His opponents all had their explanations for this apparent survival of function; they held that either Goltz's ablations were not complete or that the gray matter of the opposite hemisphere had taken over control.

Goltz's explanation was that the motor movements did not have localized representation in the cortex, and he proceeded to bilateral cortical ablations claiming no loss of movement (Fig. 61). Behind his stubborn refusal to accept cortical localization of movement lay his opinion that the controlling regions lay in the cerebellum. (This had been the view of Willis in an earlier century.) On this point Goltz wrote:

> Actions such as walking and running are not based in the cerebrum but rather in the cerebellum. How do I explain disturbances in these actions after injury to the brain, if the cerebrum is not the center for these actions. I believe, through a process of inhibition which is transmitted from a site of the wound. This inhibition gradually fades, and many functions which seemed to have been lost return, leaving only those functions disturbed which were actually controlled from the part of the brain which was destroyed. These disturbances can then not be overcome.

Goltz was also disturbed by the claim of Herman Munk[47] that there was a specific center

[47]Herman Munk (1839–1912). Über die Functionen des Grosshirns. Hirschwald, Berlin, 1881. (On the Functions of the Cerebrum.)

for sight. He had, in fact, found some loss of sight to follow nearly all his cortical extirpations: "All the animals so far described showed (he stated) disturbances of sight. . . . I maintain that one would obtain disturbances in sight by destruction of any part of the cortex."[48] It was, of course, Munk who had localized the center of vision in the occipital lobe. He made punctate lesions in the cortex and found areas for impairment of vision but was puzzled that in time the animals regained apparent vision as judged by their behavior.

Goltz was very disturbed by the failure of other workers to accept his theories and gave much space to attacking them (especially Hitzig, Munk, and Ferrier). In spite of the gathering evidence, he never changed his mind. Friedrich Goltz died in 1902, at the beginning of the 20th century, the century in which the power of subcortical regions was to be revealed, thus in part supporting his views of the role of centers other than the cortical mantle. A long obituary of Goltz with his portrait and a list of his publications was published by J. J. R. Ewald[49] in *Pflüger's Archiv.*

BIBLIOGRAPHY

Franz Josef Gall (1758–1828)

Selected Writings

Philosophishe medizinische Untersuchungen über Natur und Kunst im kranken und gesunden Zustand des Menschen. 1791. (Philosophical Medical Investigations on Nature and Art in the Sick and Well Human Condition.)

Recherches sur la Système nerveux en genéral (with J. C. Spurzheim). Schoell & Nicolle, 1809. (Researches on the General Nervous System.)

Anatomie et Physiologie du Système Nerveux en genéral, et du Cerveau en particulier, avec des Observations sur la Possibilité de reconnaître plusiers Dispositions intellectuelles et morales de l'Homme et des Animaux, par la configuration de leurs têtes. 4 vols. With atlas of 100 engraved plates. Schoell & Nicolle, Paris, 1810–19. (Gall is sole author of Vols. 3 and 4; J. C. Spurzheim, is co-author of Vols. 1 and 2.) (Anatomy and Physiology of the General Nervous System, and the Brain in Particular, with Observations on the Possibility of Recognizing many Moral and Intellectual Dispositions of Men and Animals by the Configurations of their Heads.)

Sur les Fonctions du Cerveau et sur Celles de Chacune des ses Parties (with J. C. Spurzheim). 6 vols. Ballière, Paris, 1822–25. (On the Functions of the Brain, and Each of its Parts.)

On the Functions of the Cerebellum by Drs. Gall, Vimont and Braussais, translated by George Combe. Maclachlan and Stewart, Edinburgh, 1838.

Gall, F. J., and Spurzheim, J. C. *On the Functions of the Brain and of Each of its Parts: With Observations on the Possibility of Determining the Instincts, Propensities, and Talents, or the Moral and Intellectual Dispositions of Men and Animals, by the Configuration of the Brain and Head* (with J. C. Spurzheim), translated by Winslow Lewis, Jr. 6 vols. Marsh, Capen and Lyon, Boston, 1835.

Spurzheim, J. C. *Examination of the Objections Made in Britain Against the Doctrines of Gall and Spurzheim.* Macrede, Skelly and Muchesy, Edinburgh, 1817.

Suggested Readings

Ackerknecht, E. H., and Vallois, V. *Franz Joseph Gall, Inventor of Phrenology and His Collection.* Translated from the French by Claire St. Leon. Prefaced by John Z. Bowers. Wisconsin Studies

---

[48]F. L. Goltz. *Gesammelte Abhandlungen.* Bonn, 1881.

[49]J. J. R. Ewald. Friedrich Goltz. *Pflüger's Arch. f.d. ges. Physiol.,* 93:1–64 (1903).

in Medical History. No. 1. Madison, Wisconsin, Department of History of Medicine, University of Wisconsin Medical School, 1956.

Clarke, E., and Dewhurst, K. *An Illustrated History of Brain Function*. Sanalford Publications, Oxford, 1972.

Gardener, M. *Fads and Fantasies in the Name of Science,* Dover, New York, 1957.

Jefferson, Geoffrey. *Selected Papers*. Pitman, London, 1960.

Nacquart, Jacques Baptiste. *Traité sur la nouvelle Physiologie du Cerveau*. Leopold Collin, Paris, 1808.

Riese, Walter, and Hoff, Ebbe C. A history of the doctrine of cerebral localization, *J. Hist. Med.,* 5:50–71 (1950).

Temkin, O. Gall and the phrenological movement. *Bull. Hist. Med.,* 21:275–321 (1947).

Young, R. *Mind, Brain, and Adaptation*. Clarendon Press, Oxford, 1970.

Pierre Jean Marie Flourens (1794–1867)

Selected Writings

Arch. Gen. Med., 2:321 (1823).

Recherches Expérimentales sur les Proprietés et les Fonctions du Système Nerveux dans les Animaux Vertébrés. Cervot, Paris, 1824.

Experiences sur les canaux semicirculaires de l'oreille. *Mém. Acad. Roy. Sci. (Paris),* 9:455–477 (1830).

Examen de Phrénologie. Paulin, Paris, 1843. (English translation by C. de L. Meigs, *Phrenology Examined*. Hogan and Thompson, Philadelphia, 1943.)

Note touchant l'action de l'éther sur les centres nerveux. *C. R. Acad. Sci.,* 24:340–344 (1847).

De la vie et de l'intelligence. Garnier, Paris, 1858.

Mémoire de Magendie. *C. R. Acad. Sci. (Paris),* 33, 1–17 (1858).

De la Phrénologie et des Etudes vraies sur le Cerveau. Garnier, Paris, 1863.

Suggested Readings

von Bonin, G. In: *The Founders of Neurology*. Thomas, Springfield, Ill., 1970.

Kruta, V. Pierre Jean Marie Flourens. In: *Dictionary of Scientific Biography,* Vol. 5. Scribner, New York, 1972, pp. 44–45.

Friedrich Leopold Goltz (1834–1902)

Selected Writings

Beiträge zur Lehre von den Functionen der Nervencentren des Frosches. Hirschwald, Berlin, 1869. (Contributions to the Theory of the Functions of the Nerve Centers of the Frog).

Über die physiologische Bedeutung der Bogengänge des Ohrlabyrinths. *Pflüger's Arch. f. d. ges. Physiol.,* 3:172–192 (1870). (On the physiological significance of the semi-circular canals of the labyrinth of the ear.)

Über die Aufsaugung und Fortführung von Giften nach Unterbrechung des Blutkreislaufs. *Pflüger's Arch. f. d. ges. Physiol.,* 4:147–148 (1871). (On the absorption and conveyance of poisons after interruption of the circulation of the blood.)

Über den Einfluss der Nervencentren auf die Aufsaugung. *Pflüger's Arch. f. d. ges. Physiol.,* 5:53–76 (1872). (On the influence of the nerve centers on absorption.)

Studien über die Bewegungen der Speiseröhre und des Magens des Frosches. *Pflüger's Arch. f. d. ges. Physiol.*, 6:616–642 (1872). (Studies on the movements of the alimentary canal and the stomach of the frog.)

Über die Functionen des Lendenmarks des Hundes. *Cbl. Med. Wiss.*, 12:645–647 (1874). (On the functions of the lumbar marrow of the dog.)

Ueber gefässerweiternde Nerven. *Pflüger's Arch. f. d. ges. Physiol.*, 9:174–187 (1874). (On vessel expanding nerves.)

Ueber den Einfluss des Nervensystems auf die Vorgänge während der Schwangerschaft und des Gebärakts. *Pflüger's Arch. f. d. ges. Physiol.* 9:552–565 (1874). (On the influence of the nervous system on bodily processes during pregnancy and in labor.)

Über die Functionen des Lendenmarks des Hundes. *Cbl. Med. Wiss.*, 12:645–647 (1875). (On the functions of the lumbar marrow of the dog.)

Über die Verrichtungen des Grosshirns. *Pflüger's Arch. f. d. ges. Physiol.*, 13:1–44 (1876). (On the functions of the cerebrum.)

Über die Verrichtungen des Grosshirns. *Pflüger's Arch. f. d. ges. Physiol.*, 14:412–443 (1877). (On the functions of the cerebrum.)

Ein Vorlesungsversuch mittelst des Fernsprechers (Telephons). *Pflüger's Arch. f. d. ges. Physiol.*, 16:189–190 (1878). (An attempt to hold a lecture by telephone.)

Ueber die Verrichtungen des Grosshirns. Dritte Abhandlung von Prof. Goltz zu Strassburg i. E. Unter Mitwirkung von Dr. J. v. Mering. *Pflüger's Arch. f. d. ges. Physiol.*, 20:1–54 (1879). (On the functions of the cerebrum. Third treatise by Prof. Goltz of Strasbourg.)

Über die Druckverhältnisse im Innern des Herzens. *Pflüger's Arch. f. d. ges. Physiol.*, 17:100–120 (1878). (On the pressure relationships inside the heart.)

Über die Verrichtungen des Grosshirns. *Gesammelte Abhandlungen*, Bonn, 1881. (On the Function of the Brain.)

Über die Verrichtungen des Grosshirns. Amel. Strauss, 1881. (On the Functions of the Cerebrum.)

Zur Physiologie des Grosshirns vorläufige Mittheilung. *Pflüger's Arch. f. d. ges. Physiol.*, 28:579–580 (1882). (On the physiology of the brain, preliminary report.)

Über die Verrichtungen des Grosshirns. *Pflüger's Arch. f. d. ges. Physiol.*, 34:450–505 (1884). (On the functions of the cerebrum.)

Der Hund ohne Grosshirn. Siebente Abhandlung über die Verrichtungen des Grosshirns. *Pflüger's Arch. f. d. ges. Physiol.*, 51:570–614 (1892). (The dog without a brain. Seventh treatise on the functions of the brain.)

Suggested Readings

Ewald, J. R. Friedrich Goltz. *Pflüger's Arch. f. d. ges. Physiol.*, 94:1–64 (1903).

Haymaker, W. Friedrich Goltz (1834–1902). In: *The Founders of Neurology*. Springer, New York, 1970.

Rothschuh, K. E. Friedrich Leopold Göltz. *Dictionary of Scientific Biography*. Vol. 5, 1972, Scribner, New York, pp. 462–464.

CHAPTER IX

Revelations by the Microscope

When Fritsch and Hitzig[1] published their finding that electrical stimulation of certain sites in the outer cortical layer produced specific muscle movements, their chief opponent, Ludimar Hermann,[2] maintained that the effect was due to current spread to fibers deep in the cortex. One has to remember that at the time of this argument, the neuron theory was not yet known. It had not yet been established that nerve fibers arose from the nerve cell, so that stimulation of the one would be transmitted to the other.

During the 19th century, there was a great development in microscopy that resulted in individual brain cells being visualized for the first time. The first single cell was seen in the cerebellum by Valentin[3] in 1836. The first in the motor cortex was described by Betz[4] in 1874, four years after Fritsch and Hitzig had demonstrated the evocation of motor movements by electrical stimulation. The motor and sensory lobes of the brain and the cerebellum had already been differentiated by Rolando at the beginning of the century.

Luigi Rolando (1773–1831)

Famous as Sardinia's illustrious anatomist, Luigi Rolando[5-8] was born in 1773; he had been educated in Turin at a time that coincided with the turbulent years of the Napoleonic

[1]G.T. Fritsch (1838–1927) and E. Hitzig (1838–1907). Über die elektrische Erregbarkeit des Grosshirns. *Arch. Anat. Physiol. Wiss. Med.*, 37:300–314 (1870). (On the electrical sensitivity of the cerebrum.)

[2]L. Hermann. Über elektrische Reizversuche an der Grosshirnrinde. *Pflüger's Arch. f. d. ges Physiol.*, 10:77–85 (1875). (On electrical stimulus experiments on the cerebral cortex.)

[3]Gabriel Gustav Valentin (1810–1883). Über den Verlauf und die letzten Enden der Nerven. Verhandlungen der Kaiserlichen Leopoldnish-Carolinischen Academie der Naturforscher, Breslau, 18:151–240 (1836). (On the extent and the terminals of the nerves.)

[4]V.A. Betz (1834–1894). Antomischer Nachweis zweier Gehirnzentra. *Zlb. Med. Wiss.*, 12:578–580; 595–599 (1874). (Anatomical proof of two brain centers.)

[5]L. Rolando. *Anatomie Physiologica-Comparative Disquitio in Respirationes Organa*. Turin, 1801. (Investigation of Physiological Comparative Anatomy in Respiratory Organs.)

dominance of northern Italy, times when Napoleon felt himself overextended and gave Thomas Jefferson the opportunity to make the Louisiana Purchase of a region so vast that no less than 16 of the present states of America had some part in that land area, several being wholly in it. In that year (1804), Rolando left the troubled north and accepted the Chair of Medicine at the University of Sassari, which had been founded in 1677 across the water in Sardinia. He had gained on the way experience in Florence, where another distinguished anatomist from the north (Fontana) had preceded him. Later, through the patronage of the King of Savoy, he was to attain an accessory appointment at his old university at Turin.

This distinguished man, receiving great recognition during his lifetime, left no portrait; the only likeness that exists is one made in his honor after his death. But his name lives on in all maps of the cerebral cortex as the Rolandic fissure, the clear recognition of the division of the brain into sensory and motor lobes being one of his achievements. Rolando's work was not only on the cerebrum but also on the spinal cord (on which he published a finely illustrated book) and on the cerebellum. The functions of the latter were at that time a source of argument, and Rolando was the first to declare unequivocally that its function involved muscular movement.

Born into the decade when theories of the great Bolognese anatomist Galvani were under attack, led by Volta, it is of interest to find Rolando giving serious consideration to the possible role of electricity in movements initiated by the cerebellar cortex whose layering of cells reminded him of a voltaic pile (in his publication of 1809). For the scientists whose interests were later to bring them the name of neurophysiologists, the step from intimate anatomical structure to the orchestration of function in the brain thus received a pioneer contribution from this great experimenter. He died in 1831 before his championship of the possible role of electricity received its final accolade from Du Bois-Reymond.[9]

The growth of knowledge about the gross anatomy of the cortex owed much to Rolando but there was another observation, made in the previous century, that proved a lead toward analysis of its more intimate structure. This was the observation of Francisci Gennari. In 1782 this Italian anatomist published a study[10] in which he described the laminated appearance of the cortex that could be seen even without a microscope; outstanding was a white line parallel to the cortical surface. Gennari described it as an "insole albidior" and noted that, although he found it elsewhere in the brain, it was the most conspicuous in what he described as "the internal part of the posterior lobe of the brain not far from the point in which it extends itself into the tentorium." This line, whose white appearance is due to the myelinated fibers, is now known by his name as an eponym—"the white line of Gennari."

[6]L. Rolando. *Saggio Sopra le Vera Struttura del Cervello dell'uomo e del' Animali e Sorpra le Funzioni del Sistema Nervoso*. Nella Stamperia da S.S.R.M. Privilegiata, Sassari, 1809. (Essay on the True Structure of the Human Brain and of the Animals and on the Functions of the Nervous System.)

[7]L. Rolando. *Inductions Physiologiques et Pathologiques sur les Différentes Especes d'Excitabilité et d'Excitment, sur l'Irritation, et sur les Puissances Excitantes, Debilitantes et Irritantes*. Caille et Ravier, Paris, 1822. (Physiological and Pathological Inductions on the Different Kinds of Excitability and Excitement, on the Irritation and on the Exciting Powers, Debilitating and Irritating.)

[8]L. Rolando. *Richerche Anatomiche sulla Struttura del Midollo Spinale*. Stamperia Reale, Torino, 1824. (Anatomical Research on the Structure of the Spinal Cord.)

[9]E. Du Bois-Reymond. *Untersuchungen über thierische Elektricität*. Vol. 1. Reimer, Berlin, 1848. Vol. 2, Part 1. Reimer, Berlin, 1849. (Investigations into Animal Electricity.)

[10]Francisi Gennari (1752–1792). *De peculiari Structura Cerebri Normalisque eins Morbis*. Ex Regio Typographico, Parma, 1782.

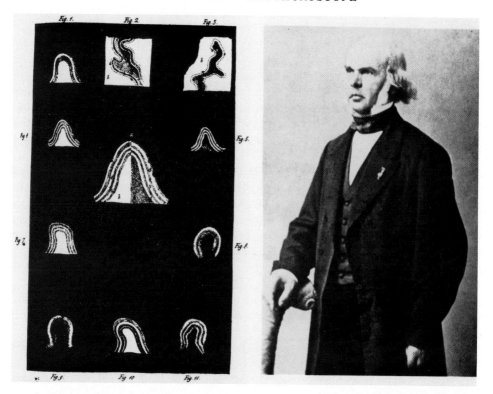

FIG. 62. Left: Drawings by Baillarger of the layers of the cortex seen with a simple microscope. Included are two sketches he reproduced to show the white line of Gennari (his Figures 2 and 3). (From: *Mem. Acad. R. Med., Paris,* 8:149–183, 1840.) **Right:** Jules Gabriel François Baillarger, alienist at the Clarenton at the age of approximately 40.

In the 19th century, one of the scientists to work most actively on this "white line" was the French psychiatrist Baillarger.

Jules Gabriel François Baillarger (1806–1891)

In the first half of the 19th century, the most persistent work on the line of Gennari was done by Baillarger[11] who recognized the name and findings of Gennari's earlier work. His illustrations are crude and hardly more detailed than those of Gennari nearly half a century before (Fig. 62). In fact, he included two of Gennari's drawings in the display he used to illustrate his communication.

His conclusions are interesting and unexpected in an anatomist. The start of his century had been rocked by the controversy between Galvani and Volta, and everyone had become familiar with pictures of Volta's pile. Baillarger compared the appearance of the layered cortex to a "plie galvanique." It is interesting that he used Volta's word "pile" with an adjective derived from the man whose findings he strove to repudiate. He wrote:[12]

[11]J. G. F. Baillarger (1806–1891). Recherches sur le structure de la couche corticale des circonvolutions du cerveau. *Mem. Acad. R. Belg.,* 1:149–183 (1840).

[12]J. G. F. Baillarger. Mémoire historique et statistique sur la maison royale de Charenton. *Ann. Hyg. Pub. Med. Leg.,* 13:5–192 (1835).

This analogy between the structure of the cerebral surface and appearance of galvanic apparatus perhaps involves another argument in favor of the two following propositions:

1. That nervous action, like that of electricity, is not due to the mass but to the surfaces.
2. The flow of nervous transmission, like that of electricity, is transmitted along surfaces.

Baillarger's contributions did not stop there, for he examined the cortex in other lobes of the brain, not only the occipital cortex where the layering is most obvious. He had help from the simple microscope that he used in his work at the psychiatric hospital of Charenton. This old 17th-century hospital (which has given its name to a station on the metro), located on the east side of Paris, had been greatly updated by Esquirol. There Baillarger had access to the brains of those who died. He had no doubt that the cortical layers were prolongations of the central white matter. He was aware that several alchemists had read rather extreme meanings into variations in these cortical layers, variations which they associated with certain abnormal conditions. Even the great Pinel[13] assigned profound meaning to the cortical layers, believing them to be pathological, thinking a single layer to be the normal condition.

Baillarger's contribution was not only to put these speculations to rest but, from his own studies, to identify six layers. It was definite, Baillarger claimed, that "the convolutions of the brain are formed by the prolongation of the central white matter covered with a layer of gray substance about a line and a half in substance."

Largely because of the inadequacy of the early microscopes, the multiple layering of the cortex had previously been missed. Occasionally, investigators found "a third layer" of gray matter but claimed that particular brain abnormal. Baillarger was determined to solve this question and described his technique in detail: He cut very thin vertical slices through the convolutions and pressed the tissue between two thin plates of glass held together with wax. With his simple microscope he found he could see no less than six layers. He examined the brains of many kinds of animals—dog, pig, horse, sheep, cat, and rabbit—but he put most care into the study of the human brain, including that of the newborn. Greatly improved microscopes led to a spreading interest in the layering of the cortex, the question raised a century before by Gennari.

Gabriel Gustav Valentin (1810–1883)

The largest cells, the first to be clearly described, were found in the cerebellum by G. G. Valentin in 1836,[14] but he noted only their cell bodies. Within three years of Valentin's report, his contemporary T. Schwann[15] published his cell theory. Valentin was born in 1810 in Breslau, a city on the Oder River. In Valentin's day this was the capital of lower Silesia but is now in Poland and named Wroclaw. He was a pupil of the great Purkyně at the University of Breslau and, after gaining his degree, moved to Berne where he spent the rest of his very active scientific life (Fig. 63). Valentin's work was outstanding, for not

[13]Philippe Pinel (1745–1826). *Traité médico-philosophique sur l'Alientionmentale ou la Manie*. Caille et Ravier, Paris, 1801.

[14]G. G. Valentin. Über den Verlauf und die letzen Enden der Nerven. Verhandlungen der Kaiserlichen Leopoldnisch-Carolinischacademie der Naturforscher, Breslau, 18:151–240 (1836). (On the Development and Endings of the Nerves.)

[15]Theodor Schwann (1810–1882). *Mikroskopische Untersuchungen über die Übereinstimmung in der Structur und dem Wachstum der Thiere und Pflanzen*. Reimer, Berlin, 1839. (Microscopic Investigations of the Relationship of the Structure and Growth of Animals and Plants.)

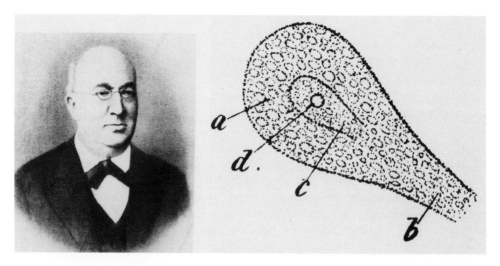

FIG. 63. Left: Gabriel Gustav Valentin (1810–1883). Professor of Physiology and Zootomy at Berne. **Right:** Valentin's drawing of a human cerebellar cell that was later to be named for Purkyně. (*a*) = parenchyma (*b*) tail-like extension (*c*) = nucleus (*d*) = nucleolus. (*Nova Acta Acad. Leopoldina, Breslau,* 18:151–240, 1836.) (From: V. Kruta. *Jan Purkyně. (1787–1869) Physiologist.* Academia, Praha, 1969.)

only did he identify the cell body (he used the word "Kugel"), but he described the nucleus within it as well as the nucleolus. He also observed the fibers but did not regard them as extensions of the cell, believing them not to be actually connected, an opinion that would be utterly destroyed by the work of his contemporary Robert Remak.

Valentin's concept was that the fibers he found coming up to the cortical surface turned in loops ("Endumbiegungschlingen") to go down to lower layers, a view also accepted by Kölikker,[16] the Swiss histologist several years later, for Remak's clear demonstration that the fiber connected with the cell body was not immediately accepted.

Valentin had described the cerebellar cell as having an oblong shape. He said (in translation):

> Often one end is slightly pointed, often even elongated into a small tail-like appendix. In particular, the latter configuration could well lead us to assume that this elongation continues as an individual, organic nerve fiber. Apart merely from the great differences between the content of the primary fibers and the paraenchyma of the globules, the fact alone that one occasionally finds such tailed globule completely enclosed by its sheath of cellular tissue must bring to mind this assumption which hitherto was based only on the doctrine of physics.

Valentin's own description of this first examination of these single cells reads (in translation):[17]

> In the yellow substance of the cerebellum, for example, we sometimes succeed in obtaining cross-sections which indicate sufficiently the position of these peculiar bodies. Then we see that they stand in rows, their rounded ends to the inside toward the white matter, their tail-like extensions, on the other hand, are directed to the surface, toward the cortex, as

[16]Rudolf Albert von Kölliker (1817–1905). *Mikroskopische Anatomie oder Gewebelehre des Menschen.* Vol. 2, Part 1. Engelmann, Leipzig, 1850. (Microscopic Anatomy of the Theory of Human Tissue.)

[17]G. G. Valentin. Über den Verlauf und die letzten Enden der Nerven. Verhandlungen der Kaiserlichen Leopoldnisch-Carolinischen Academie der Naturforscher, Breslau, 18:31–240 (1836). (On the extent and terminals of Nerves.)

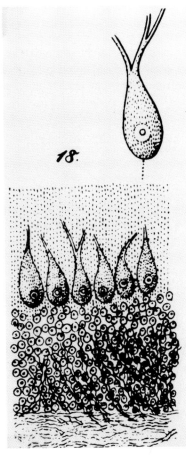

FIG. 64. **Left:** Jan Evangelista Purkyně (1787–1869). The great Czech anatomist. (Photograph by courtesy of the late Professor V. Kruta). **Right:** One of cerebellar cells that he reported to the Congress of Physicians and Scientists held in Prague in 1837. It is these cerebellar cells that later were to bear his name. (Über die Struktur der menschlichen und Säugethierzähne nach eigenen in Gesellschaft mit Dr. Fränkel angestellten Untersuchungen. *Übens. d. Arbeiten u. Veranderungen d. Schles. Gesell. f. vaterl. Kultur, Breslau,* 2:86–87, 1836.)

Purkyně was the first to observe. In usable cross-sections of the cerebellum of the horse, I have seen that all of these bodies are so placed that the rows which follow each other alternate. Thus each rounded end of the bodies of one row fits between the two tail-shaped extensions of two bodies lying side by side in the row which immediately precedes it and is directed toward the white matter. In this way, these bodies, in their placement in cross-section form a spiral line projection onto a flat surface.

In 1837 the cerebellar cell was demonstrated with great clarity by the great Czech scientist Jan Evangelista Purkyně.[18] He published drawings of these large cells now named for him (Fig. 64). He showed that these large tailed cells formed a definite layer in the cortex. He presented his findings first at the International Congress of Physicians and Scientists meeting in Prague in 1837.

[18]J. E. Purkyne. Über die Struktur der menschlichen und Säugethierzähne nach eigenen in Gesellschaft mit Dr. Fränkel angestellten Untersuchungen. *Arb. Ver. Schles. Gesell. Kult.,* Breslau, 55:86–87 (1836). (On the structure of human and mammal teeth, based on experiments carried out with Dr. Fränkel.)

In the eruption of interest in electrical stimulation following the publication of the experiments of Fritsch and Hitzig, Valentin himself wrote on electrical stimulation.[19,20] Valentin was living in the period when scientists were beginning to reject the teachings of vitalism. However, he still clung to some of its tenets, a stance that brought scorn from Du Bois-Reymond. In a volume entitled *Reden,*[21] Du Bois-Reymond wrote a long polemic against vitalism in which he deplored the lack of mathematical measurements by physiologists, and he specifically mentioned Valentin. His purpose was (as said in translation):

> . . . to introduce mathematical procedures to the physiologists who, at the present time, are partly prejudiced against this kind of problem treatment in their field, partly still remain in the dark concerning the essence of the physical-mathematical method, regardless of how much they have been talking about it lately. How could it be otherwise, since they acquired for the most part only a morphological, medical, at most a chemical training, and their knowledge of that method is only all too often limited to what they absorbed from Valentin's textbook.

Valentin continued a very productive life, outliving by 14 years his friend and teacher Purkyně, for whom the cell is named. He died in 1883 in a Europe profoundly changed since his birth 73 years before.

Vladimir Alekseyevich Betz (1831–1894)

The sighting of the nerve cell in the brain was a great step forward. As early as 1833, Ehrenberg,[22] an anatomist trained in Leipzig and working in Berlin, described cortical slices as seen with a simple microscope in unstained material. To the modern anatomist, his conclusions are bizarre. He held that the fibers he found in the white matter of the cortex were continuous with those of the spinal cord and even those from the periphery.

The Betz cell of the cortex and the Purkyně cell of the cerebellum were the largest and therefore the first to be described. The latter was named for the Ukrainian anatomist Betz by the Viennese H. Obersteiner[23] in his doctoral thesis, he name being preserved to this day. Unfortunately, Betz himself published no illustration of the cells he described.

Vladimir Alekseyevich Betz[24] was a native of Kiev, the great city on the banks of the Dneiper as it flows down to the Black Sea. During his lifetime he was to see the lands to the west of his city change hands several times following wars and treaties that dismembered the Ukraine, and in the reign of Czar Nicholas I, attempts were made to destroy its language. The city of Kiev, with its beautiful gold-domed cathedral, is also the site of a fine university, and it was there that Betz did the research that left his name on anatomy. For a neurophysiologist, this was a most important demonstration.

[19]G. G. Valentin. Die Interferenzen elektrischer Erregungen. *Pflüger's Arch. f. d. ges. Physiol.,* 7:458–496 (1873). (The Interferences of Electrical Stimuli.)

[20]G. G. Valentin. Einige Bemerkungen über elektrische Tetanisation des Nerven und der Muskeln. *Pflüger's Arch. f. d. ges. Physiol.,* 11:481–501 (1875). (Some Remarks on the Electrical Tetanization of Nerve and Muscle.)

[21]E. Du Bois-Reymond. *Reden.* Veit, Leipzig, 1886–1887. (Speeches.)

[22]Christian Gottfried Ehrenberg (1795–1876). Nothwendigkeit einer feineren mechanischen Zehrlegung des Gehirns und der Nerven vor der chemischen dargestellt aus Beobachtungen von G. G. Ehrenberg. *Ann. Phys.,* 28:449–473 (1833). (The Necessity of a Fine Mechanical Dissection of the Brain and Nerves over a Chemical Analysis Based on Observations.)

[23]H. Obersteiner. *Anleitung beim Studium des Baues der nervösen Centralorgane.* Deuticke, Vienna, 1896. (Introduction to a Study of the Structure of the Central Nervous Organs.)

[24]V. A. Betz. Antomischer Nachweis zweier Gehrinzentra. *Zbl. Med. Wiss.* 12:478–480, 494–499 (1874). (Anatomical evidence for two brain centers.)

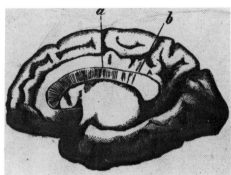

FIG. 65. **Left:** Vladimir Alekseyevich Betz (1831–1894). From a photograph taken when he was working in Vienna. (Courtesy of Professor P. G. Kostyuk.) **Right:** Betz's description of his figure of the human brain: "In the figure this lobe is precisely defined by two straight lines, *a* and *b*. Since it is inside and in front of the Sulcus Rolando or centralis, I will designate it the 'Lobulus paracentralis.'" (*Cbl. Med. Wiss.*, 12:556–569, 1874.)

Betz was a prominent Ukrainian anatomist, teacher, and public figure. As a student he had graduated in 1860 from the Medical Faculty at Kiev University and was sent to Vienna to do scientific work with the anatomist Hirtle and physiologists Brücke[25] and Ludwig.[26] There he performed investigations on the determination of sugar in urine by Brücke's method and studied the mechanism of hepatic circulation. In 1863 he defended his doctoral thesis "On the Mechanism of Circulation in Liver." In 1868 Betz was appointed to the Chair of Anatomy at the Anatomical Theatre at Kiev University. Later, he published his work "Some Comments on Microscopic Structure of the Adrenals" (1864) where at first he used the method of serial sections through the whole organ and gave descriptions of arteries, veins, lymphatic vessels, and nerves in the adrenals. He discovered the chromaffin reaction in adrenals.

For the neurophysiologist, the importance of Betz was that he was a founder of the modern doctrine about architectonics of the cerebral cortex based on macro- and microscopic studies of the cortical surface (Fig. 65). In the work entitled (in English) "Two Centers in the Cortical Layer of Human Brain" (1874), Betz described giant pyramidal cells of brain cortex, which are now called Betz cells. He linked his discovery with the experiments on electrical stimulation of the motor centers of the brain. In the work "On Details of the Cerebral Cortex Structure in Man" (1882), Betz gave the principle of architectonic division of the brain cortex into the main zones adopted at present. The studies of architectonics allowed him to publish an atlas of the human brain (1890).

[25]Ernst Wilhelm von Brücke (1819–1892).

[26]Carl Ludwig (1816–1895).

In two articles published in 1874,[27] Betz described his findings as follows: "The fissure of Rolando divides the outer surface of the brain into two parts: anterior in which are the large pyramidal cells and a posterior part (including the temporal lobes in which the nuclear layers predominate . . .)."

Betz found this neuronal pattern in human brains and in those of many animals he examined—chimpanzees, baboons, and green monkeys. He also found them in the brains of idiots, which led him to write:[28]

> It is up to the scientists and psychiatrists to research more thoroughly these centers in the brain and especially anteriorly . . . this could possibly throw light on the sudden occurrence of temporary paralysis of cerebral origin; on eclampsia of pregnancy; or epileptic convulsions, hysterical laughter, and finally it could, in a greater area, explain the appearance of partial or total aphasia.

In a later paper published in 1881,[29] Betz wrote: "My atlas of the surface of the human brain will treat in more detail what I have just said, as well as the structure of the convolutions." Betz died just before the end of the century and is buried in Kiev.

Robert Remak (1815–1865)

Robert Remak was responsible for the clear illustrations that exist today of the connection of the axon with the cell body.[30] Remak was born in the then Polish city of Posnan, in the year of Napoleon's final defeat. To study medicine, he moved to Berlin where he stayed for the rest of his life, first as a pupil of Johannes Müller. He graduated in medicine in 1838, eventually becoming Professor at the University of Berlin. His devotion to histology led him to some of the most basic discoveries in the nervous system. Among these was the clear demonstration that the nerve fiber was derived directly from the nerve cell; for the neurophysiologist, this was a demonstration of the most fundamental importance.[31] The "tail-like appendix" that Valentin described (Fig. 66) is very clearly seen in his illustration, but Remak's drawings of his preparation from the spinal cord of the ox made the connection indubitable. These were indeed axons, though his name for them was "organic fibers." As a pupil of Johannes Müller, Remak had some difficulty getting acceptance of his finding but having the occasion to demonstrate these cells directly to Purkyně, the agreement of the master ensued acceptance that these "tail-like" growths were indeed excrescenes from the cell bodies. But even in his concept, there was no suggestion that these protruding fibers actually contacted other cells in a manner that could produce interaction within the cortex.

Remak made very clear distinctions between what we now call myelinated fibers and the non-myelinated, which he was the first to describe. His name for the myelinated nerve was "organic fiber" and for the non-myelinated he used the name "primitive fiber." He wrote:

[27]V. A. Betz. Antomischer Nachweis zweier Gehirnzentra. *Zbl. Med. Wiss.*, 12:578–580, 595–599 (1874). (Anatomical evidence for two brain centers.)

[28]V. A. Betz. Antomischer Nachweis zweier Gehirnzentra. *Zbl. Med. Wiss.*, 12:578–580, 595–599 (1874). (Anatomical Evidence of Two Brain Centers.)

[29]V. A. Betz. Quelque mots sur la structure du l'écorce cérébrale. *Rev. Anthr.* 4:427–438 (1881). (Some Remarks on the Structure of the Cerebral Cortex.)

[30]R. Remak. Anatomische Beobachtungen über das Gehirn, das Rückenmark und die Nervenwurzeln. *Arch. Anat. Physiol. Wiss. Med.*, 506–522 (1841). (Anatomical observations of the brain, the spinal cord, and nerve roots.)

[31]R. Remak. Neurologische Erläuterungen. *Arch. Anat. Physiol. Med.*, 463–472 (1844). (Neurological commentaries.)

 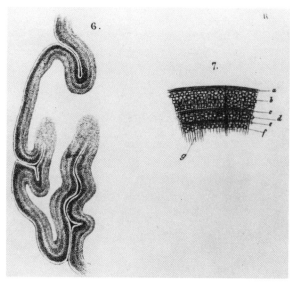

FIG. 66. Left: Robert Remak (1815–1865). **Right:** Remak's own description: "The sixth figure of Plate XII illustrates a slightly enlarged cross-section of the posterior lobe of the cerebrum of a sheep. It shows that the gray matter is composed of 6 layers—and that a white layer is on the outermost surface followed by gray and white ones alternatively." (From: R. Remak. Neurologische Erläuterungen. *Arch. Anat. Physiol.*, 506–522, 1844.)

> They are not tubular, that is surrounded by a sheath but naked transparent, almost gelatinous and much finer than most of the primitive tubes. Almost invariably they have longitudinal lines upon their surface and they readily divide into very small fibers. In their course they are frequently provided with oval nodules, and they are covered with certain small, oval or rounded but more rarely irregular, corpuscles with one or more nuclei which are about the same as the nuclei of the ganglion globules.

His allusion here to "very small fibers" is to what are now called neurofibrils.

Remak went on to note that all fibers did not streak up toward the cortical surface, for he found many which he called "crossing fibers." He described them as follows:

> The primary fibers which radiate from the central white matter towards the surface of the gyri are crossed by primary fibers in their course through the layers of the grey cortex, and pass through it in a direction parallel to the surface of the gyri, and, with respect to their primary fibers surrounding the gyri at their surface. These latter which, in contrast to those radiating from the central white matter, might be called crossing fibers, proceed so sparsely through the grey layers of the cortex that they escape observation almost entirely.

Remak's illustrations[32] are really the first in the field. They were published in 1844 together with further descriptions. They are mostly from the brains of sheep. The clarification of the processes of the cell body that we now call axons was expanded by O. F. K. Deiters who differentiated the dendrite from the axon.[33]

Deiters (who died so young) had, when working in Bonn (the birthplace of Beethoven),

[32]R. Remak. Anatomische Beobachtungen über das Gehirn, das Rückenmark und die Nervenwurzeln. *Arch. Anat. Physiol. Wiss. Med.*, 507–522 (1841). (Anatomical observations on the brain, the spinal cord and the nerve cells.)

[33]Otto Friedrich Karl Deiters (1834–1863). *Untersuchungen über Gehirn und Rückenmark des Menschen und der Saugerthiere.* Braunschweig, 1865. (Investigations of the Brain and Spinal Cord of Man and Mammals.)

 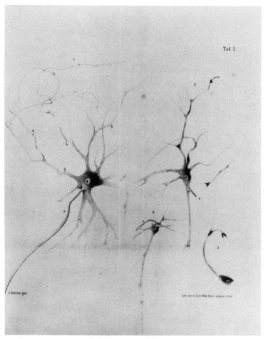

FIG. 67.   Left: Otto Friedrich Karl Deiters (1834–1863). **Right:** One of his illustrations of nerve cells and dendrites in the spinal cord. His Fig. 1 *(left)* is labeled "a large ganglion cell with appendages." The axon is labeled "a." His Fig. 2 shows sensory cells with appendages, in other words, dendrites. These are from the dorsal horn of the spinal cord. (From: Otto Friedrich Karl Deiters. *Untersuchungen über Gehirn und Rückenmark des Menschen und der Säugerthiere.* Braunschweig, 1865.) (Investigations on the Brain and Spinal Cord of Humans and Mammals.)

described the dendrites and the characteristics that distinguished them from axons (Fig. 67). He was insistent that the dendrites (which he called photoplasmic processes) were not immature axis cylinders or in any way related to the function of axis cylinders exiting from the parent nerve cell, or ganglion cell, as he called it.

Deiters died at the young age of 29; his work on the nerve cell was to be followed and expanded by the great company of histologists that dominated the first half of the 19th century. But his name is preserved for the vestibular nucleus in the pons.

W. Bevan Lewis (1847–1929)

In England, W. Bevan Lewis, writing from the West Riding Lunatic Asylum (the insane asylum that also was the workplace of David Ferrier), published two important papers. One was in the inaugural volume of *Brain*[34] in 1878 and a second in the *Proceedings of the Royal Society*[35] in the same year. These were articles in which he described, for the first time in man, the giant pyramidal cells of the precentral gyrus. He declared his belief that these cells had "motor significance," a conclusion that, in fact, Betz himself had reached from their counterparts in the lower animals (Fig. 68).

[34]W. B. Lewis. On the comparative structure of the cortex cerebri, *Brain,* 1:79–86 (1878).

[35]W. B. Lewis and H. Clarke. The cortical lamination of the motor area of the brain. *Proc. Trans. R. Soc.,* 27:38–49 (1878).

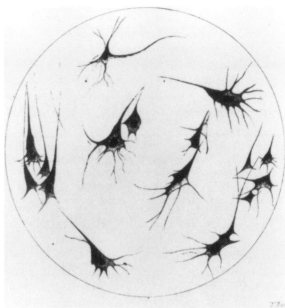

FIG. 68. **Left:** W. Bevan Lewis (1847–1929). A physician at the West Riding Lunatic Asylum who, with the achromatic microscope designed by J. J. Lister, demonstrated for the first time the cortical cells of man. **Right:** Lewis' picture of human cortical cells from the precentral gyrus. (From: W. B. Lewis and H. Clarke. The cortical lamination of the motor area of the brain. *Proc. Trans. R. Soc.*, 27:38–49, 1878.)

The West Riding Lunatic Asylum was in a remote country district of Yorkshire in England, the nearest city being Leeds. Yet, from this asylum came outstanding work in histology, not only from Lewis[36] but from his mentor Herbert Major,[37] who himself had published a volume on the histology of the brain. It was perhaps the ready access to the brains of the deceased insane that drew so many investigators to that outlying asylum. It was there that Clifford Allbutt[38] introduced electrical stimulation of the head in patients he described as having "acute primary dementia" (he used a constant current). For neurophysiologists, the outstanding scientist to come later from the same hospital was David Ferrier. The microscope that Lewis used had been designed for him by Joseph Jackson Lister. The early microscopes were neither achromatic nor were they free of spherical aberration, defects that led to unfortunate mistakes of interpretation, including in the 17th century Malpighi's[39] belief that what he saw in the cortex were "minute glands." It was nearly 200 years later that these technical difficulties were conquered, the main contributor being Joseph Jackson Lister, the father of a famous physician, Lord Lister. Together with the physician Thomas

---

[36]W. B. Lewis. On the histology of the great sciatic nerve in general paralysis of the insane. *West Riding Lunatic Asylum Medical Reports*, 5:85–104 (1875).

[37]H. C. Major. Observations on the histology of the morbid brain. *West Riding Lunatic Asylum Medical Reports*, 4:223–239 (1874).

[38]Clifford Allbutt. The electrical treatment of the insane. *West Riding Lunatic Asylum Medical Reports*, 2:203–232 (1872).

[39]Marcello Malpighi (1675–1715). De cerebre cortice. *In De Viscerium Structura Exercitatio Anatomica*. Montius, Bologna, 1666.

Hodgkin,[40] Lister published in 1827 a description of the miscroscope he designed tha. was achromatic and had a correction for spherical aberration. Only after this had been achieved could the exploration of the intimate connections of the cerebral cortex begin.

Lister's great improvements consisted of making the "object-glasses" by joining together a plain concave flint lens with a convex one, using transparent cement to bind them together. The agent he used was Canada Balsum and the result was to correct for spherical aberration. With his colleague Hodgkin, he most surely saw the nerve cells of the cortex, though they doubted their findings, for their description reads:[41]

> If there is any organized animal substance which seems more likely than another to consist of globular particles, it is undoubtedly that of the brain. Our examination of it has been but slight; but we have noticed that when a portion of it, however fresh, is sufficiently extended to allow to its being viewed in the microscope, one sees instead of globules a multitude of very small particles, which are most irregular in shape and size, and are probably more dependent on the disintegration than on the organization of substance.

In Lewis' use of this microscope, we find the third psychiatrist in search of the architectonics of the cerebral cortex. He became director of the West Riding Lunatic Asylum and later Professor of Psychiatry at Leeds. Lewis and his colleague Clarke, working in the same institution as Ferrier, were undoubtedly influenced by the latter's extensive studies of motor function, for their histological work focused on the precentral cortex. They were also working after Fritsch and Hitzig's[42] demonstration of the motor cortex by electrical stimulation, a revolutionary discovery that opened a new era of research on the brain.

Theodor Meynert (1833–1892)

The new microscopes were soon being used in many centers where the cell structure of the cortex was of major interest. One of these users was the great Viennese psychiatrist Theodor Meynert, whose classic textbook was entitled *Psychiatrie*.[43] This book contained illustrations of the precentral cortex that, as in the case of Baillarger, were essentially from brains of the insane.

Meynert was born in 1833 in Dresden, then in Bohemia, but his family moved to Vienna and there he received his education and spent the rest of his life. After obtaining his doctoral degree in 1861, he continued to advance in his profession until he was appointed Professor and Director of the Psychiatric Clinic. His depiction of the cortical layers was published in 1862. By naked eye he could detect three layers, but with his microscope he could differentiate five layers in the precentral cortex and eight in the occipital cortex (Fig. 69). He held that the large pyramids (the giant cells of Betz) merged into the same layer as the granular cells above them, an opinion later to be challenged.

Meynert's work opened the initial intense attack on the subject of the myelarchitectonics of the cortex—the arcuate fibers, the corpus callosum, the anterior commissura, the cerebel-

[40]T. Hodgkin and J. J. Lister. Notice of some microscopic observations of the blood and animal tissue. *Phil. Mag.*, 2:130–138 (1827).

[41]J. J. Lister (1786–1869). On the improvement of achromatic compound microscopes. *Phil. Trans. roy. Soc.*, 120:187–200 (1830). (See also J. J Lister. On the limit to defining power in vision with the unassisted eye, the telescope, and the microscope. *J. roy. Micr. Soc.*, 33:34–35, 1913.)

[42]G. T. Fritsch and E. Hitzig. Über die elektrische Erregbeit des Grosshirns. *Arch. Anat. Physiol. Wiss. Med.*, 37:300–332 (1870). (On the electrical sensitivity of the brain.)

[43]T. Meynert. *Psychiatrie: Klinik der Erkrankungen des Vorderhirns* (Erste Hälfte). Braumüller, Vienna, 1884. (Psychiatry: Clinical Instruction of the Diseases of the Forebrain.)

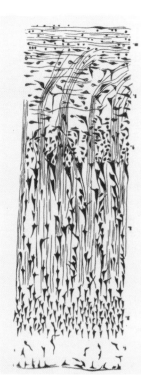  

FIG. 69. **Left:** Meynert's illustration of the six-layered cortex he identified in the human brain. **Center:** Theodor Meynert (1833–1892). **Right:** His drawing of the eight-layered occipital cortex: (1) neuroglia, (2) pyramids, (3) external granular layer, (4) external bare layer with secretory cells, (5) middle granular layer, (6) inner bare layer intergranular layer with solitary cells, (7) inner granular layer, (8) spindle-shaped cells, (9) white matter. (From: T. Meynert *Psychiatrie.* Braumuller, Vienna, 1884.)

lar peduncles—in other words, the destination and function of what he called the ''conducting tracts'' of his association and projection system. He used the latter term for the fiber tracts running from the cortex to the thalamus and for connections via the brainstem to the spinal cord, as well as for connections from the retina to the occipital lobe. It is in an address[44] given to the Ninth International Congress of Medicine in Berlin that Meynert's most philosophical analysis of the findings of interaction among brain regions was expressed.

Meynert had interested himself not only in the cells of the cortex but in the mechanisms necessary for interaction;[45,46] he had identified association pathways, structures that were to bear his name as the ''association fibers of Meynert.'' Meynert's interest in the distribution and function of the fiber tracts was a great step forward in the understanding of the cortex and led to the realization of its relationship to, for example, the basal ganglia (a name intro-

[44]T. Meynert. Der Bau der Gross-hirnrinde und sine örtlichen Verschiedenheiten, nebst einem pathologisch-anatomischen corollarium. Vierteljahrschr. *Psychiatrie,* 1:77–93, 125–217 (1867); 1:381–403 (1868); 2:88–113 (1868). (The structure of the cerebral cortex and its local variations together with a pathological-anatomical corollarium.)

[45]T. Meynert. Vom Gehirne der säugethiere. *Handbuch der Lehre von den Geweven des Menschen und der Thiere,* 2:694–808. Translated by the New Sydenham Society, 1870–1893 (edited by Stricker). (On the brains of mammals.)

[46]T. Meynert. Über das Zusammenwirken der Gehirntheile. Tenth International Medical Congress, Vol. 1. Hirschwald, Berlin, 1891, pp. 173–190. (On the Interaction of Parts of the Brain.)

duced by Gall). He should receive credit for the sensory pathways to the cortex, including thalamocortical projections only guessed at by earlier workers. He called these tracts "Stiele" and distinguished four such tracts: an anterior one to the frontal lobe, one from the anterior thalamus to what is now called the cingulate gyrus, and most clearly the connection between the occipital lobe and the pulvinar. This work foreshadowed that of the Swedish neurologist S. E. Henschen,[47] who at the end of the 19th century published his concepts, derived from autopsies, of the visual pathways to the calcarine cortex, ideas that were received with skepticism by Horsley, Hitzig, von Monokow, and other prominent scientists.

The second major problem facing the histologists of the cerebral cortex was: How do the neural components interact? Granted, eventually, that nerve fibers derived their origin from the nerve cell, there remained the question as to whether or not anastomoses formed between fibers from different nerve cells. The idea of some degree of anatomosis died hard. Even Meynert, in his famous book on the brains of mammals, when announcing the postulate which he called "the law of isolated conduction," stated that he found "its morphological expression in the fact that the axes of the nerve cells appear to be elongated in the direction of the nerve fiber with which they are continuous." But he made some exceptions and stated that "even in the gray masses, which doubtless constitute paths for transverse conduction by means of anastomoses, the law of isolated conduction holds good, though only conditionally." He returned to this consideration of anastomoses when discussing the work of Arndt[48] and of Besser[49] who, he said, demonstrated them as occurring between the processes of cells in the cortex. Meynert later claimed to have seen them. "The fusiform body," Meynert wrote, "gives off laterally also from five to seven processes which, like the apical process, form demonstrable anastomoses." Later, he wrote: "On occasion I distinctly saw the division of a medullary fibre from the processes of two cells."

Meynert died in the last decade of the century, having made major contributions not only to psychiatry but to histology. He has left his name on a nucleus in the forebrain near the optic chiasm—the nucleus vasilis of Meynert, a region of acetylcholine-using neurons at one time implicated in Alzheimer's Disease.

At this time the question remained whether or not a distinction had to be made between vertebrates (as studied by Meynert) and invertebrates. One may take an example from the work of one of his pupils who was also to make his career in a totally different field— Sigmund Freud. In a paper written in 1882, Freud gave a camera lucida drawing of an apparent anastomosis of nerve fibers in tissue from the river crayfish (Fig. 70), the same animal used so much by Remak.[50]

Although Meynert continued to work in his field until the 1890s, he never adopted the new technique of localization by electrical stimulation, introduced so ingeniously by Fritsch and Hitzig in 1870. But during and following the period under review technical advances were to come once more to the aid of the histologist. In the last decades of the 19th century, these technical developments led to a tremendous expansion of work on the structure of the cortex and to the rise of the Spanish school spearheaded by Santiago Ramón y Cajal.

[47]S. E. Henschen. On the visual path and centre. *Brain,* 16:170–180 (1893).

[48]R. Arndt. Studien über die Architectonik der Grosshirnrinde. *Schutze's Arch. Anat.,* 3; 4:441–476 (1867). (Studies on the structure of the cerebral cortex.)

[49]L. Besser. Zur Histogenese der nervösen Elementartheile in den Centralorganen des neugeborenen Menschen. *Arch. Pathol. Anat. Physiol. Klin. Med.,* 36:305–334 (1866). (On the histogenesis of the elemental parts of the nerves in the central organs of the new born human being.)

[50]S. Freud (1856–1939). Über den Bau der Nervenfasern und Nervenzellen beim Flusskrebs. *Sitz. Akad. Wiss. Wien.,* 85:9–46 (1882). (On the structure of the nerve fibers and nerve cells in the river crayfish.)

 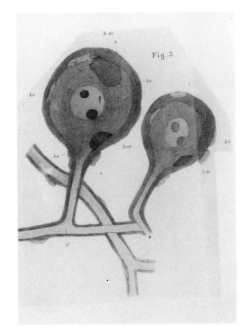

FIG. 70. Left: Sigmund Freud (1856–1939). **Right:** The illustration he published from his work with crayfish. He held that this was an example of anastamosis of nerve fibers, a concept disproved by Cajal. (From S. Freud. Über den Bau der Nervenfasern und Nervenzellen beim Flusskrebs. *Sitz Akad. Wiss. Wien.*, 85:9–46, 1882.) (On the structure of nerve fibers and nerve cells in the river crayfish.)

Santiago Ramón y Cajal (1852–1934)

Santiago Ramón y Cajal was born in the village of Petilla de Aragon in northern Spain, and his father, a surgeon, sent him to medical school in Zaragoza where he graduated in 1877. After a period as an army surgeon in Cuba, he moved many times, from Zaragoza to Valencia to Barcelona, always pursuing his work in histology but publishing in journals little known to the rest of Europe. He achieved full recognition in Madrid where he founded the Laboratorio de Investigationes Biologicas at the University of Madrid.

Cajal's earlier publications in Spanish were in *Boletins, Gaceta,* and *Revista,* which were published in very few copies and remain rarities outside Spain.[51] The greater number of them appeared in the last decade of the 19th century and the first of the 20th century, including his major work, the three volumes of which in the Spanish edition are dated 1894–1904. These, however, were not generally distributed until translated into French (as two volumes).[52] Scientists all over the world are indebted to the many translators who have made these writings available also in other languages. Cajal added immensely to the knowledge of the intimate structure of the cerebral cortex (Fig. 71). To mention just one example, one may cite his analysis of the short-axoned cells of the first layers, the layer noted by Meynert to be so poor in cells and thought by Golgi to be solely occupied by neuroglia. Using several different staining procedures, Cajal identified many short-axoned cells in this first plexiform layer, and not all were immature cortex. These studies of Cajal's mark the

[51]S. Ramón y Cajal. Las celulas de cilindro-eje corto de la capa molecular de cerebro. *Rev. Trim. Microger.,* 2:5–7 (1897).

[52]S. Ramón y Cajal. *Histologie du Système Nerveux de l'Homme et des Vertébrés.* 2 vols. Maloine, Paris, 1911.

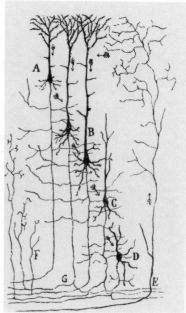

FIG. 71. Left: Santiago Ramón y Cajal (1852–1934). Master anatomist of the cerebral cortex. **Right:** His own drawing of the afferent inflow to the mammalian cortex. (From: S. Ramón y Cajal. *Histologie du Système Nerveaux de l'Homme et de Vertébrés.* Maloine, Paris, 1909–1911).

development of interest moving from the solely vertically oriented distribution of neurons in the cortex to their horizontal interactions and to the whole question of tangential spread of fibers.

Cajal had long been working with silver reduction methods but was jubilant when he first tried Golgi's technique of osmic dichromate fixation followed by silver. He exclaimed (in the French translation):

> Spectacle inattendu! Sur un fond jaune d'une translucidité parfaite, apparaissent, clairsemés des filaments noirs, lissés et minces, ou épineux et épais, des corps noirs, triangulaires, étoilés, fusiformes! On dirait des dessins à l'encre de Chine sur un papier transparent du Japon.

Cajal's very many contributions to the knowledge of the intimate structure of the brain—knowledge so essential before its functions could be understood—were made known mostly in the century following the one currently under review, for he lived long on into the 20th century.

In the 19th century, under review here, he received full recognition by the scientific world in the form of the Nobel Prize. This he shared with Camillo Golgi, who was still tied to his belief in reticular networks within the cortex. Cajal's work had gradually become more widely known in Europe; he was now using gold as well as the Golgi stain. In 1906 he was elected a Foreign Fellow of the Royal Society and began to travel more widely, especially after the end of World War I. In the last year of his life, he lectured to students at London University, leaving an indelible memory on at least one of the student audience. Cajal died in 1934; his antagonist had died before him.

Camillo Golgi (1843–1926)

Camillo Golgi, a Lombard, was born in Corteno in the still divided Italy; his youth covered the turbulent years of Garibaldi's uprisings and the eventual unification of Italy in 1861 under a single flag. Educated at the University of Parvia, he spent the first years of his professional life in pathology, under the influence of the great Italian pathologist Cesare Lombroso, and it was not until he was nearly 30 that his interest turned to the field that made his name—the intimate structure of the nervous system.[53-56] For the understanding of cortical function, his outstanding contribution was the identification of the Type I and Type II neurons now named for him—the first, a motor cell with its downward influence, and the second, the Golgi Type II sensory in action and wholly contained in its cortical layer. Golgi with his superior staining method was able to develop further his distinction, however, always clinging to the mistaken concept that these excrescences from the cell body formed a network.

The differential between axon and dendrite is fundamental for the understanding of the neurophysiology that is based on anatomy, and Golgi's further development of cortical networks tends, in our memories, to overshadow the unfortunate attack he made on Cajal on the occasion of their sharing the Nobel Prize in 1906 (Fig. 72). The issue was then known as the "neuron theory," or in Golgi's own words: "This theory, which I thought ought to be briefly summarized, should not be considered as an essential part of the neuron theory. In fact it only expresses one interpretation of nerve function."

Later in his speech, Golgi brought the attack directly against his fellow prizewinner, saying:

> I do not think that I need prolong the discussion of the above any longer to achieve the purpose I intend. I shall therefore confine myself to saying that, while I admire the brilliancy of the doctrine which is a worthy product of the high intellect of my illustrious Spanish colleague, I cannot agree with him on some points of an anatomical nature which are, for the theory, of fundamental importance, for example, that the peripheral branch of spinal ganglion cells must be identified with a protoplasmic process, since one must consider the myelin sheath as an absolutely secondary event, for it is only necessitated by the length of the process. Similarly, I cannot accept as a good argument in support of the theory the statement which, however, is its starting point, that says the processes of the cells of the molecular layer of the cerebellum terminate by forming endings of the bodies of the cells of Purkyně, for I have verified that the fibers coming from the nerve process of the cells of the molecular layer only pass near the cells of Purkyně to proceed into the rich and characteristic network existing in the granular layer.

Golgi, who received many honors in his own country, died in 1926 at the age of 83, Cajal outliving him by another eight years.

As the 19th century came to an end, the neurophysiologists in their studies of the function of the nervous system now found themselves with information about its basic structure. At the opening of the 20th century, not even the nerve cell itself had been seen; but by the final decade the complex neuronal networks of the brain were beginning to reveal their power.

[53]C. Golgi. Sulla sostanzia grigia del cervello. *Gaz. Ital. Lombardia*, 6:244–246 (1873).

[54]C. Golgi. Sulla sostanza grigia del cervello. *Reale Inst. Lombardia.* Rep. II. 7:69 (1874).

[55]C. Golgi. *Untersuchungen über den feineren Bau des centralen und peripherischen Nervensystems.* Fisher, Jena, 1894. (Investigations on the Finer Structure of the Central and Peripheral Nervous System.)

[56]C. Golgi. Intoro alla struttura delle cellule nervose della corteccia cerebrale. *Verh. Anat. Ges.*, 14:164–176 (1900).

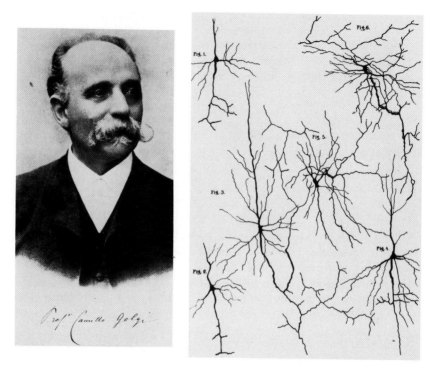

FIG. 72. Left: Camillo Golgi (1843–1926). Co-winner with Cajal of the Nobel Prize in 1906. **Right:** Drawings by Golgi: Fig. 1 and Fig. 2: Ganglion cells Type 2 in cortex of dog; Fig. 3 and Fig. 4: Large ganglion cells from the cat. The bifurcation in the Type 2 cell convinced Golgi of the reticular nature of the cortex so strongly denied by Cajal. (From: C. Golgi. *Untersuchungen über den feineren Bau des centralen und peripherischen Nervensystems.* Fischer, Jena, 1894.) (Investigations on the Finer Structure of the Central and Peripheral Nervous System.)

BIBLIOGRAPHY

Luigi Rolando (1773–1831)

Selected Writings

Anatomie Physiologica-Comparative Disquisitio in Respirationes Organa. Turin, 1801. (Investigation of Physiological Comparative Anatomy in Respiratory Organs.)

Saggio sopra la vera struttura del cervello dell' uomo e delgl' animali e sopra le funzioni del sistema nervoso. Nella Stamperia da S.S.R.M. Privilegiata, Sassari, 1809. (Essay on the True Structure of the Human Brain and of the Animals and on the Functions of the Nervous System.)

Inductions Physiologiques et Pathologiques sur les différentes Espèces d'Excitabilité et d'Excitement, sur l'Irritation, et sur les Puissances Excitantes, Debilitantes et Irritantes. Caille et Ravier, Paris, 1822. (Physiological and Pathological Inductions on the Different Kinds of Excitability and Excitement, on the Irritation and on the Exciting Powers, Debilitating and Irritating.)

Expériences sur les fonctions du système nerveux. *J. Physiol. Exp. Pathol.,* 3:95 (1823).

Richerche Anatomisch sulla Struttura del Midollo Spinale. Stamperia Reale, Torino, 1824. (Anatomical Research on the Structure of the Spinal Cord.)

Osservazioni sul cervelletto. *Mem. Accad. Sci. Turino,* 29:163 (1825).

Jules Gabriel François Baillarger (1806–1891)

Selected Writings

Mémoire historique et statistique sur la Maison Royale de Charenton. *Ann. Hyg. Méd.* 13:5–192 (1835).

Recherches sur la structure de la couche, corticale des circoncolutions du cerveau. *Mém. Acad. Roy. Belg.,* 1:149–483 (1840). (English translation by G. von Bonin. *The Cerebral Cortex.* Thomas, Springfield, 1960.)

De l'étendue de la surface du cerveau et de ses rapports avec l'intelligence. *Gaz. Hop.* 18:179 (1854).

Gabriel Gustav Valentin (1810–1883)

Selected Writings

Über die Dicke der varikösen Fäden in dem Gehirne und dem Rückenmarke des Menschen. *Arch. Anat. Physiol. Wiss. Med.,* 401–409 (1834). (On the thickness of various fibers in the brain and spinal cord of man.)

Über den Verlauf und die letzten Enden der Nerven. Verhandlungen der Kaiserlichen Leopoldnisch Carolinischeacademie der Naturforscher, Breslau, 18:31–240 (1836). (On the Development and Endings of Nerves.)

Die Interferenzen elektrischer Erregungen. *Pflüger's Arch. f. d. ges. Physiol.,* 7:485–496 (1873). (The interferences of electrical stimuli.)

Einige Bemerkungen über elektrische Tetanisation der Nerven und Muskeln. *Pflüger's Arch. f. d. ges. Physiol.,* 11:481–501 (1875). (Some remarks on the electrical tetanization of nerve and muscle.)

Einiges über Ermüdungscurven quergestreifter Muskelfasern. *Pflüger's Arch. f. d. ges. Physiol.,* 29:257–285 (1882). (Remarks on fatigue curves of striated muscle fibers.)

Vladimir Alekseyevich Betz (1834–1894)

Selected Writings

Anatomischer Nachweis zweier Gehirncentren. *Zbl. Med. Wiss.,* 12:578–580, 595–599 (1874). (Anatomical evidence for two brain centers.)

Über die feinere Strucktur der Grosshirnrinde des Menschen. *Zbl. Med. Wiss.,* 19:209–213 (1874). (On the finer structure of the cerebral cortex in man.)

Quelque mots sur la structure du l'écorce cérébrale. *Rev. Anthr.,* 4:427–438 (1881). (Some remarks on the structure of the cerebral cortex.)

Jan Evangelista Purkyně (1778–1869)

Selected Writings

Beobachtungen und Versuche zur Physiologie der Sinne. Calve, Prague, 1823. (Observations and Experiments on the Physiology of the Senses.)

Der microtomische Quetscher, ein bei microscopishen Untersuchungen unenbehrliches Instrument. *Arch. Anat. Physiol. Wiss. Med.,* 385–390 (1834). (The microtomic squasher, an indispensable instrument for microscopic investigations.)

Über Flimmerbewegungen im Gehirn. *Arch. Anat. Physiol. Wiss. Med.,* 289–290 (1836). (On ciliary action in the brain.)

Bericht über die Versammlung deutscher Naturforscher und Ärtze in Prag. *Anat. Physiol. Ver.,* 3:177–180 (1837). (Report on the conference of German scientists and doctors in Prague.)

Entdeckung continuirlicher durch Wimperhaare erzeugter Flimmerbewegungen, als eines allgemeinen Phänomens in den Klassen der Amphibien, Vögel und Säugethiere. *Arch. Anat. Physiol. Wiss. Med.,* 391–400 (1834). (Discovery of continuous flickering movements produced by lashes, as a general phenomenon in the classes of amphibians, birds and mammals.)

Robert Remak (1815–1865)

Selected Writings

Vorläufige Mittheilungen microscopischer Beobachtungen über den innern Bau der Cerebrospinalnerven und über die Entwickelung ihrer Formelemente. *Arch. Anat. Physiol. Wiss. Med.,* 145–161 (1836). (Preliminary report of microscopic observations of the inner structure of the cerebral spinal nerves and the development of their formal elements.)

Observations anatomicae et microscopicae de systematis nervosi structura. Beroline, sumtibus et formis Reimerianis. Reimer, Berlin, 1838. (Discovery of the non-medullated nerve fibers ["fibers of Remak"].)

Anatomische Beobachtungen über das Gehirn, das Rückenmark und die Nervenwurzeln. *Arch. Anat. Physiol. Wiss. Med.,* 506–522 (1841). (Anatomical observations of the brain, the spinal cord and nerve roots.)

Neurologische Erläuterungen. *Arch. Anat. Physiol. Wiss. Med.,* 463–472 (1844). (Neurological Commentaries.)

Über den Entwicklungsplan der Wirbelthiere. *Arch. Anat. Physiol. Wiss. Med.,* 374–375 (1854). (On the Development Plan of Vertebrates.)

Über Theilung thierischer Zellen. *Arch. Anat. Physiol. Wiss. Med.,* 376 (1854). (On the Division of Animal Cells.)

Über methodische Electrisirung gelähmter Muskeln. Hirschwald, Berlin, 1855. (On Methodical Electrification of Paralyzed Muscles.)

Galvantherapie der Nerven- und Muskelkrankheiten. Hirschwald, Berlin, 1855. (Galvanic Therapy of Nerve and Muscle Illnesses.)

Otto Friedrich Karl Deiters (1834–1863)

Selected Writings

Untersuchungen über Gehirn und Rückenmark des Menschen und der Säugerthiere. Braunschweig, Berlin, 1865. (Investigations of the Brain and Spinal Cord of Man and Mammals.)

W. Bevan Lewis (1847–1929)

Selected Writings

On the comparative structure of the cortex cerebri. *Brain,* 1:79–86 (1878).

The cortical lamination of the motor area of the brain. *Porc. Trans. R. Soc.,* 27:38–49 (1878). (With H. Clarke.)

Methods of preparing, demonstrating, and examining cerebral structure in health and disease. *Brain,* 3:314–336 (1880–1881).

Methods of preparing, demonstrating, and examining cerebral structure in health and disease. *Brain,* 3:502–515 (1880–1881).

Theodor Meynert (1833–1892)

Selected Writings

Der Bau der Gross-Hirnrinde und seine örtlichen Verschiedenheiten, nebst einem pathologisch-anatomischen Corrollarium. Vierteljahrschr. *Psychiatrie,* Vol. 1, pp. 77–93, 125–217, 381–402, (1867); Vol. 2, pp. 88–113 (1868). (The structure of the cerebral cortex and its local variations, together with a pathological-anatomical corollarium.)

Vom Gehirn der Säugethiere, in Stricker. *Handbuch der Lehre von den Geweben des Menschen und der Thiere,* 2:694–808 (1872). (On the brains of mammals. Translated by the New Sydenham Society.)

Psychiatrie: Klinik der Erkrankungen des Vorderhirns begründet auf dessen Bau, Leistungen und Ernährung. Braumüller, Vienna, 1884. (Psychiatry: Clinical instruction of ailments of the forebrain, based on its structure, functions, and nourishment.)

Gehirn und Gesittung. Vortrag in der Versammlung der Naturforscher und Ärtze in Cöln. Sept. 1888. Braumüller, Vienna, 1889. (Brains and manners. Speech at the Conference of scientists and doctors in Cologne.)

Sammlung von populär-wissenschaftlichen Vorträgen über den Bau und die Leistungen des Gehirns. Braümuller, Vienna, Leipzig, 1892. (Collection of popular scientific speeches on the structure and function of the brain.)

Santiago Ramón y Cajal (1852–1934)

Selected Writings

Estructura de los centros nerviosos de los aves. *Rev. Trimest. Histol.,* 1:305–315 (1888). (Structure of neural centers in birds.)

Estructura del cerebelo. *Gac. Catalana,* 11:449–457 (1888). (Structure of the cerebellum.)

Sur la structure de l'écorce cérébrale de quelques mammifères. *Cellule,* 7:123–176 (1891).

Observaciones anatómicas sobre la corteza cerebral y asta de Ammon. *Actas de la Sociedad Española de Historía Natural.* Second series, Vol. I. Session of December 1892. (Anatomical observations on the cerebral cortex and Ammon's horn.)

La fine structure des centres nerveux. *Proc. R. Soc. London,* 55:444–468 (1894). (Fine structure of neural centers.)

Structure et connexions des neurones. Le Prix Nobel en 1906. Norstedt & Söner, Stockholm, 1906. (Structure and Connections of Neurons.)

Über die feinere Struktur des Ammonshornes, translated by A. von Kölliker. *Z. Wissen Zool.,* 56:613–663 (1893). (On the fine structure of Ammon's horn.)

Histologie du système nerveux de l'homme et des vertébrés, translated from Spanish by Dr. L. Azovlay. 2 vols. Maloine, Paris. 1909–11.

Estudios sobre la degeneración y regeneración del sistema nervioso. Madrid. 2 vols. 1912–1914. (Studies on Degeneration and Regeneration in the Nervous System.)

Recuerdos de mi vida. 3rd ed. Alianza University, Madrid, 1923. (Memoirs from My Life.)

Degeneration and Regeneration of the Nervous System, translated by R. M. May. 2 vols. Milford, London, 1928.

Neuronismo o reticularismo? Las pruebas objectivas de la unidad anatomica, de las celulas nerviosas. *Archos. Neurobiol.,* 13:217–291, 579–646 (1933). (Neuronism or reticularism? The objective evidence for the anatomical unity of nerve cells.)

Studies of the cerebral cortex (limbic structure), translated by L. M. Kraft. Lloyd-Luke, London, 1955.

Camillo Golgi (1843–1926)

Selected Writings

Untersuchungen über den feineren Bau des Centralen und peripherischen Nervensystems. *Gaz. Med. Ital. Lombardia*, 6:244–246 (1873). (Investigations on the fine structure of the central and peripheral nervous systems.)

Sulla fina struttura dei bulbi olfactoria. *Riv. Sper. Freniatria Med. Legal.*, 1:66–78 (1875). (On the fine structure of the olfactory bulb.)

Origine del Tractus olfactorius e struttura dei lobi olfattorii dell uomo e di altri mammiferi. *Rend. R. Inst. Lomb.*, 20:216, 1882; *Arch. Ital. Biol.*, 2, 1882. (Origins of the olfactory tract and structure of the olfactory lobes of man and other mammals.)

Sulla fina anatomia delgi organi centrali del sistema nervoso. *Rivist. Psichiatr.*, p. 165 (1882). (On the fine structure of central organs in the nervous system.)

Reserches sur l'histologie des centres nerveux. *Arch. Ital. Biol.*, 3:285–317 (1883). (Research on the Histology of Nervous Centers.)

Sulla fina anatomie delgi organi centrali del sistema nervoso. Hoepli, Milan, 1886. (On the Fine Structure of Central Organs of the Nervous System.)

Untersuchungen über den feineren Bau des centralen und peripherischen Nervensystems. Fisher, Jena, 1894. (Investigations on the Fine Structure of Central and Peripheral Nervous Systems.)

Suggested Readings on Neuroanatomists Working in the 19th Century

Ackerknecht, E. H. Mediziner und Zellenlehre Gesenerus. 25:188–184 (1968). (Doctors and the cell theory.)

Bailey, P., and von Bonin, G. *The Isocortex of Man.* University of Illinois Press, Urbana, 1951.

Brazier, M. A. B. Historical development of our knowledge of cortical investigation. In: *Cortical Investigation,* edited by F. Reinosi-Saurez and C. Ajmone Marsan. Raven Press, New York, 1984.

Brazier, M. A. B. Architectonics of the cerebral cortex. Research in the 19th century. In: *Architectonics of the Cerebral Cortex,* edited by M. A. B. Brazier and H. Petsche. Raven Press, New York, 1978.

Campbell, A. W. *Histological Studies on the Localisation of Cerebral Function.* Cambridge University Press, Cambridge, 1905.

Clarke, E., and O'Malley, D. *The Human Brain and Spinal Cord.* University of California Press, Berkeley, 1968.

Fulton, F. A note on Francesci Gennari. The early history of cytoarchitectural studies of the cerebral cortex. *Bull. Hist. Med.*, 5:895–913 (1937).

Grisolia, S., Guerri, C., Samson, F., Norton, S., and Reinosi-Suarez, F. (Eds.). *Ramón y Cajal's Contribution to the Neurosciences.* Elsevier, New York, 1983.

Haymaker, W., and Schiller, F. *The Founders of Neurology.* Thomas, Springfield, 1958.

Kirsch, B. *Forgotten Leaders in Modern Medicine: Valentin, Gouby, Remak, Auerbach.* American Philosophical Society, Philadelphia, 1954.

Kruta, V. *J. E. Purkyně (1787–1869) Physiologist.* Academia Publishing House of the Czech Academy of Science, Prague, 1969.

Saunders, J. B. de C. M. Review of *Forgotten Leaders in Modern Medicine: Valentin, Gouby, Remak, Auerbach* by Bruno Kisch. *ISIS*, 46:383–385 (1955).

van der Loos, H. The history of the neuron. In: H. Hyden, *The Neuron.* Elsevier, New York, 1967.

Walker, A. E. The development of the concept of cerebral localization in the nineteenth century. *Bull. Hist. Med.*, 31:99–121 (1957).

CHAPTER X

A New Discovery in Brain Physiology

The surge of interest in the relation of specific brain loci to the operation of the rest of the body was being pursued by ablation techniques (Flourens,[1] Goltz[2]) and by clinical signs (Jackson[3]). A totally new approach was introduced in 1870 by a report[4] claiming that electrical stimulation of certain areas of the cortex produced both discrete and complex motor movements on the opposite side of the body. No longer could Flourens' denial of the role of the cerebral cortex be accepted. The two experimenters responsible for this breakthrough were Gustav Theodor Fritsch and Edouard Hitzig. The two young men in their early thirties were "Privatdocents" in Berlin when they carried out the work that established unequivocally the existence of the motor cortex.

The history-making discovery by Fritsch and Hitzig (Fig. 73) was that electrical stimulation of specific areas of the dog's cortex produced discrete muscle movements—a great advance on ablation techniques, for the cortex was left undamaged. They were also able to delineate the specific area of the cortex that was motor and the others that evoked no muscular response. Thus, with great clarity, the motor cortex responsible for contralateral movements was defined. These two young physicians, on writing their paper, reported that they designed the experiment on the knowledge that electrical stimulation of the nerves and of the spinal cord produced movements, and consequently they questioned why should not

[1]P. Flourens (1794–1867). *Recherches Expérimentales sur les Propriétés et les Fonctions du Système Nerveux dans les Animaux Vertébrés*. Clevot, Paris, 1824.

[2]F. L. Goltz (1834–1902). Der Hund ohne Grosshirn. *Pflüger's Arch. f. d. ges. Physiol.*, 51:570–614 (1892). (The dog without a brain.)

[3]J. Hughlings Jackson. The Croonian lectures. On the evolution and dissolution of the nervous system. *Proc. R. Soc. London*, 3–5 (1884).

[4]G. T. Fritsch and E. Hitzig. Über die elektrische Erregbarkeit des Grosshirns. *Arch. Anat. Physiol. Wiss. Med.* 300–314 (1870). (On the electrical excitability of the cerebrum.)

FIG. 73. Pioneers in the electrical stimulation of the cerebral cortex. **Left:** Gustav Theodor Fritsch (1838–1907), Professor of Physiology and Anatomy, Berlin. **Right:** Edouard Hitzig (1838–1927), Professor of Psychiatry, Zurich.

the brain do the same? They therefore defied the idea held since the beginning of the century that the brain did not respond to stimulation.

Flourens[5] had concluded that the cerebral lobes could take over tasks of parts of the brain that had been amputated and that there was no specific seat either for particular abilities or for particular perceptions. He concluded also that after partial ablation a remaining part of the hemispheres could achieve the full use of all functions. Flourens concluded his presentation by saying, "A point excited in the nervous system excites all others. A point denervated, denervates them all. . . . Unity is the great principle, it is everywhere, it dominates all. The nervous system therefore forms one unitary system."

Among the claims that Flourens had made was that he had opened the skull of a dog and had systematically pricked the cortex from back to front producing muscle twitchings except in one place. On moving more rostrally to the brainstem, he obtained contractions. From his results he concluded that the cerebral hemispheres were not involved in muscle movement, only in perception and mental activity.

For lack of facilities for animal experimentation at the Medical Institute, these two recently qualified physicians set up their experiments in Hitzig's home. They opened their report saying that "these experiments began with observations one of us had made on the movements of voluntary muscles evoked by direct stimulation of the central organ in man." The investigator (according to his later publications) was Hitzig, who had written a brief report.[6]

[5]P. Flourens. *Recherches Expérimentales sur les Properiétés et les Fonctions du Système Nerveux dans les Animaux Vertébrés*. Clevot, Paris, 1824.

[6]E. Hitzig. *Klin. Woch.* 7:137–138 (1870).

Fritsch and Hitzig in their paper described the physical procedure of removing the top of the skull and then the dura. Some of their experiments were done with "narcotics," some without—in spite of the pain induced by manipulating the dura. The electrical equipment they used was a series of Daniel cells. The electrodes were made of very fine platinum wire with molded ends to prevent injury. The description of their findings reads (in translation):

> One part of the convexity of the cerebrum of the dog is motor and another is not. . . .
> The motor area is generally more anterior; the nonmotor area is more posterior. In stimulating the motor area electrically, one elicits combined muscular contractions in the opposite half of the body. By using very weak current, one can localize these contractions exactly in narrowly delimited groups of muscle. If more intense current is used, other muscles—and indeed, also corresponding muscles in the same half of the body—immediately take part in the reaction as a result of the stimulation of the same or neighboring places. The possibility of isolated stimulation of a limited group of muscles is therefore limited to weak current on very small areas. For brevity's sake, we call these areas "centers." Minute shifting of the electrodes, to be sure, generally sets the same extremities in movement. But if, for example, extension is the first reaction, shifting the electrodes may produce flexion or rotation. We found that the part of the surface of the brain that lies between the centers lacked the capacity to be stimulated by the method described or by using minimal electrical intensity. If we increased either the distance between the electrodes or the intensity of the current, then convulsive movements could be brought forth; but these muscular contractions took place throughout the body, so that we could not once distinguish whether they were on one or both sides.
> In dogs, the location of the centers (which will soon be described) is very consistent. The exact demonstration of this fact at first met with several difficulties, which we avoided by first seeking the place where the weakest electrical stimulus elicited the strongest contraction of the appropriate muscle group. Then, as a marker, we struck a pin into the brain of the still-living animal between the two electrodes. After the brain had been removed we compared the points thus marked with those from earlier experiments that had been preserved in alcohol. How consistently the same centers are placed may best be seen in the fact that we repeatedly succeeded in finding the desired center in the middle of a single trephined hole without further opening the skull. When the dura was removed, the muscles dependent on that area contracted with the same regularity as if the entire hemisphere had been opened.

They found the center (Fig. 74) for stimulating the neck muscles ($\triangle$) in the prefrontal gyrus. At the edge of the postfrontal gyrus was found the center for extensors and abductors of the foreleg at the end of the frontal fissure ($+$). The areas of the centers for the flexion and rotation of the foreleg lay posterior to this and nearer the coronal fissure ($\dagger$). The center for the rear leg ($\#$) was also in the postfrontal gyrus, nearer the middle and more to the rear than that for the foreleg. The facial nerve ($\diamond$) was controlled from the middle portion of the suprasylvian gyrus. This area is often larger than 0.5 square centimeters and stretches down and forward from the main bend of the Sylvian fissure.

These two young (then unknown) doctors met outraged opposition from the establishment. For example, Burdon-Sanderson,[7] Wayneflete Professor of Physiology at Oxford, attacked their findings in a paper published by the Royal Society. Burdon-Sanderson had a personal interest in the effect of electrical stimulation on tissues, working himself mostly on the heart. Burdon-Sanderson, in attacking the results of Fritsch and Hitzig, declared these to be caused by spread of the current to the striatum, which he maintained was the motor center. The issue was even further obscured by his much publicized experiment in this field. Burdon-Sanderson's erroneous conclusion, which was published in the presti-

---

[7]J. C. Burdon-Sanderson. Notes on the excitation of the surface of the cerebral hemispheres by induced currents. *Proc. R. Soc.*, 22P:368–370 (1873).

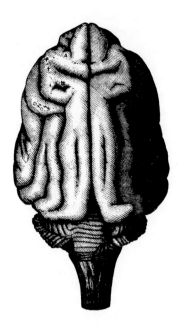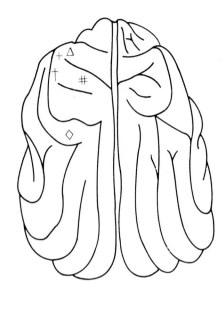

FIG. 74. **Left:** The original illustration of the responding areas of the dog's cortex, published by Fritsch and Hitzig. (From: *Archiv. Anat. Physiol. Wiss. Med.,* 37:300–332, 1870). **Right:** A drawing to make clear the labelling of cortical sites indicated by Fritsch and Hitzig.
Δ: Stimulation produced twitching of neck mucles.
+ : Abduction of foreleg.
†: Flexion of foreleg.
#: Movement of rear leg.
◇ : Facial twitching.

gious journal *Proceedings of the Royal Society,* did not convince Hitzig, who continued his research in support of his and Fritsch's conclusions.

The experiments of Fritsch and Hitzig evoked interest in one of the most prominent electrophysiologists of that time—Ludimar Hermann.[8] He described experiments specifically designed to test whether the results claimed by these two young doctors were due to stimulation of the cortex itself or from some deeply lying region. Hermann wrote (in translation):

> When these experiments were made public, no one could fail to be astonished that the vast majority of the most experienced and painstaking of the experimenters of the past should have not only overlooked such an easily measurable sensitivity of the cerebral cortex, but have actually disputed it, and that, according to the authors, the sensitivity is limited to an electrical stimulus, which is otherwise unheard of. There had to be the suspicion that the authors obtained their positive results because of the strength of electrical current they were using, never before attempted, and that this high electrical current had set into action motor apparatus deeper within the brain. In fact, in an organ containing an incredibly fine net of apparatus in a compact mass, an electrical stimulus must appear most dangerous, since it cannot be limited to one particular area.

Fritsch and Hitzig had discussed this possibility, writing as follows: "Since the brain matter possesses a high resistance, since other conductive parts were not in the vicinity, and since the electrodes were placed close together . . . the current density even very close to the points of stimulation could only be minimal."

[8]L. Hermann. Über elektrische Reizversuche an der Grosshirnrinde. *Pflüger's Arch. f. d. ges. Physiol.,* 10:77–85 (1875). (On electrical stimulation experiments on the cerebral cortex.)

Hermann set out to test the claims in his own laboratory in Zurich. His procedure was to search the cortical surface (in dogs) for a locus that evoked contraction of the rear leg in response to electrical stimulation. An interesting detail of these 19th-century experiments was that he used an 18th-century method of detecting spread of current on the surface, namely, he used frog's thigh muscles. Hermann then went on to cauterize the surface area that had given a response to stimulation. On repeating approximately the same current strength to the ''burnt'' area, he again got leg contractions. From his own—surely rather crude—experiments Hermann concluded that Fritsch and Hitzig's claim for localization in the cortical layer was unjustified. He wrote:

> We are far from underestimating the worth of newly discovered facts; but in the experiments of Fritsch and Hitzig we can only recognize as fact, that electrical stimulation has certain motor effects when electrodes are placed on certain gyri. The conclusion of the authors and many others, that at those positions motor centers lie, is completely unjustified. Quite apart from the general proposition that one can never recognize a motor center by stimulation experiments, the objection that the results of stimulation could come from deeper parts has not been disproved.

In summary, Hermann suggested that the functions of the cortical layer are psychic and only deeper regions are motor in function. He concluded his report with an unequivocal statement: ''I close with the assertion that the experiments of Fritsch and Hitzig, as interesting as they are, do not entitle us to draw any conclusions as to the function of the cerebral cortex.''

Seventy-nine years after Galvani had shown the world that peripheral nerves responded to electrical stimulation, proof was given that so did the cortical cells. Only five years later, an even more startling fact was demonstrated, namely, that these cortical cells themselves emitted electricity.[9]

Unnoticed by both the protagonists and the antagonists was that A. I. Tysheshetsky,[10] a student of Sechenov's in Moscow, presented his thesis to the Moscow Surgical Academy in May of that same year (1870). His work was on the central nervous system of amphibians and the production of muscular movements by stimulation of their brains.

Gustav Theodor Fritsch (1838–1927)

Gustav Theodor Fritsch was essentially a comparative anatomist greatly influenced by Du Bois-Reymond. In the pattern of his times, he carried out research on the electric fish, publishing the anatomy of several orders.[11]

Although qualified in medicine, Fritsch's interests took him into many fields, including travels to South Africa to study anthropology and yet another expedition to watch an eclipse to follow his interest in astronomy. It was on return from these excursions that he worked with Hitzig. Their famous paper appeared in the year of the Franco-Prussian War (1870). The war caused Fritsch to drop his scientific work and go into the army. It was the period of Bismarck's war on France, and the Germans were already in Paris. Fritsch published very frequently, but after this interruption by the war, he no longer contributed to neurophysiology, although after he returned from the war he was appointed Professor of Comparative Anatomy in Berlin. However, he never lost his interest in the electric fish and again pub-

[9]R. Caton. The electric currents of the brain. *Br. Med. J.*, 2:278 (1875).

[10]A. I. Tysheshetsky (1828–1878). Thesis. Moscow Surgical Academy. Moscow, 1870.

[11]G. T. Fritsch. *Untersuchungen über den feineren Bau des Fischgehirns.* Guttman'sche Buchhandlung (C. Ensler), Berlin, 1878. (Investigations of the Finer Structure of the Fish Brain.)

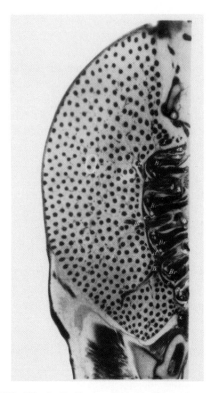 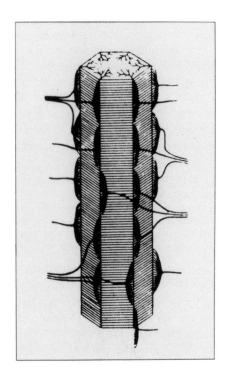

FIG. 75. Left: Gustav Fritsch's illustrations of the electrical fish. (From: Bericht über die Fortsetzung der Untersuchungen an elektrischen Fischen. Beiträge zur Embryologie von Torpedo. *Sitzungsberichte der Königlich Preussischen Akademie der Wissenschaften zu Berlin,* 1883, pp. 205–209.) **Right:** Fritsch's sketch of how he envisaged the charge of the electric fish as a voltaic pile. (From: G. Fritsch, *Die elektrischen Fische.* Veit, Leipzig, 1890.)

lished [12,13] in this field, moving into Du Bois-Reymond's Institute to pursue his anatomical work.

Fritsch published frequently on the electric fish, contributing two sections to Karl Sachs' large volume on the subject.[14] Du Bois-Reymond also contributed to this work. In 1887 and 1890 Fritsch published two large, beautifully illustrated volumes,[15] again on marine animals (Fig. 75). Fritsch outlived Du Bois-Reymond by nearly 30 years; when the latter retired, Fritsch moved his interest from electrophysiology to human anatomy. In 1899 he published a book[16] on that subject directed toward anthropologists and profusely illustrated by his own photographs. It followed another treatise in anatomy in which he studied human

[12]G. T. Fritsch. Berichte über eine Reise zur Untersuchung der in den Museen Englands und Hollands vorhandenen Torpedineen. *Archiv. Anat. Physiol.,* 70–73 (1884).

[13]G. T. Fritsch. Bericht über die Fortsetzung der Untersuchungen der elektrischen Fische. *Arch. Anat. Physiol.,* 73–78 (1884). (Report on the continuation of the investigation of electric fish.)

[14]Karl Sachs. *Untersuchungen am Zitteraal Gymnotus electricus, nach seinem Tode bearbeitet von Emil du Bois-Reymond mit zwei Abhandlungen von Gustav Fritsch.* Veit, Leipzig, 1881. (Investigation of the Electric Eel *Gymnotus electricus* Posthumously Revised by Emil du Bois-Reymond with Two Essays by Gustav Fritsch.)

[15]G. T. Fritsch. *Die Elektrischen Fische. Nach neuen Untersuchungen, anatomisch-zoologisch dargestellt.* Leipzig, 1887–1890.

[16]G. T. Fritsch. *Die Gestalt des Menschen.* Neff, Stuttgart, 1899. (The Human Form.)

hair types,[17] including those of the American Indian.[18] Fritsch's proflific and varied life ended in 1927.

Edouard Hitzig (1838–1907)

Edouard Hitzig led a very different life from that of Fritsch. Also a Berliner, he gained his doctorate there and devoted his life primarily to medical practice, first in Berlin and later in Zurich where he became a psychiatrist. He remained in psychiatry all his long life, with professorships first in Zurich, then in Halle. After the partnership with Fritsch had ended, Hitzig continued some research on the effect of cortical stimulation in the dog and monkey, refining the areas for evocation of movements. He was always very polite in his publications, although his findings disproved the theories of generalized functions sponsored by Goltz and Flourens.

In 1870, the year in which he and his colleague Fritsch published their epoch-making paper, Hitzig (while it was still in press) described their work in a speech to the Berlin Medical Society.[19] He was polite but firm in refuting those that misinterpreted the findings in their experiments. In his speech he said:

> Gentlemen! You will know that the peripheral nerves reply to all influences which change their condition at a certain speed, with their specific energy. This property of sensitivity is agreed upon by all physiologists.
>
> Only with regard to the central nervous system are there differing opinions. Almost the only universally accepted opinion is that of the insensitivity of the brain's hemispheres. Even those experimenters who claim to have caused movements or feelings emanating from other central sections were agreed that the brain reacted to no stimulus.
>
> Then some basal parts of the brain were attributed sensitivity by several researchers. As for the spinal cord, opinions are still very diverse.

Hitzig then went on to describe to the Society the experiments he had done with Fritsch, subtly suggesting that this followed earlier work by himself:

> Following this point of view, I undertook a new investigation on the electrical sensitivity of the brain hemispheres, together with Dr. Fritsch, an assistant at the Berlin Institute of Anatomy. We have achieved a number of positive results, but cannot as yet term our work as completed.
>
> Our main result has been to demonstrate that in the brain hemispheres of the dog there are circumscript seats of innervation which are susceptible to electrical arousal. Groups of muscles can be set in motion by the use of weak current, the motor reaction can be restricted to individual parts of the body. We were thus able to gradually determine the centers for a large part of the voluntary musculature. If strong currents are used then very widespread muscle spasms occur.
>
> The motor centers which we determined all lie, as you can see in Dr. Fritsch's drawing, quite far to the front. The rear part of the brain is insensitive for even the strongest currents.
>
> The movements which are caused usually consist of a simple spasm, but when the stimulus is carried out with tetanizing induction currents, then an initial maximum of contraction is achieved, which decreases.

[17]G. T. Fritsch. Das menschliche Haar als Rassenmerkmal. *Verh. Berliner Ges. Anthrop.* In *Zeit. Ethnol. Anthrop. Urgeschicte,* 1885. (Human hair as racial characteristic.)

[18]G. T. Fritsch. Die Lehre von der Einheit der amerikanischen Eingeborenenrassen untersucht an der Haarbeschaffenheit. *Ver. Amerik. Cong.* Berlin, 1888. (The Theory of the Unity of the American Indigenous Peoples, Investigated by Means of Hair Types.)

[19]E. Hitzig. Über die elektrische Erregbarkheit des Grosshirns. *Klin. Woch.,* 7:137–138 (1870). (On the electrical activity of the brain.)

Hitzig continued to publish quite frequently, always defending the response of the cortex to electrical stimulation as well as less elaborate experiments on the peripheral nerve.[20]

Out of therapeutic interest, Hitzig once more put to the test the fact already noted by Du Bois-Reymond, Pflüger, and Munk, namely, that the current flow falling vertically on the longitudinal axis of a nerve is ineffective. He laid the nerves of two frog thighs next to each other in such a way that they touched each other only for a short length. He then passed current through both, by placing an anode on one thigh, a cathode on another. Higher currents were always necessary than for simple longitudinal passage of current. The closure spasm of the cathode stimulus did not always correspond to the opening spasm of the anode stimulus. The size of the electrodes and many other influences led to many differences.

His most prized opportunity came when he was invited to give the Hughlings Jackson Lecture in London. This was published in its entirety in *Brain* in 1900.[21] In this lecture he reviewed the development of ideas on cerebral localization dealing rather heavily with Goltz but praising, naturally, Jackson. He did, however, mention Ferrier, saying in a footnote:

> A remark is here necessary against *Ferrier*. On page 225 of the second edition of his work, "The Functions of the Brain," he has thought it necessary to observe, against my word "excitable," that there is no reason to assume that any parts of the cortex are not excitable. This observation is a stroke into the air, for he will find that on page 19 of my book, "Untersuchungen über das Gehirn," in the footnote, I have defined the expression "unerregbar" (not suitable) in the following words: "We call here, without any prejudgment, all those parts 'not excitable' from which we can produce no twitchings."

After the initial classic paper in 1870, Hitzig never worked again with Fritsch. He died in 1907, Fritsch outliving him by 20 years.

John Hughlings Jackson (1835–1911)

In the 19th century, a great deal of neurophysiological research stemmed from findings in the clinic. This was especially so in brain research. One whose clinical observations were seminal was John Hughlings Jackson, who had been born plain John and adopted the name of Hughlings.

Jackson, a Yorkshireman born and educated in the north of England, became interested in the brain early, for he worked with Thomas Laycock.[22] Laycock, a neurologist greatly concerned with the brain and problems of consciousness, developed the idea of reflexes in the brain, thus anticipating Sechenov. The interest of his pupil Jackson also lay in the nervous system, and this led to his being suggested by Brown-Séquard[23] to a post at the newly established hospital in Queen Square; he was the fifth physician to be appointed (in 1862) to this new hospital, which was devoted to the problem of epilepsy and named the National Hospital for the Paralysed and Epileptic. The line of Jackson's future work was now clear; he would devote his life to the study of convulsive disorders.

---

[20]E. Hitzig. Über quere Durchströmung der Froschnerven. *Pflüger's Arch. f. d. ges. Physiol.*, 7:263–273, 1873. (*Cbl. Med. Wiss.*, 12:263–273, 1873.) (On passing transverse current through frog nerves.)

[21]E. Hitzig. Hughlings Jackson and the cortical motor centres in the light of physiological research. *Brain*, 28:545–581 (1900).

[22]Thomas Laycock (1812–1876). On the reflex function of the brain. *Br. For. Med. Rev.*, 19:298 (1845).

[23]C. E. Brown-Séquard (1817–1894).

Jackson's first five papers[24] were published by Crichton Browne in the *Reports of the West Riding Lunatic Asylum*. This hospital near Wakefield in the north of England produced a remarkable series of papers all published in the *Reports*, a journal forum created by the director, Crichton Browne.[25] He not only drew to his group an outstanding series of research workers concerned with the brain, but he also founded the series of *Reports* from this hospital that appeared for seven years (1871–1876) and included a gamut of studies, many of which were on the brain, from Clifford Allbutt's early attempts at electrical shock treatment for the mentally ill[26] to Jackson's studies on epilepsy. Those were followed by Ferrier's explorations of electrical stimulation of the cortex.[27,28] This rich source of neurological papers ceased when Browne left the West Riding Lunatic Asylum, but he was undaunted and moved to the founding of an even more prestigious outlet for papers on nervous diseases; in 1878, together with a group of neurologists, he founded the journal *Brain*.

Jackson often had second thoughts about what he had published, and in the archives of the National Hospital is one of those early papers that he had presented, over which were many handwritten corrections.[29] Those are of interest because they reflect the opinion of the 59-year-old experienced epileptologist since the statements were made 19 years earlier. On page 110 of his comments he added in his handwriting: ''I cross out the word 'mental' as we have to distinguish the psychical from the physical. Psychical states *may attend* the nervous activities of the highest cerebral centres which activities lead to the so-called automatic actions.'' The underlining is Jackson's and the comment is signed ''J H. J.''[30] Although his clinical work gave so much inspiration to those seeking cortical localization and although he had confirmed the speech center declared by Pierre Paul Broca[31] at the Bicêtre (Fig. 76), Jackson himself was more interested in an approach suggested by the works of Herbert Spencer.[32] In his Croonian lecture[33] he said:

> The doctrine of Evolution daily gains new adherents. It is not simply synonymous with Darwinism. Herbert Spencer applies it to all orders of phenomena. His application of it to the nervous system is most important for medical men. I have long thought that we shall

---

[24]Hughlings Jackson's first five papers contributed to the *West Riding Lunatic Asylum Medical Reports* were: (1) Observations on the localisation of movements in the cerebral hemispheres as revealed by cases of convulsion, chorea and 'aphasia,' 3:175–195 (1873); On the anatomical, physiological, and pathological investigation of epilepsies, 3:315–349 (1873); On a case of recovery from double optic neuritis, 4:24–29 (1874); On temporary mental disorders after epileptic paroxysms, 5:188–197 (1875); On epilepsies and on the after effects on epileptic discharges (Todd and Robertson's hypothesis), 6:266–309 (1876).

[25]J. Crichton Browne. A plea for the minute study of mania. *Brain*, 3:347–362 (1880–1881).

[26]T. Clifford Allbutt (1836–1925). The electrical treatment of the insane. *West Riding Lunatic Asylum Reports*, 2:203–222 (1872).

[27]D. Ferrier. Experimental research in cerebral physiology and pathology. *West Riding Lunatic Asylum Medical Reports*, 3:30–96 (1873).

[28]D. Ferrier. Pathological illustration of brain function. *West Riding Lunatic Asylum Medical Reports*, 4:30–62 (1874).

[29]The author is grateful to the National Hospital for a photocopy of this paper corrected in J. Hughlings Jackson's handwriting.

[30]J. Hughlings Jackson. On temporary mental disorders after epileptic paroxysms. *West Riding Lunatic Asylum Medical Reports*, 5:105–129 (1875).

[31]Pierre Paul Broca. Remarques sur le siège de la faculté de language articulé, suivies d'une observation d'amphémie (perte de la parole). *Bull. Soc. Anat. Paris*, 36:330–337 (1861).

[32]Herbert Spencer (1820–1903).

[33]J. Hughlings Jackson. Croonian lecture. On the evolution and dissolution of the nervous system. *Proc. R. Soc.*, 3–5 (1884).

FIG. 76. Two explorers of localization in the brain. **Left:** Pierre Paul Broca (1824–1880) whose clarification of the speech center was achieved by clinical observation with findings at autopsy. **Right:** John Hughlings Jackson whose studies of convulsive seizures led his followers to electrical search for the motor cortex.

be very much helped in our investigations of diseases of the nervous system by considering them as reversals of Evolution, that is, as Dissolutions. Dissolution is a term I take from Spencer as a name for the reverse of the process of Evolution.

The localization of speech disabilities (Fig. 77) was a subject of disagreement between Jackson and Broca. Broca's term ''aphemia'' was designated by him to mean a loss of the ability to articulate—in other words, a motor loss not attributable to a single hemisphere. He differentiated this from ''amnesia verbale,'' defined as a loss of memory, and it was this that he put in the left hemisphere.

Hughlings Jackson[34] expressed his view as follows: ''I must here say that I believe less in some of the views propounded by Broca than I did, although I think the scientific world is under vast obligation to him for giving precision to an important inquiry.'' Jackson was still skeptical that all language function lay in a single hemisphere.[35] Even in his comments on Broca, his interpretations were his own. He wrote:[36]

[34]J. Hughlings Jackson. Notes on the physiology and pathology of language. *Med. Times Gaz.*, 1:659–662 (1866).

[35]At this time the main contribution to the localization of speech and of articulation came from France by a combination of clinical observation and postmortem anatomy. See the following papers:

Marcel Dax. Lésions de la moitié gauche de l'encéphale coincident, l'oubli des signes de la pensée. *Gaz. Hebdom. Med. Chir.*, 2:259 (1865).
Ernst Aubertin. Considérations sur les localisations et, en particulier sur le siège de la faculté du language articulé. *Gaz. Hebd. Med. Chir.*, 10:318, 348, 397, 455 (1863).
Pierre Marie (1853–1940). Revision de la question de l'aphasie; la troisième circonvolution frontale gauche ne joue aucun rôle spécial dans la fonction du language. *Sem. Méd.*, 26:241–247 (1906). [Marie disputed Broca's claim that the third frontal convolution is the speech's center. He classified aphasia into three groups: anarthria (defects of articulation), Broca's (motor) aphasia, and Wernicke's (sensory) aphasia.]

[36]J. Hughlings Jackson. On the anatomical and physiological localisation in the brain. *Lancet*, 1:84–85 (1837).

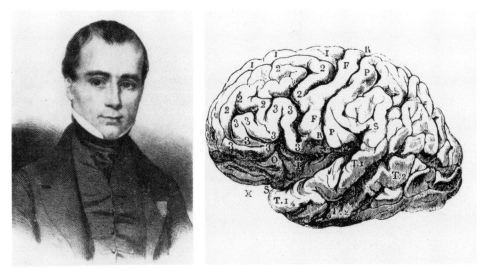

FIG. 77. Left: Ernst Aubertin (1825–??). **Right:** His picture of the human brain with the cortical regions he claimed to be the speech centers. (From: E. Aubertin *Gaz. Hebd. Med. Chir.,* 10, 397, 1863.)

> M. Broca says, that aphemia may be considered to be either the loss of a purely mental faculty, or merely a kind of locomotive ataxy. It seems to me, that a loss of guiding power in the articulatory apparatus, is at least a great part of the defect in the confused talking of which I am now speaking. It certainly exists without any obvious paralysis of the lips, tongue, or palate. . . .

Jackson's name is honored for the clinical state called Jacksonian epilepsy. His contribution was the major one, although the condition had been described before by the French physician Bravais.[37] It was in the middle of Jackson's career that Fritsch and Hitzig published their demonstration that certain movements could be elicited by electrical stimulation of certain areas of the cortex. This confirmed Jackson's proposal that there was a motor area.

Jackson himself was not a neurophysiologist, but his detailed descriptions of clinical signs were extremely suggestive to experimenters. He maintained his appointment at the National Hospital while also acting as physician to the London Hospital. He lived into the 20th century, dying in 1911.

John Burdon-Sanderson (1828–1905)

Trained in medicine at the University of Edinburgh, John Burdon-Sanderson qualified in 1851 and then went to Paris, ostensibly to study chemistry, but he also attended some lectures by Claude Bernard, which instilled in him an interest in physiology. He returned to London in 1852 to practice medicine and later to lecture at St. Mary's Hospital. After several different experiences in medicine, he finally took the decision to concentrate on physiology. This concentration brought him a reward in the form of the Chair of Physiology at University College, London, that had recently been vacated by Michael Foster, the first holder. This professorship was a first step toward a series of distinguished appointments: the Jodrell professorship in London, the Wayneflete Chair in Physiology at Oxford (of

---

[37]L. F. Bravais. Recherches sur les symptomes et le traitement de l'épilepsie hémiplégique. Thèse No. 118, Paris, 1827.

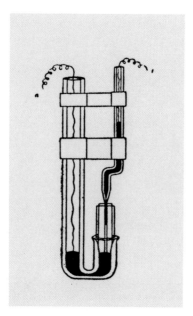

FIG. 78. Left: The electrometer built by Frederick Page for use in the electrophysiological experiments of John Burdon-Sanderson. **Right:** John Burdon-Sanderson (1828–1905).

which he was the first holder), and then in 1895 he received one of the Regents professorships that had been founded by Henry VIII.

During his first years of research, he focused on pathology, but he was almost inevitably caught up by the prevailing interest in electrophysiology. For the neurophysiologist, his first contribution is a paper he wrote on the speed of conduction of the nerve impulse in the muscle. Writing, as he was, from the prestigious Wayneflete Chair, his paper was readily accepted by the Royal Society for publication in its *Proceedings*.[38,39] Being uncertain in electronics, he collaborated with an electrical engineer named Frederick Page, and with him he published other papers in this field, which were mostly concerned with the currents of the heart.[40] With a capillary electrometer designed by Page, he made multiple measurements of the T wave (Fig. 78).

One of his contributions most useful to the neurophysiologist was the volume[41] he masterminded of translations from the great German scientists—Du Bois-Reymond, Hermann, Hering and others. Burdon-Sanderson became very powerful with his editorship of the *Proceedings of the Royal Society*, but he was not readily open to new ideas. His personal weakness in electrophysiology did not deter him from denying the results of others. He attacked Fritsch and Hitzig's report as incorrect, and some years later he would not encourage Richard Caton[42] in his discovery of the brain's own electricity, for he had a conviction that cortical cells themselves were not electrically involved (an opinion he passed on to

[38]J. Burdon-Sanderson and F. J. M. Page. Some experiments as to the influence of the surrounding temperature on the discharge of carbonic acid in the dog. *J. Physiol.*, 2:228–234 (1879–1880).

[39]J. Burdon-Sanderson and F. J. M. Page. On the time-relations of the excitory process in the ventricle of the heart of the frog. *J. Physiol.*, 2:384–455 (1879–1880).

[40]J. Burdon-Sanderson and F. J. M. Page. On the time-relations of the excitory process in the ventricle of the heart of the frog. *J. Physiol.*, 2:384–435 (1879–1880).

[41]J. Burdon-Sanderson. *Transactions of Foreign Biological Memoirs*. Frowde, London, 1877.

[42]R. Caton. The electrical currents of the brain. *Br. Med. J.*, 2:278 (1875).

Victor Horsley). Caton had hoped to give a report to the Royal Society but met with little encouragement from Burdon-Sanderson who was at that time (1875) vice-president of the society. Burdon-Sanderson made his attack on the conclusion reached by Fritsch and Hitzig, namely, that the cerebral cortex was electrically excitable. In a much publicized paper printed in the *Proceedings of the Royal Society,* he wrote:[43]

> If that part of the surface of the hemisphere which comprises the active spots is severed from the deeper spots by a nearly horizontal incision made with a thin-bladed knife, and the instrument is at once withdrawn without dislocation of the severed part, and the excitation of the active spots there upon repeated, the result is the same as when the surface of the injured organ is acted upon.

Comments from Burdon-Sanderson were not to be dismissed lightly because of his distinguished appointment as the Jordell Professor of Physiology at the University of London and his membership in the Royal Society. His alternative proposal was that the movements observed by Fritsch and Hitzig were secondary to sensation and that the direct effect of the stimulation was sensory rather than motor. The "sensation" reached the brain, an "idea" then intervened and excited the basal ganglia to activate their motor function.

Burdon-Sanderson may, perhaps, be regarded as a forerunner of the movement that was to develop into stereotaxic methods for localization of nervous structures in the depths of the brain. So impressed was he by his experiments that produced movements on stimulation of the underlying white matter after removal of the cortical layers that he became even more convinced that the motor centers lay in the basal ganglia. He wrote:

> In a brain hardened in alcohol, a needle is plunged vertically, i.e., at right angles to the surface, from the active spot for retraction of the opposite ear, reaches the posterior part of the *corpus striatum* at a depth of from 10 to 12 millimeters. If a horizontal incision is made in the living brain, at this depth, and is met by two others, of which one is directed antero-posteriorly and the other transversely, and the part comprised within the incisions removed, a surface of brain is exposed in the deepest part of the wound which corresponds to the outer and upper part of the corpus striatum. If now the electrodes are applied to this surface, the movements . . . are produced in the same way as before, but more distinctly; the active spots are quite strictly localised. . . .

Burdon-Sanderson's conclusion was as follows:

> From these facts it appears that the superficial convolutions do not contain organs which are essential to the production of the combination of muscular movements now in question. They further make it probable that the centres for such movements are to be found in the masses of grey matter which lie in the floor and outer wall of each lateral ventricle is true.

Burdon-Sanderson's attack[44] on Fritsch and Hitzig was not his only challenge to the growing evidence for cortical excitability. In 1876 he wrote a long commentary[45] on the work reported by Hermann[46-48] in 1877–1878. (Burdon-Sanderson says Hermann sent him

[43]J. Burdon-Sanderson. Notes on the excitation of the surface of the cerebral hemispheres by induced currents. *Proc. R. Soc.,* 22, 368–370 (1878).

[44]J. Burdon-Sanderson. Notes on the excitation of the surface of the cerebral hemispheres by induced currents. *Proc. R. Soc.,* 22P:368–370 (1873).

[45]J. Burdon-Sanderson. A report of Prof. L. Hermann's recent research on the electro-motive properties of muscle. *J. Physiol.,* 1, 196–212 (1876).

[46]L. Hermann. Untersuchungen über die Entwicklung des Muskelstroms. *Pflüger's Arch. f. d. ges. Physiol.,* 15:191–232 (1877). (Research into the Development of the Muscle Current.)

[47]L. Hermann. Versuche mit dem Fall-Rheotom über die Erregungsschwankung des Muskels. *Pflüger's Arch. f. d. ges. Physiol.,* 15:233–245 (1877). (Experiments with the Fall-rheotome on the stimular variation of the muscle.)

[48]L. Hermann. Untersuchungen über die Actionsströme des Muskels. *Pflüger's Arch. f. d. ges. Physiol.,* 16:191–262 (1878). (Investigations into the Action Currents of the Muscle.)

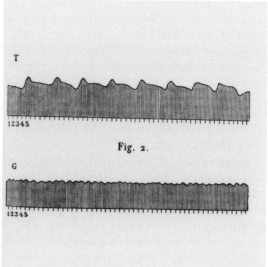

FIG. 79. Left: Étienne Jules Marey (1830–1904). Investigator of recording techniques and introducer of photography into the laboratory. A distinguished institute was named for him but is unfortunately no longer existent. (Portrait by courtesy of Professor D. Albe-Fessard.) **Right:** Two recordings made with an electrometer. *Above:* Electrocardiagram of a tortoise. *Below:* Electrocardigram of a frog. (From: E. J. Marey and G. Lippmann. Des variations électriques des muscles et du coeur en particulier étudiées au moyen de l'électrometre de M. Lippmann. C. R. *Acad. Sci.* 82:957–977, 1876.)

37 pages of reports.) This article of Burdon-Sanderson's did not, in fact, bring any new knowledge or experimental results to the problem. The problem was the difference in opinion between Du Bois-Reymond and Hermann. Burdon-Sanderson's report was on Hermann, and he restricted his discussion to muscle current and not the nerve. Du Bois-Reymond held that muscles and nerves contained electromotive particles suspended in an inactive medium. On stimulation the electromotive force of these particles diminishes (negative variation).

One of Hermann's arguments was that the current flow caused by a cut should flow for a long time if the fiber were filled (as Du Bois-Reymond claimed) with multiple electromotive elements. Hermann held that the flow of current diminished rapidly and was contributed to only by those cells which had been injured by the cut.

The capillary electrometer designed for him by Page was soon eclipsed by the model designed by Marey and Lippmann,[49–54] known popularly as the Lippmann electrometer. These instruments reflected oscillations of the electrical input by fluctuations of the surface of mercury (Fig. 79). Electronics soon introduced techniques with faster response.

Etienne Jules Marey, the distinguished Professor of Medicine at the Collège de France, was most ingenious in his development of instrumentation for measuring physiological

[49]Etienne Jules Marey (1830–1904) and Gabriel Lippmann.

[50]E. J. Marey. Inscription photographique des indications de l'électromètre de Lippmann. *C. R. Acad. Sci.*, 83: 278–280 (1876).

[51]E. J. Marey. *Recherches sur le Pouls.* Thunot, Paris, 1860.

[52]E. J. Marey. Loi qui préside à la fréquence des battements du coeur. *C. R. Acad. Sci.*, 53:95–98 (1861).

[53]E. J. Marey. *La Circulation du Sang à l'Etat Physiologique.* Masson, Paris, 1881.

[54]E. J. Marey. *Le Mouvement.* Masson, Paris, 1894.

events. He had contributed much to the techniques for measuring blood pressure, and in 1884 he applied these techniques to neurophysiology. Marey was a great innovator, having constructed a camera in 1887, and this innovation brought photography into the physiology laboratory. He developed his photographic technique to study muscle movements by serial photographs. This was a contribution to the beginning years of cinematography.

Marey had been born in the wine country in 1830 and went to Paris to qualify in medicine. Although interested in mechanics, he soon began to contribute to the recording and measuring of physiological events. He increased the number of exposures he could make to 60 per second, and with this advancement he achieved photography of the flight of birds.

But an attack was coming to the capillary electrometer. Einthoven designed the string galvanometer, and this was to become the preferred method of measurement in electrophysiology with the development of the cathode ray oscillography. Einthoven was one of the outstanding critics of the capillary electrometer,[55-57] which he had himself tried to improve before designing the string galvanometer. He received the Nobel Prize for his invention in 1903.

There were those, however, who had used the capillary electrometer extensively before. This was especially so in England where Waller[58] continued to use it successfully in his exploration of the electrical fields of the heart. Another was the English physiologist G. T. Burch,[59-63] who published three papers in the *Proceedings of the Royal Society*.

David Ferrier (1843–1928)

The outstanding results of Fritsch and Hitzig,[64] published in 1870, changed the whole approach to research on the localization of brain function. There had been the fantasies of Gall,[65] the denials of Flourens[66] and of Goltz,[67] as well as the attempts to deduce such information from clinical signs. One of the most industrious investigators, David Ferrier, re-

[55]Wilhelm Einthoven (1860–1927). Hete meter von snell Wisselende Potential versuchillen mit Behulp von Lippmann Capillar Electrometer. *Pflüger's Arch. f. d. ges. Physiol.*, 56:528–541 (1894). (The measurement of rapidly fluctuating potentials with the help of Lippmann's electrometer.)

[56]W. Einthoven and M. A. B. Geluk. Die Registrirung des Herztons. *Pflüger's Arch. f. d. ges. Physiol.*, 57:617–639 (1894). (The registering of the heartbeat.)

[57]W. Einthoven. Ein neues Galvanometer. *Ann. Physik.*, 4:1069–1071 (1903). (A new galvanometer.)

[58]Augusts Volney Waller (1816–1870).

[59]G. T. Burch. On a method of determining the value of rapid variations of potential by means of the capillary electrometer. *Proc. R. Soc.*, 48:89–93 (1890).

[60]G. T. Burch. On the calibration of the capillary electrometer. *Proc. R. Soc.*, 59:18–24 (1895–1896).

[61]G. T. Burch. Contributions to a theory of the capillary electrometer. *Proc. R. Soc.*, 71:102–105 (1902).

[62]G. T. Burch and Francis Gotch. The electrical response of nerve to two stimuli. *J. Physiol.*, 24:410–426 (1899).

[63]G. T. Burch and Francis Gotch. The electrical response of nerve to a single stimulus investigated with the capillary electrometer. *Proc. R. Soc.*, 63:300–303 (1899).

[64]G. T. Fritsch (1838–1891) and E. Hitzig (1838–1907). Über die elektrische Erregbarkeit des Grosshirns. *Arch. Anat. Physiol. Wiss. Med.*, 300–314 (1870). (On the Electrical Excitability of the Cerebrum.)

[65]F. J. Gall and J. C. Spurzheim. *Anatomie et Physiologie du Système Nerveux en Général et du cerveau en Particulier et des Animaux, par la Configuration de leur Têtes.* Paris, Schoell, 1810–1819. (Vols. I & II by Gall and Spurzheim; Vols. III & IV by Gall.)

[66]P. Flourens (1794–1867). *Recherches expérimentales sur les Propriétés et les Fonctions du Système Nerveux dans les Animaux Vertébrés.* Clevot, Paris, 1824.

[67]F. L. Goltz (1834–1902). Über die Verrichtungen des Grosshirns. *Pflüger's Arch. f. d. ges. Physiol.*, 13:1–44 (1876). (On the functions of the cerebrum.)

ported that it was the observations made by the clinical neurologists that inspired his research. He gave especial recognition to J. Hughlings Jackson[68] to whom he dedicated his famous book *The Functions of the Brain*.[69] In later years, he was himself to receive a similar distinction when Sherrington dedicated to Ferrier his well-known book *The Integration of the Nervous System*.

Ferrier, a Scotsman with a medical degree from Edinburgh, was schooled at the universities of Aberdeen and of Heidelberg. He eventually moved to London where, after a series of junior positions, he received appointments at King's College Hospital in 1874 and the National Hospital, Queen Square, in 1880. In Ferrier's day, London's Queen Square had just opened as the National Hospital for the Paralysed and Epileptic. Before this, Ferrier had had a period in pathology at the West Riding Lunatic Asylum in Yorkshire, a unit particularly strong in pathology. It was the same hospital where W. Bevan Lewis had been the first to demonstrate the cortical cells of humans.[70] This was, of course, a demonstration of great importance for all those using electrical stimulation of the cortex. As already noted, the initial criticism for Fritsch and Hitzig's results was that they were owing to the spread of current to deeper regions. That the cells themselves had fibers that projected downward had not yet been revealed, and the "neuron doctrine" still lay in the future. It was from the West Riding Lunatic Asylum in its medical reports that Ferrier[71] published his first results on cortical localization. His goal was to prove in the pathological laboratory the conclusions drawn from the clinical signs. He was working under a distinguished director, Crichton Browne.[72,73] In this first paper, Ferrier described experiments with birds, moving on to cats and dogs and finally to monkeys. Ferrier's results did indeed confirm Hughlings Jackson's conclusions derived from epilepsy but disproved Meynert's idea that the thalamus was solely motor in function. He was also able to relate the movement of the eyes to cerebellar function.

Ferrier's stimulus to embark on the method of electrical stimulation to seek support for the clinical concepts was directly evoked by the 1870 report of Fritsch and Hitzig, though Ferrier's first attempt to stimulate the cortex had been by chemicals, that is, by injection of chromic acid. When this did not work, he turned to electrical stimulation. Ferrier followed his 1873 paper[74] the next year with one describing experiments on more animals, publishing again from the West Riding Lunatic Asylum.[75] Although he owed his use of electrical stimulation to Fritsch and Hitzig, he was ungracious in failing to make adequate recognition. Ferrier had influential friends, Burdon-Sanderson and Jackson being supporters, and this gave him access to publication in the *Proceedings of the Royal Society*.[76] They

[68]J. Hughlings Jackson (1835–1911). On the anatomical, physiological, and pathological investigations of the epilepsies. *West Riding Lunatic Asylum Medical Reports*, 3:315–339 (1873).

[69]David Ferrier (1843–1928). *The Functions of the Brain*. Smith Elder, London, 1876.

[70]W. B. Lewis (1847–1929). On the comparative structure of the cortex cerebri. *Brain*, 1:79–86 (1878).

[71]David Ferrier. Experimental research in cerebral physiology and pathology. *West Riding Lunatic Asylum Medical Reports*, 3:30–96 (1873).

[72]J. Crichton Browne. On the weight of the brain and its component parts in the insane. *Brain*, 2:42–67 (1879–1880).

[73]J. Crichton Browne. A plea for the minute study of mania. *Brain*, 3:347–362 (1880–1881).

[74]D. Ferrier. Experimental research in cerebral physiology and pathology. *West Riding Lunatic Asylum Medical Reports*, 3:30–96 (1873).

[75]D. Ferrier. Pathological illustrations of brain function. *West Riding Lunatic Asylum Medical Reports*, 4:30–62 (1874).

[76]D. Ferrier. The localization of function in the brain (MS), communicated by J. Burdon-Sanderson, 5 March 1874; Archives of the Royal Society, A.P. 56.2. *Proc. R. Soc.*, 22:229–232 (1874).

were uncomfortable about Ferrier's lack of adequate acknowledgment to the young pioneers in Berlin and even called in the prestigious Thomas Huxley to try to persuade Ferrier to make a more generous recognition. Ferrier, however, did not comply and merely removed his description of work on dogs (as used by Fritsch and Hitzig), publishing only his results on monkeys.[77] Understandably, one of those who complained was Hitzig.[78]

In 1875 Ferrier reported what he had found on monkeys he had studied after surgical removal of various parts of the brain. In one[79] of his reports the description runs as follows:

> The experiments show conclusively that an animal deprived of its frontal lobes retains all its powers of voluntary motion unimpaired, and that it continues to see, hear, smell, and taste, and to perceive and localise tactile impressions as before. It retains its instincts of self-preservation, retains its appetites, and continues to seek its food. It is also capable of exhibiting various emotions. The result, therefore, is almost negative, and the removal of a part of the brain which gives no external response to electric stimulation excercises no striking positive effect; and yet the facts seem to warrant the conclusion that a decided change is produced in the animal's character and disposition. For this operation I selected the most active, lively, and intelligent animals I could obtain. To one seeing the animals after removal of their frontal lobes little effect might be perceptible, and beyond some dullness and inactivity they might seem fairly up to the average of monkey intelligence. They seemed to me, after having studied their character carefully before and after the operation, to have undergone a great change. While conscious of sensory impressions, and retaining voluntary power, they, instead of being actively interested in their surroundings, ceased to exhibit any interest in aught beyond their own immediate sensations, paid no attention to, or looked vacantly and indifferently at, what formerly would have excited intense curiosity, sat stupidly quiet or went to sleep, varying this with restless and purposeless wanderings to and fro, and generally appeared to have lost the faculty of intelligent and attentive observation.

In 1881 an event of great historic meaning for neurophysiologists was held in London. This was the International Medical Congress at which Goltz demonstrated the dogs from whom he had removed cortical tissue and at which Ferrier also demonstrated one of his animals. Ferrier's report brought enthusiastic support from his audience, which set him firmly on the road of his future career in research. Goltz, in contrast, failed to convince the audience. At this congress, Ferrier demonstrated two monkeys: one in whom, some weeks earlier, he had destroyed the superior temporal convolution bilaterally; the other in whom he had destroyed the left motor area. Both monkeys had survived well, both could move around, but the first one was deaf. Neither of these demonstrations caused Goltz to change his mind.

Ferrier was familiar with Goltz's work and critical of it. In a book he published in 1878,[80] Ferrier wrote:

> According to Goltz it is not so much the position as the extent of the injury on which the phenomena of cortical lesions depend. . . . Instead of laying bare a distinct region in the brain, and accurately limiting his destrictive lesion, he merely trephines in the temporal region and destroys the cerebral substance by squirting it out with a strong stream of water. . . . While Goltz's description of the phenomena themselves resulting from this procedure may be accepted without question, his theory that the effects of cortical lesions depend more on their extent than on their position must, I think, be unhesitatingly rejected.

---

[77]Ferrier, a devotee of J. Hughlings Jackson, was perhaps offended that Fritsch and Hitzig had not been impressed in their search by the clinical results of Jackson. Ferrier wrote that the localization of the convolutions was merely "indicated" by the work of Fritsch and Hitzig.

[78]E. Hitzig. Hughlings Jackson and the cortical motor centers in the light of physiological research. *Brain*, 23:545–581 (1900).

[79]D. Ferrier. Experiments on the brain of monkeys. *Proc. R. Soc.* 23:409–430 (1875).

[80]D. Ferrier. *The Localization of Cerebral Disease.* Smith Elder, London, 1878.

Ferrier had come into the field of cerebral localization in an era when the clues came from observation of brain damage seen in the clinic and from ablation experiments performed in the laboratory. Ferrier used both these sources and, as noted, on learning of Fritsch and Hitzig's classic work, he too turned to electrical stimulation. His technique varied from theirs in two ways: he used the monkey instead of the dog and faradic instead of galvanic current.

For the neurophysiologist, it is clear that the observations from clinical neurology were extremely suggestive for the design of future experiments. For Ferrier, this was especially focused on Jackson's observations of focal epilepsy and the leads it could give for the motor system. In other systems there were the important results of Bouillard,[81] Aubertin,[82] and Broca[83,84] for the localization of the speech center.

All the experiments described so far could give information only about motor function; there was no clue to the localization of sensory function in the brain and, of course, no clue to the function of speech, for extirpation experiments on animals could give no clue to the cortical representation of speech. This had to come from clinical observation with studies at autopsy. Gall had placed language in the anterior lobes, and the first clinical reports seemed to confirm this. In fact, the great surge of work aiming to establish localized centers in the human brain began with the speech center. In his studies of encephalitis, Bouillard, a pupil of Magendie and later Professor of Medicine, had reasoned that the anterior lobes of the brain were necessary for speech and went on to observe that other focal lesions of the brain caused localized impairment of muscular movement. The cause of cerebral localization was taken up by his son-in-law, Aubertin, who predicted that a lesion would be found in the anterior lobes of an aphasic patient who was at that time in the hospital of Bicêtre under the surgeon Pierre Paul Broca. Autopsy confirmed Aubertin's prediction, pinpointing the lesion in the left anterior lobe. The next aphasic patient on Broca's service was found at autopsy to have an even more discrete lesion—in what is known to this day as Broca's area. The name of Aubertin has been forgotten, as has Broca's term ''aphemia'' for aphasia.

We have only one illustration from Fritsch and Hitzig's classic paper, and it shows five specific loci in the cortex of the dog for evoking discrete contractions of certain muscle groups (Fig. 80). Ferrier, taking up their lead, defused 15 loci in the monkey. Ferrier had no opportunity to test for localization of function in the human cortex but, perhaps rather surprisingly, took a sketch of the cortex of man and pasted numbers indicating locations of function drawing his guidance from what he had found in the monkey as reported in his first publication from the West Riding Lunatic Asylum. Perhaps the impact of Darwin's *The Descent of Man,* published in 1871, made this transference of data from monkey to man more acceptable. (A similar presumption that it was justifiable to present the human cortex and display on it the findings from the monkey's brain was followed by Sherrington in his work with Gründbaum in the early part of the next century.)[85]

[81]Jean Baptiste Bouillard (1796–1881). *Traité Clinique et Physiologique de l'Encéphalite ou Inflammation du Cerveau.* Ballière, Paris, 1825.

[82]Ernst Aubertin (1825–   ). Consideration sur les localisations cérébrale, et en particular sur le siège de la faculté du language articulé. *Gaz. Hebd. Med. Chir.,* 10:318, 348, 397, 455 (1863).

[83]Pierre Paul Broca (1824–1880). Perte de parole, ramollissement chronique et destruction du lobe antérieur gauche du cerveau. *Bull. Soc. Anthropol. Paris,* 2:235–238, 301–321 (1861).

[84]P. Broca. Rémarques sur le siège de la faculté de language articulé, suivies d'une observation d'aphémie. *Bull. Soc. Anat. Paris,* 6:330–357 (1801).

[85]C. S. Sherrington and A. S. P. Grundbaum. Observations on the physiology of the cerebral cortex of the anthropoid apes. *Proc. R. Soc.,* 72:152–155 (1903).

 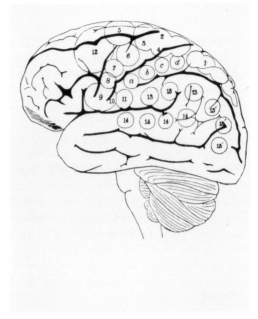

FIG. 80. Left: David Ferrier (1843–1928). **Right:** Ferrier's map of the human brain with numbered labels attached according to the results he obtained in the monkey. (1) Center for movements of the opposite leg and foot as are concerned in locomotion; (2,3,4) Centers for various complex movements of arms and legs, as in climbing and swimming; (5) Center for extension forward of arm and hand, as in putting forth hand to touch something in front; (6) Center for movements of hand and forearm, in which biceps are particularly engaged, supination of hand and flexion of forearm; (7,8) Centers for the elevators and depressors of the angles of the mouth, respectively; (9,10) Center for movement of lips and tongue, as in articulation; disease causes aphasia—generally known as "Broca's convolution"; (11) Center of the platysma, retraction of angle of mouth; (12) Center for lateral movement of the head and eyes, with elevation of eyelids and dilation of the pupil; (a,b,c,d) Centers for movement of the head and wrist; (13,13^1) Center of vision, supramarginal lobule and angular gyrus; (14) Center of hearing, superior tempro-sphenoidal convolution. The center of smell is situated in the subiculum cornu ammonius. The center of taste is near the center of smell but not defined. The center of touch is in the hippocampal region.

Modern students of cerebral localization note errors introduced by his method of presuming the cortex of man to resemble that of the monkey. For example, Ferrier's numbers 13 and 13′ purport (erroneously) to be the centers for vision. (Numbers 1–12 represent motor function.)

In electrical stimulation of the gyri around the Rolandic fissure, Ferrier correctly identified the order of responses from superior to inferior foot, leg, arm, angle of mouth, lips, tongue, and anterior to the precentral gyrus the center for lateral movements of the head and eyes. He recognized the greater complexity of the convolutions of the human brain, caused by the development of numerous secondary and tertiary gyri (the great expansion of the human neocortex). In his own words:

> An exact correspondence can scarcely be supposed to exist, inasmuch as the movements of
> the arm and hands are more complex and differentiated than those of the monkey; while,
> on the other hand, there is nothing in man to correspond with the prehensile movements of
> the lower limbs and tail in the monkey.

Nevertheless, he pasted the map of the monkey's responses on the outline of the human brain. It is well into the 20th century before a map is made directly from the human cortex from which these movements can be obtained.[86]

Ferrier's book *The Functions of the Brain* was denounced for inaccuracies in two reviews published in the authoritative journal *Nature*.[87] The reviews were written by George Henry Lewes. Lewes' name has faded from scientific literature, but at the time of Ferrier's work, Lewes was a prominent writer. He published an authoritative study of Goethe, and his book on the nervous system was in general use.

George Henry Lewes, whose name has almost faded from scientific memory, was then a respected authority, for his book *Physiology of Common Life*[88] was the best account of the nervous system available at the time. Lewes, better remembered for his liaison with George Eliot, wrote a pungent attack, both on Ferrier and on the original Fritsch-Hitzig concepts of functional localization. Lewes did not spare his fire. He deplored "the increasingly popular but thoroughly unphysiological conception of Localisation." He proclaimed that "we should marvel to witness so many eminent investigators cheering each other on the wild-goose chase of a function localised in a cerebral convolution." Just because stimulation of a cortical area evoked a movement, that did not, in Lewes' opinion, prove it to be a motor center. "We do not," he wrote, "consider the centre of laughter to be located in the sole of the foot, because tickling the sole causes laughter."

Ferrier died in 1928 having seen his work have great influence on clinical neurology and especially on neurosurgery.[89] His findings also spurred further research by such workers as Munk, Beevor, and Horsley, but some early attempts of cortical stimulation in a human patient were disastrous. He had many influential friends, especially John Burdon-Sanderson, and he received many honors, including knighthood.

Victor Alexander Haden Horsley (1857–1916)

One result of the explosion of interest in the electrical excitability of the cortex was its clear meaning for neurosurgery. One of the most prominent neurosurgeons of the 19th century to realize the possibilities of this procedure was Victor Horsley. As a neurosurgeon, he is well reported in the history of his field, but he also has some place in neurophysiology.

Born to privileged family in London in 1857, Horsley took his medical degree at University College, London. His interests were weaned away from clinical medicine and problems of the endocrines to the cerebral cortex by the influence of his professor, Schäfer, and others more closely involved with the nervous system. In 1886 he obtained an appointment at the National Hospital, Queen Square; there he came even closer to neurophysiologists and he also became more important as a neurosurgeon. It was at this time that he collaborated with the anatomist R.H. Clarke in the design of the stereotaxic instrument, which has been used widely since then for localizing the exact site for placing electrodes within the brain (Fig. 81).[90]

The development of his scientific life reflected the concern of this time with the problem (and explanation) of epileptic attacks, as seen in his first paper written at the age of 23 in

[86]O. Foerster and W. Penfield. The structural basis of traumatic epilepsy and results of radical operation. *Brain*, 52:99–119 (1930).

[87]G. H. Lewes. Ferrier on the brain. Book review. *Nature*. Nov. 23, and Nov. 30, 1876.

[88]G. H. Lewes. *The Physiology of Common Life*. 2 vols. Blackwood, Edinburgh, 1859–1860.

[89]R. Barthalow. Experimental investigations into the functions of the human brain. *Am. J. Med. Sci.*, 67:305–313 (1874).

[90]V. Horsley and R. H. Clarke. The structure and functions of the cerebellum examined by a new method. *Brain*, 31:1–80 (1908).

FIG. 81.   Left: R. H. Clarke, anatomist, who with Victor Horsley developed a stereotaxic instrument designed for recording from inside the brain. **Right:** One view of the stereotaxic instrument published just after the turn of the century. (V. Horsley and R. H. Clarke. The structure and functions of the cerebellum examined by a new method. *Brain,* 31:1–80, 1908.)

collaboration with Charlton Bastian,[91] physician at the National Hospital and member of the Royal Society known for his work on aphasia. His concern with epilepsy clearly led to his interest in the experiments of Ferrier and his followers. Not himself a neurophysiologist, he relied heavily on colleagues, one of the most prominent was Charles Edward Beevor. Beevor, medically qualified from University College, had some training in Vienna with Obersteiner and later in Leipzig and Berlin. Although appointed physician to the National Hospital, he devoted most of his time to his laboratories at University College. There he was joined by Victor Horsley, and their long series of research reports began (Fig. 82).

Beevor himself made a detailed study of the impairment of muscles in cerebral palsy.[92] He also developed a staining technique for detecting and following the precise morphology of arteries in the brain.[93]

With Horsley a long series of experiments started. These intense studies of the brains of monkeys[93,94] were all directed to the eventual application of these techniques to the human patient suffering from epilepsy. By 1886 Horsley's experience with animal experimentation and his natural dexterity were tested in his first brain surgery in man. This operation was

[91]H. C. Bastian and V. Horsley. Arrest of development in the left upper limb in association with an extremely small right ascending parietal convolution. *Brain,* 3:113–116 (1880–1881).

[92]V. Horsley and C. E. Beevor. A minute analysis (experimental) of the various movements produced by stimulating in the monkey different regions of the cortical center for the upper limb as defined by Professor Ferrier. *Phil. Trans. R. Soc.,* 178:153–167 (1886).

[93]V. Horsley and C. E. Beevor. On the causes of the fibers of the cingulum and the posterior parts of the corpus callosum and fornix in the marmoset monkey. *Phil. Trans. R. Soc.,* 182 (1892).

[94]C. E. Beevor. On Professor Hamilton's theory concerning the corpus callosum. *Brain,* 9:63–73 (1886–1887).

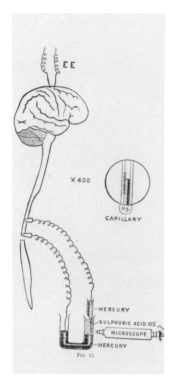 

FIG. 82. Left: Horsley's diagram of his experiments with C. E. Beevor to record cortical responses to peripheral stimulation. (From: *Transaction of the Congress of American Physicians and Surgeons,* 1:340–350, 1888.) **Right:** Victor Alexander Haden Horsley (1857–1916).

also the first cranial surgery done at the National Hospital. It is of interest that both Jackson, the epileptologist who had sparked so much research, and Ferrier, who had laid the background of surgical research in monkeys, were present at this operation.

Horsley had been appointed to the Brown Institution, a research unit of University College (demolished unfortunately by bombs in World War II). Horsley began research affiliations with more physiologists, including Francis Gotch (Fig. 83). Gotch had followed Richard Caton as Professor of Physiology at Liverpool but had never acknowledged his work. When the controversy broke out in the *Centralblatt*[95] in 1891 over the claims to discovery of the brain's own electricity, Gotch and Horsley made a claim for priority of the findings reported by Beck and Fleischl von Marxow. They did not mention Caton and, in fact, were averse to the concept that cortical cells could themselves give electrical activity, for they related all such phenomena to the underlying white matter.

Horsley's researches were not only on the cerebrum but also on the cerebellum. With Lowenthal he published the results of electrical stimulation of the cerebellum in animals and its effect on decerebrate rigidity.

Horsley's brilliance was as a surgeon, and he received great recognition from his field (there are several fine biographies describing his surgical expertise). His work in the laboratory was clearly directed toward its use in the treatment of the human patient. Many honors came his way, including knighthood.

[95]F. Gotch and V. Horsley. Ueber den Gebrauch der Elektricität für die Localisirung der Erregungserscheinungen im Centralnervensystem. *Cbl. Physiol.,* 4:22:649–651 (1890). (On the use of electricity in localizing arousal phenomena in the central nervous system.)

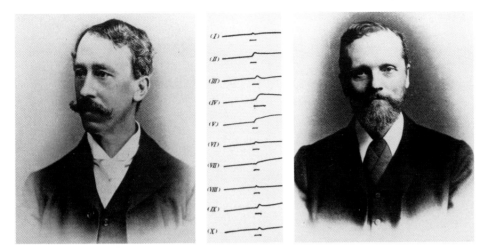

FIG. 83. Two physiologists who used the capillary electrometer to record nerve potentials. **Left:** Francis Gotch (1853–1913). **Right:** G. T. Burch. They were among the first to publish photographs of nerve potentials in the spinal cord evoked by stimulation. (From: G. T. Burch and F. Gotch. *J. Physiol.*, 24 410–426, 1899, and *Proc. R. Soc.* 63:300–303, 1899.)

Horsley's brilliant career was in surgery, not only of the brain but also of the spinal cord, but he ceased to publish in this field after 1890. He threw his energies into community projects, especially the temperance movement against alcoholic drinks. When World War I broke out, although already 59, he joined the army. On the ship on its way to the campaign in the Dardanelles, he developed a fever and died. He is buried in Mesopotamia.

Horsley, the great innovative surgeon, was covered with honors including knighthood. His success in neurosurgery was due to his seven years of laboratory research in collaboration with neurophysiologists, not only Beevor and Gotch but also with his professor at University College, Schäfer (Fig. 84), and with Semon and Spencer. His work is a brilliant

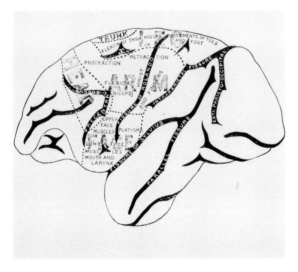

FIG. 84. **Left:** Schäfer and Horsley's diagram of the cortex of the monkey over which they have written the localizations that they found (From: *Phil. Trans. R. Soc.*, 179B:1–45, 1888). **Right:** Charles Edward Beevor (1854–1908). One of the several experimentalists with whom Horsley worked.

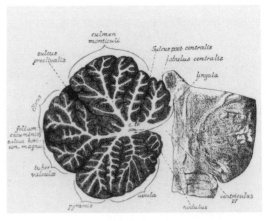

FIG. 85. **Left:** Luigi Luciani (1840–1921). The great Italian neurophysiologist. **Right:** Luciani's special interest was in the cerebellum and this is one of his diagrams of its connections. (From: L. Luciani. *Fisiologia del'Homo*. 5 vols. Le Monnier, Societa Edit Libraria, Milan, 1912–1913.)

example of the application of fundamental research to the clinical field. His research work falls in the century under review; his surgical triumphs came in the 20th century. Horsley never accepted the concept of the electrical activity of the cortical cells themselves, thus following Burdon-Sanderson in his disbelief.

The field of cortical localization in the brain was dominated by the English school, three of whom were knighted, all of them with access to the major publication outlets by virtue of their membership in the Royal Society and their seats on the editorial boards of *Philosophical Transactions* and *Brain*. But they were not alone in the field. In Rumania this field interested the distinguished physiologist Marinesco[96,97] and the prominent Italian physiologist Luigi Luciani.[98] Luciani, born in Ascola Picello in 1840, was a product of the universities of Bologna and Naples, and he also had some training with Ludwig in Leipzig. After posts in Siena and Florence, he was in 1892 appointed to the post in Rome. He had a deep interest in cerebral localization but concentrated particularly on the cerebellum[99] (Fig. 85). His works on this subject are famous.

Hermann Munk (1839–1912)

Since the first spark ignited by Fritsch and Hitzig, the growth of interest in cortical localization came early from England. Greatly encouraged by the clinical neurologists and especially by Jackson in his research for the mechanisms of epileptic convulsions, the efforts of the neurophysiologists focused particularly on the motor systems. It was among the more psychologically minded that the localization of sensory systems was of interest. One of

[96]G. Marinesco. *Oeuvres Chosies*, pp. 376–385. [*Acad. Rep. Pol. Rumania* (1963).]

[97]G. Marinesco and J. Nicolesco. Contribution à l'étude de la pathologie diencéphalo-mésencéphalopontine. A propos des tuvercules intéressand les formations de ces régions et plus spécialement le noyau rouge. *Z. Neurol. Psychiatr.*, 157:81–91 (1897).

[98]Luigi Luciani (1840–1921). *Il Cervelleto. Novistudi di fisiologie normale e pathelogia*. Monnier, Florence, 1891.

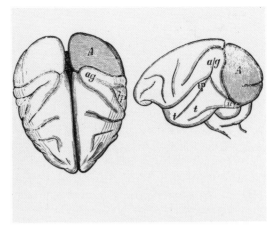

FIG. 86. **Left:** Hermann Munk (1839–1912). Critic of Ferrier, especially for the latter's localizing of the visual centers in the angular gyrus. **Right:** Munk's diagrams of the monkey's brain with the letter *A* placed to indicate the visual center. (From: H. Munk *Über die Functionen der Grosshirnrinde.* Hirschwald, Berlin, 1881.)

these was Hermann Munk. Munk was born in Posen in 1839, a city then in East Prussia. He was educated in Berlin as a student of Johannes Müller and in Göttingen, graduating just before the outbreak in 1870 of the Franco-Prussian War. Like his friend Gustav Fritsch, he went off to serve in the army but, unlike Fritsch, he returned after the war to the laboratory to the work that led to his intense interest in cortical localization. He succeeded in gaining a position at the Physiological Institute of the Veterinary School in Berlin, and there his work began.

Munk's researches essentially concentrated on localization of cortical areas that had received little attention from the English school. He gave especial attention to the visual system, though other sensory systems (acoustic, somatosensory) were also studied. His work on the visual centers threw much doubt on Ferrier's claims, for Ferrier had placed vision in the angular gyrus of man. Munk introduced a distinction between peripheral blindness and mental blindness. His work was mostly on dogs, though some was on monkeys (Fig. 86), and he reported his results[100] as follows (the references to figures are to his 1881 publication):

> To examine the brain of the monkey was not only prompted by the question how far the insight gained was valid for the human brain, I had a special occasion. In my first communication on the physiology of the cortex which I made in March of last year I did not say anything about Ferrier's work on the monkey because there was nothing good to be said about it. But asked in the subsequent session I had to say that Mr. Ferrier's declarations that the visual center in the monkey was situated in the angular gyrus . . . right underneath in the gyrus temporo-sphenoidalis superior (tp) the acoustic center; in the lower part of the temporal lobe (tt) the center for smell and taste; in the gyrus hippocampi and hippocampus major (inside of u) the center for touch; finally in the occipital lobe (A) the center for hunger, that all these statements and what followed from them as far as the character and restitution of disturbances set by the operation are concerned, are worthless and gratuitous constructions since the operated animals were examined by Mr. Ferrier in quite an insufficient

[99]L. Luciani. *Das Kleinhirn.* Leipzig, 1893. (The Cerebellum.)

[100]H. Munk. *Über die Funktionen der Grosshirnrinde.* Hirschwald, Berlin, 1881. (On the Functions of the Cortex.)

manner and only at the time of general depression of brain function. If I have gone too far in this statement which is based on a general survey of Mr. Ferrier's experiments it was up to me to restore the injury, the sooner the better. However, as the experiments show now I have said at that time rather too little than too much, Mr. Ferrier had not made one correct guess, all his statements have turned out to be wrong.

 The visual center of the monkey is the cortex of his occipital lobe A. Only injuries of this cortex lead to disturbances of the visual sense and only such disturbances follow lesions of this cortex.

The result was that the center for vision in the monkey was now firmly recognized as being in the occipital cortex. Munk's major publication that pulled together some of his earlier reports was his book *Über die Functionen der Grosshirnrinde*. In this book he reported the findings that led to his identification of the projection areas not only for sight but also for hearing and feeling in dogs and monkeys. He relied principally on ablation techniques rather than electrical stimulation.

Munk was, in fact, not by any chance the first to look for representation of the visual system in the brain. One of the little known was Panizza[101] in Pavia who had placed it in the posterior lobe; another was Salomen Eberhard Henschen, the Swede who pinpointed vision in the calcarine fissure. Henschen had been working on the connections from the eye to the brain,[102] a subject also at that time being pursued by V. M. Bechterev.[103]

Munk did not restrict himself to the visual system but claimed to find touch represented in the hippocampus and intelligence in the frontal lobe. He is perhaps remembered best for his differentiation between "mental" blindness and cortical blindness. He also made this distinction for hearing. He identified an area in the cortex of the dog of which in translation he says, "Removal . . . produces mental deafness after extirpation of both sides. The dog still hears—his ears pick up at a sound—but he no longer understands commands which he formerly comprehended."

By a great number of skilled, highly localized excisions, Munk was able to give very detailed analyses of the cortical representation for the sensations derived from various loci on the dog's body. It was from these experiments that he reached his views on the location of intelligence in the frontal lobe. He said (in translation):

> I assume that intelligence is the sum result of all sensation concepts, and therefore any injury to the cerebral cortex damages Intelligence. In describing mental blindness, deafness and paralysis to you, I have therefore been describing a partial loss of intelligence.

Munk died in 1912 having received little recognition from the other workers on cortical localization.

BIBLIOGRAPHY

Gustav Theodor Fritsch (1838–1927)

Selected Writings

Zur vergleichenden Anatomie der Amphibienherzen. *Arch. Anat. Physiol.* (1869). (On the comparative anatomy of the hearts of amphibians.)

[101]B. Panizza. *Ricerchi Sperimentali supra i nervieco*. Pavia, 1834.

[102]Salomen Eberhard Henschen (1847–1930). On the visual path and center. *Brain*, 16:170–180 (1893).

[103]V. M. Bechterev. Ueber den Verlauf der die Pupille verengenden Nervenfasern im Gehirn und über die Localisation eines Centrums für die Iris und Contraction der Augenmuskeln. *Pflüger's Arch. f. d. ges. Physiol.*, 31:60–87 (1883). (On the path in the brain of those nerve fibers which cause the pupil to narrow, and on the localization of a center for the iris and for contraction of the eye muscles.)

Abnorme Muskelbündel der Achselhöhle. *Arch. Anat. Physiol. Wiss. Med.*, 361–371 (1869). (Abnormal muscle bundles of the arm pit.)

Über die elektrische Erregbarkeit des Grosshirns. *Arch. Anat. Physiol. Med. Wiss.*, 300–322 (1870) (with E. Hitzig). (On the electrical sensitivity of the brain.)

Untersuchungen des Zitteraals. *Arch. Physiol.*, 61–75 (1882). (Investigations on the electrical eel, electrophorus electricus.)

Berichte über die Fortsetzung der Untersuchungen an elektrischen Fischen. Beiträge zur Embryologie von Torpedo. *Sitzungsberichte der Königlich Prussischen Akademie der Wissenschaften zu Berlin*, 8:205–209 (1883). (Report on the contributions on investigations on electrical fish. Contributions to the embryology of the torpedo.)

Berichte über eine Reise zur Untersuchung der in den Museen Englands und Hollands vorhandenen Torpedineen. *Arch. Anat. Physiol.* 70–73 (1884). (Report on a journey to study the electrical rays to be found in the museums of England and Holland.)

Berichte über die Fortsetzung der Untersuchungen von elektrischen Fischen. *Arch. Anat. Physiol.*, 73–78 (1884). (Report on the continuation of the investigations of electric fish.)

Das menschliche Haar als Rassenmerkmal. *Verh. Berliner Gellsch. Anthrop.* In *Zeit. Enthnol. Anthrop. Urgeschite*, 1885. (Human hair as evidence of race.)

Die Parasiten des Zitterwelses. *Sitz. Akad. Wiss.* Berlin, 1886. (The parasites of the electric catfish.)

Zur Anatomie der Bilharzia haematobia. *Arch. Mikro. Anat.*, 1888. (On the Anatomy of Bilharzia Haematobia.)

Ueber einige bemerkenswerthe Elemente des Centralnervensystems von Lophius piscatorius L. *Arch. Mikro. Anat.*, 27:18–31 (1886). (On Several Notable Elements of the Central Nervous System of Lophins Piscatorius.)

Die elektrischen Fische. Nach neuen Untersuchungen anatomisch-zoologisch dargestellt. 2 vols. Illustrated with 32 large litographed plates and 33 woodcuts in the text. Veit, Leipzig, 1898–1890. (The electric fish. An anatomical-zoological presentation based on new investigations.)

Die Lehre von der Einheit der amerikanischen Eingeborenenrassen untersucht an der Haarbeschaffenheit. *Verh. Amerik. Cong.*, Berlin, 1888. (The theory of the unity of native Americans investigated according to hair type.)

Die Gestalt des Menschen. Neff, Stuttgart, 1899. (The Human Form.)

Aegyptische Volkstypen der Jetztzeit. C. Kreidel, Wiesbaden, 1904. (Egyptian peoples of the present day.)

Über Bau und Bedeutung der Area centralis des Menschen. Reimer, Berlin, 1908. (On the structure and significance of the area centralis in man.)

Edouard Hitzig (1838–1907)

Selected Writings

Beiträge zur Kenntnis der Peripheren Lähmung des Facilis. *Klin. Woch.*, 6:18–24 (1869). (Contributions to Knowledge of Peripheral Paralysis of the Face.) (Speech held at the sitting of the Berlin Medical Society, Nov. 18, 1868.)

Über die galvanischen Schwindelempfindungen und eine neue Methode galvanischer Reizung der Augenmuskeln. *Verh. Berl. Ges.* vom. 19 Jan. 1870, in *Klin. Woch.* No. 11, 1870, *Eine ausführliche Bearbeitung wird demnächst erfolgen.* (On Galvanic Sensations of Faintness, and a New Method of Galvanic Stimulation of the Eye Muscles. Meeting of the Berlin Medical Society, Jan. 19, 1870. In the Berlin *Clinical Weekly* No. 11, 1870. *An extensive treatment will follow shortly.*)

Ueber die beim Galvanisiren des Kopfes entstehenden Störungen der Muskelinnervation und der Vorstellungen vom Verhalten im Raume. *Arch. Anat. Physiol. Wiss. Med.*, 716–770 (1871). (On disturbances of muscle innervation and conceptions of spatial behavior which occur during galvanization of the head.)

Weitere Untersuchungen zur Physiologie des Gehirns. *Arch. Anat. Wiss. Med.*, 771–772 (1871). (Futher investigations on the physiology of the brain.)

Über quere Durchströmung der Froschnerven. *Pflügers Arch. f. d. ges Physiol.*, 7:263–273 (1873) (*Cbl. Med. Wiss.*, 12:263–273, 1873). (On passing transverse current through frog nerves.)

Untersuchungen über das Gehirn. Abhandlungen physiologischen und pathologischen Inhalts. Hirschwald, Berlin, 1874. (*Cbl. Med. Wiss.*, 12:821–824, 1874.) (Investigations of the Brain. Report on Physiological and Pathological Content.)

Untersuchungen über das Gehirn. *Cbl. Med. Wiss.*, 12:548–549 (1874). (Investigations of the brain.)

Ueber die Resultate der electrischen Untersuchung der Hirnrinde eines Affen. *Med. Trans.* 2L:375–376 (1874). (On the Results of the Electrical Investigation of the Cerebral Cortex of an Ape (translation). *Klin. Woch.* No. 6, 1874.)

Untersuchungen über das Gehirn. *Arch. Anat. Physiol. Med. Wiss.* 343–381 (1876). (Investigations of the Brain.)

Untersuchungen über das Gehirn. *Arch Anat. Physiol. Med.*, 692–711 (1876). (Investigations of the Brain.)

Hughlings Jackson and the cortical motor centers in the light of physiological research. *Brain*, 23: 545–581 (1900).

Suggested Readings

Bailey, P., and von Bonin, G. *The Isocortex of Man.* University of Ill. Press, Urbana, 1951.

Grundfest, H. *The Different Careers of Gustav Fritsch (1838–1927).*

Herrick, C. I. Neurologists and Neurological Laboratories, No. F. Professor Gustav Fritsch. *J. Comp. Neurol.*, 2:84–88 (1892).

Walker, A. E. Stimulation and ablation, their role in the history of cerebral physiology. *J. Neurophysiol.*, 20:435–448 (1957).

Young, R. M. *Mind, Brain and Adaptation in the Nineteenth Century.* Clarendon Press, Oxford, 1970.

John Hughlings Jackson (1835–1911)

Selected Writings

Unilateral epileptiform seizures, attended by temporary defect of sight. *Med. Times Gaz.*, 1:588–589 (1863).

Loss of speech. *Clinical Lecture Reports (London Hospital)* 1:388–471 (1864).

Observations on defects of sight in brain disease. *Ophthal. Hosp. Rep.*, 4:1–9; 185–186; 389–446; 5:51–78, 251–306 (1863–1865).

Notes on the physiology and pathology of language. *Med. Times Gaz.*, 1:659–662 (1866).

On a case of muscular atrophy, with disease of the spinal and medulla oblongata. *Med.-chir. Trans.*, 50:489–496 (1867) (with J. A. L. Clarke.)

A study of convulsions. *Trans. St. Andrews Med. Grad. Assoc.*, 3:162–204 (1870).

On a case of paralysis of the tongue from haemorrhage in the medulla oblongata. *Lancet*, 2:271–273 (1872).

On the anatomical and physiological localisation in the brain. *Lancet*, 1:84–85 (1873).

Observations on the localisation of movements in the cerebral hemispheres, as revealed by cases of convulsion, chorea and 'aphasia.' *West Riding Lunatic Asylum Medical Reports*, 3:175–195 (1873).

On the anatomical, physiological, and pathological investigation of epilepsies. *West Riding Lunatic Asylum Medical Reports,* 3:315–349 (1873).

On a case of recovery from double optic neuritis. *West Riding Lunatic Asylum Medical Reports,* 4: 24–29 (1874).

On temporary mental disorders after epileptic paroxysms. *West Riding Lunatic Asylum Medical Reports,* 5:188–197 (1875).

Case of hemikinesis. *Br. Med. J.,* 1:773–774 (1875).

On epilepsies and on the after effects of epileptic discharges (Todd and Robertson's hypothesis). *West Riding Lunatic Asylum Medical Reports,* 6:266–309 (1876).

Auditory vertigo. *Brain,* 2:29–38 (1879–1880).

On affections of speech from disease of the brain. *Brain,* 2:203–222 (1879–1880).

On affections of speech from disease of the brain. II. *Brain,* 2:323–356 (1879–1880).

On right or left-sided spasm at the onset of epileptic paroxysms, and on crude sensation warnings, and elaborate mental states. *Brain,* 3:192–206 (1880–1881).

The Croonian lecture. On the evolution and dissolution of the nervous system. *Proc. R. Soc.,* 3–5 (1884).

On temporary paralysis after epileptiform and epileptic seizures: a contribution to the study of dissolution of the nervous system. *Brain,* 3:433–451 (1880–1881).

A contribution to the comparative study of convulsions. *Brain,* 9:1 (1886).

Case of tumour of the right temporo-sphenoral lobe bearing on the localisation of the sense of smell and on the interpretation of a particular variety of epilepsy. *Brain,* 12:346–357 (1889–1890) (with C. Beevor.)

On convulsive seizures. *Br. Med. J.,* 2:703–707 (1890).

Remarks on the relations of different divisions of the central nervous system to one another and to parts of the body. *Br. Med. J.,* 65–69 (1898).

Suggested Readings

Brain, W. R. Hughlings Jackson's ideas of consciousness in the light of today. In: *The Brain and its Functions,* edited by F. L. N. Poynter. Thomas, Springfield, 1958.

Haymaker, W. and Schiller, F. (Eds.). *The Founders of Neurology.* Thomas, Springfield, 1963.

Hernstein, R. and Borny, E. (Eds.). *A Source Book in the History of Psychology.* Harvard University Press, Cambridge, 1965.

Joynt, R. J. The Great Confrontation: The meeting between Broca and Jackson. In: *Historical Aspects of the Neurosciences,* edited by F. C. Rose and W. F. Bynum. Raven Press, New York, 1982.

Melville, I. D. The medical treatment of epilepsy: a historical review. In: *Historical Aspects of the Neurosciences,* edited by F. C. Rose and W. F. Bynum. Raven Press, New York, 1982.

Riese, Walter. *A History of Neurology.* MD Publications, New York, 1959.

Selected Writings of John Hughlings Jackson, edited by J. Taylor, 2 vols. Hodder & Stoughton, London, 1931.

Pierre Paul Broca (1824–1880)

Selected Writings

Remarques sur le siège de la faculté de language articulé, suives d'une observation d'amphémie (perte de la parole). *Bull. Soc. Anat. Paris,* 36:330–337 (1861).

Nouvelle observation d'amphémie produite par une lesion de la moitié posterieure des deuxième et troisième circonvolutions frontales gauche. *Bull. Soc. Anat. Paris*, 36:398 (1861).

Perte de la parole; ramollissement chronique et destruction partielle du lobe antérieur gauche du cerveau. *Bull. Soc. Anthrop. Paris*, 2:235–238 (1861).

Localisation des fonctions cérébrales. Siège du language articulé. *Bull. Soc. Anthrop. Paris*, 4:200 (1863).

Sur le siège de la faculté du language articulé. *Bull. Soc. Anthrop. Paris*, 6:377 (1865).

Nouveau cas d'amphémie traumatique. *Bull. Soc. Anthrop. Paris*, 1:396 (1866).

Sur le siège de la faculté du language articulé. *Trib. Méd.*, 74:265 (1869).

Mémoires d'Anthropologie. 3 vols. Reinwald, Paris, 1871–1877.

Mémoires sur le Cerveau de l'Homme et des Primates. Reinwald, Paris, 1888.

Luigi Luciani (1840–1921)

Selected Writings

Sulla patogenesi dell'epilepsia. *Riv. Sperim. Freniatria.* 1878, 1881. (On the Pathogenesis of Epilepsy.)

Benham, F. L. Sulle Funzioni del Cervello. Richerche Sperimentali dei Professor Luigi Luciani e Augusto Tamburini. Seconda Comunicazione: Centri Posco-Sensori Corticali. Reggio—Emilia. 1879. (Critical Digests and Notices of Books.) *Brain*, 2:234–256 (1879–1880). (On the functions of the brain. The psycho-sensory cortical centers. Research experiments of Professors Luigi Luciani and Augusto Tamburini.)

Localizone funzionali del cervello. Vallar du Naples, 1885. (with Seppilli.) (Localization of Function in the Cerebellum.)

Die Functions—Localisation auf der Grosshirnrinde, 1886. (with Seppilli.) (On the Functional Localization of the Cerebrum.)

Cervelletto—Novistudi di fisiologia normale e patelogica. Florence, 1891. (The Cerebellum: Physiology and Pathology.)

Fisiologia del'Homo. Milan, 1912–1913. (The Physiology of Man.)

Human Physiology. (English translation by F. A. Welby.) Macmillan, London, 1915.

John Burdon-Sanderson (1828–1905)

Selected Writings

Notes on the excitation of the surface of the cerebral hemispheres by induced currents. *Proc. R. Soc.*, 22P:368–370 (1873).

On the mechanical effects and on the electrical distribution consequent in the excitation of the leaf Dionaea Muscipula. *Proc. R. Soc.*, 25:411–434 (1875).

A report of Prof. L. Hermann's recent researches on the electromotive properties of muscle. *J. Physiol.*, 1:196–212 (1876).

Note on the electromotive properties in muscle. *Proc. R. Soc.*, 25:435–439 (1876).

Correction to note of Dec. 6. *Proc. R. Soc.*, 26:332–333 (1877).

Experimental results relating to the rhythmical and excitatory motions of the ventricle of the heart of the frog, and the electrical phenomena which accompany them. *Proc. R. Soc.*, 27:410 (1878). (with F. J. M. Page.)

On the time relations of the excitatory process in the ventricle of the heart of the frog. *J. Physiol.* 2:384–435 (1879–1880). (with F. J. M. Page.)

On the electrical phenomena of the excitatory process in the heart of the frog and of the tortoise as investigated photographically. *J. Physiol.*, 4:327–338 (1883–1884) (with F. J. M. Page.)

Photographic determination of the time-relations of the changes which take place in muscle during the period of so-called 'latent stimulation.' *Proc. R. Soc.*, 48:14–19 (1890). (Measurement, by means of photography of the speed of the nervous impulse.)

The electrical response to stimulation of muscle and its relation to the mechanical response. *J. Physiol.*, 18:117–159 (1895).

Etienne Jules Marey (1830–1904)

Selected Writings

Physiologie Expérimentale. Trauveau du Laboratoire de M. Marey. Masson, Paris, 1866.

Animal Mechanism, a Treatise on Terrestrial and Aerial Locomotion. Appleton, New York, 1874.

Des variations électriques des muscles et du coeur en particulier étudiés au moyen de l'électromètre de M. Lippmann. *C. R. Acad. Sci.*, 82:957–977 (1876).

Inscription photographique des indications d'électrometre de Lippmann. *C. R. Acad. Sci.*, 83:278–280 (1876).

Méthode Graphique. Masson, Paris, 1878.

La Méthode graphique dans les sciences expérimentales et principalement en physiologie et en médecine. Masson, Paris, 1885.

International Physiological Congress (Minutes). *J. Physiol.*, 23 (Suppl.): 6–7 (1898).

Obituary. *C. R. Acad. Sci.*, 138:1185–1187 (1904).

Wilhelm Einthoven (1860–1927)

Selected Writings

Ein neues Galvanometer. *Ann. Physik.*, 4:1069–1071 (1903). (An new galvanometer.)

Hete meter von snel Wisselende Potential verschillen mit Behulp von Lippmann Capillar Electrometer. *Pflüger's Arch. f. d. ges Physiol.*, 56:528–541 (1894). (The measuring of rapidly fluctuating potential with the help of Lippmann's capillary electrometer.)

Die Registrirung des Herztons. *Pflüger's Arch. f. d. ges. Physiol.*, 57:617–639 (1894). (with M. A. B. Geluk.) (The registering of the heart tone.)

David Ferrier (1843–1928)

Selected Writings

Experimental research in cerebral physiology and pathology. *West Riding Lunatic Asylum Medical Reports*, 3:30–96 (1873).

The localization of function in the brain (MS) communicated by J. B. Sanderson, March 5, 1874; Archives of the Royal Society, A.P. 56.2; abstract. *Proc. R. Soc.*, 22:229–232 (1874).

Pathological illustrations of brain function. *West Riding Lunatic Asylum Medical Reports*, 4:30–62 (1874).

Experiments on the brain of monkeys, No. 1. *Proc. R. Soc.*, 23:409–430 (1875).

The Croonian lecture. Experiments on the brain of monkeys, No. 2. *Phil. Trans. roy. Soc.*, 165: 433–488 (1875); abstract. *Proc. R. Soc.*, 23:431–432 (1875).

The Function of the Brain. Smith Elder, London, 1876.

The Localization of Cerebral Disease. Smith Elder, London, 1878.

Vomiting in connection with cerebral disease. *Brain,* 2:223–233 (1879–1880).

Crural monoplegia—limited cortical lesions of opposite hemisphere. *Brain,* 3:128–131 (1880–1881).

An experimental research upon cerebro-cortical afferents and efferents. *Phil. Trans. roy. Soc.,* 190: 1–44 (1898). (with W. A. Turner.)

The Regional Diagnosis of Cerebral Disease. In: *A System of Medicine,* edited by C. Abbott and H. D. Robertson. Smith Elder, London, 1911.

Suggested Readings

Jefferson, G. The prodromes to cortical localization. In: Sir Geoffrey Jefferson. *Selected Papers.* Thomas, Springfield, 1960.

Koshtouants, K. S. The history of the problem of brain cortex excitability, Actes du 8 ème Congres Internat. d'Hist. des Sciences, 1956.

Sherrington, C. S. Sir David Ferrier. In: *Dictionary of National Biography.* Scribner, London, 1930.

Spillane, J. D. A memorable decade in the history of neurology, 1874–1884. *Br. Med. J.,* 2:701– 706 (1974).

Viets, H. R. West Riding, 1871–1876. *Bull. Hist. Med.,* 6:427–487 (1938).

Walker, A. E. *A History of Neurological Surgery.* Williams & Wilkins, 1951.

Walker, A. E. Stimulation and ablation. Their role in the history of cerebral physiology. *J. Neurophysiol.,* 20:435–448 (1957).

Young, R. M. *Mind, Brain and Adaptation,* Clarendon Press, Oxford, 1970.

Victor Horsley (1857–1916)

Selected Writings

The motor centers of the brain, and the mechanism of the will. Royal Institution Lecture. *Br. Med. J.,* 1:111–115 (1885).

Recherches Expérimentales sur l'Ecorce Cérébrale des Singes, démontrées par une expérience actuelle devant la société de Biologie de Paris. London, H. K. Lewis (undated). (with Charles E. Beevor.)

A minute analysis of the various movements produced by stimulating in the monkey different regions of the cortical center for the upper limb as defined by Professor Ferrier. *Phil. Trans. roy. Soc.* 178:153–167 (1886–1887). (with Charles E. Beevor.)

On the relation between the posterior columns of the spinal cord and the excito-motor area of the cortex, with especial reference to Prof. Schiff's views on the subject. *Brain,* 9:42–62 (1886–1887).

A further and final criticism of Prof. Schiff's experimental demonstration of the relation which he believes to exist between the posterior columns of the area of the spinal cord and the excitable area of the cortex. *Brain,* 9:311–328 (1886–1887).

A record of experiments upon the functions of the cerebral cortex. *Phil. Trans. roy. Soc.,* 179B: 1–45 (1888). (with E. A. Schäfer.)

A further minute analysis by electrical stimulation of the so-called motor region of the cortex cerebri in the monkey (Macacus sinicus). *Phil. Trans. roy. Soc.,* 179B:205–256 (1888). (with Charles E. Beevor.)

An experimental investigation into the arrangement of the exitable fibers of the internal capsule of the bonnet monkey. *Phil. Trans. roy. Soc.,* 181:49–88 (1890) (with Charles E. Beevor.)

A record of the results obtained by electrical excitation of the so-called motor cortex and internal capsule in an orang-outang (Simia satyrus). *Phil. Trans. roy. Soc.,* 181B:129–158 (1890). (with C. E. Beevor.)

An experimental investigation into the arrangement of the excitable fibers of the internal capsule of the bonnet monkey. *Phil. Trans. roy. Soc.*, 181:187–211 (1890). (with C. E. Beevor.)

Ueber den Gebrauch der Elektricität für die Localisirung der Erregungserscheinungen im Centralnervensystem. *Cbl. Physiol.*, 4:649–651 (1891). (with F. Gotch.) (On the Use of Electricity for Localizing Arousal Phenomena in the Central Nervous System.)

On the changes produced in the circulation and respiration by increase of the intra-cranial pressure or tension. *Phil. Trans. roy. Soc.*, 182B:201–254 (1891). (with H. Spencer.)

On the mammalian nervous system, its functions, and their localisation determined by an electrical method. Croonian lecture for 1891. *Phil. Trans. roy. Soc.*, 182B:267–526 (1891). (with F. Gotch.)

The Structure and Functions of the Brain and Spinal Cord. Fullerian lectures for 1891. Griffin, London, 1892.

A further minute analysis by electrical stimulation of the so-called motor region (facial area) of the cortex cerebri in the monkey. *Phil. Trans. roy. Soc.*, 185:39–81 (1894). (with C. E. Beevor.)

An experimental investigation of the central motor innervation of the larynx. *Phil. Trans. roy. Soc.*, Series B:187–211 (1896). (with F. Semon.)

The structure and functions of the cerebellum examined by a new method. *Brain*, 31:1–80 (1908). (with R. H. Clarke.)

Suggested Readings

Lyons, J. B. *The Citizen Surgeon: A Biography of Sir Victor Horsley*. Dawnay, London, 1966.

Miles, C. P. Cerebral localization. *Trans. Congres. Amer. Phys. Surg.*, 1:184–284 (1888).

Paget, S. *Sir Victor Horsley. A Study of His Life and His Work*. Constable, London, 1919.

Walker, A. E. The development of the concept of cerebral localization in the nineteenth century. *Bull. Hist. Med.*, 31:99–121 (1957).

Hermann Munk (1839–1912)

Selected Writings

Nachweiss des Muskelstromes am enthäuteten Frosche ohne Aetzung der Haut. *Arch. Anat. Physiol. Wiss. Med.*, 649–653 (1869). (Proof of muscle current in an unskinned frog without cauterization of the skin.)

Ueber die Harnstoffbildung in der Leber, ein experimentaller Beitrag zur Frage der Harnstoffuntersuchung in Blut und Parenchymen. *Arch. Anat. Physiol. Wiss. Med.*, 100–112 (1875). (On the formation of urea in the liver, an experimental contribution to the question of the investigation of urea in blood and parenchyma.)

Die elektrischen und Bewegungs-erscheinungen am Blatte der dionaea muscipula. *Arch. Anat. Physiol. Wiss. Med.*, 167–203 (1876). (On the electrical phenomena during stimulus.)

Ueber cerebrale Epilepsie. *Arch. Anat. Physiol. Wiss. Med.*, 169–170 (1876). (On cerebral epilepsy.)

Zur Physiologie des Grosshirnrinde. *Klin. Wiss.* der 14:505–506 (1877). (On the physiology of the cerebral cortex.)

Über die Funktionen der Grosshirnrinde. Hirschwald, Berlin, 1881. (On the Functions of the Cerebral Cortex.)

Ueber Grosshirn-Exstirpation beim Kaninchen. *Arch. Anat. Physiol. Wiss. Med.*, 470–566 (1884). (On brain extirpation in the rabbit.)

Of the visual area of the cerebral cortex, and its relation to eye movements (trans. F. W. Mott). *Brain*, 13:45–67 (1890).

Suggested Readings

Schiller, F. Hermann Munk. In: *Founders of Neurology,* edited by Webb Haymaker. Thomas, Springfield, Ill., 1970.

von Bonin, G. *The Cerebral Cortex.* Thomas, Springfield, Ill., 1960.

Clarke, E., and O'Malley, C. D. *The Human Brain and Spinal Cord.* University of California Press, Berkeley, 1968.

CHAPTER XI

The Brain Yields Its Electricity

Richard Caton (1842–1926)

Among those who studied Du Bois-Reymond's experiments and repeated them in his own laboratory was a young lecturer at the Royal Infirmary School of Medicine in Liverpool—Richard Caton. At this time there was no university as such at Liverpool, although a medical school had been founded there in 1834. It was on the staff of this institution that Caton held an appointment as Lecturer in Physiology at the period of his work in electrophysiology.

He was born in 1842, the son of a physician practicing in the town of Scarborough in Yorkshire. In 1863 a small legacy enabled him to attend medical school at the University of Edinburgh. In 1867 he gained his M.B., Ch.M., and in 1868 he went to Liverpool, obtaining an appointment at the Children's Infirmary as practicing physician specializing in diseases of the heart.

He was not satisfied with merely a qualifying degree and immediately began research for a thesis. He chose the migration of blood cells and their relation to the formation of pus; this research, the subject of his thesis,[1] was published the following year in the *Quarterly Journal of Microscopic Science*.[2] Caton had considerable artistic ability, and a more detailed paper is illustrated by his own drawings.[3] In the same year he obtained a lectureship in Comparative Anatomy and Zoology at the Royal Infirmary School of Medicine, and in 1872 he became a Lecturer in Physiology. During this period Caton began to work on the electrical activity of the nervous system. In February 1875, he presented some of this work at a meeting of the Medical Institution at Liverpool in a paper entitled "On the electrical

---

[1]R. Caton. On the best methods of studying transparent vascular tissues in living animals, *J. Micr. Sci.,* 10:236–241, New Ser. (1870).

The original research on the works of Richard Caton was published in 1861 by Pitman Medical Publishing Co., London. Material used from that book is by courtesy of Pitman.

[2]R. Caton. Migration theory. *Quart. J. Micr. Sci.,* 10:417, New Series (1870).

[3]R. Caton. Contributions to the cell-migration theory. *J. Anat. Physiol.,* 5:35–47 (1871).

relations of muscle and nerve.''[4] At the same time he gave a demonstration of the effects he was describing and explained his apparatus which consisted of a Thomson reflecting galvanometer and Du Bois-Reymond's non-polarizable electrodes.

As this was many years before the invention of the vacuum tube and of the electronic amplifier, Caton had to use optical magnification in order to demonstrate to his audience the small changes in current that he evoked in his nerve-muscle preparations. His method was described as follows: ''A graduated scale, some eight or nine feet in length, being placed on one of the walls of the theatre, a beam of light from an osyhydrogen lamp was thrown on to the mirror of the galvanometer, and thence reflected to the scale.''

At this meeting Caton demonstrated muscle currents evoked by stimulation of the nerve as well as those from direct stimulation of the frog's triceps muscle. He showed the diminution of current when the muscle contracted and used Du Bois-Reymond's term ''negative variation'' to describe it. He also demonstrated heart currents in frogs. His final experiment was one which showed how the passage of a constant current could increase or decrease excitability in the nerve.

Those were, of course, demonstrations of the electrophysiological phenomena that had been discovered and elucidated earlier in the century by Matteucci in Pisa and by Du Bois-Reymond's school in Berlin. But at the time of this meeting, Caton was already extending this investigative approach to the brain. In order to understand how these experiments fitted into the physiology of the time, it should be noted that this was a period of intense interest in localization of function in the cortex. In the case of the motor systems, a new technique had been added to that of the ablation experiments of Goltz[5] and others. The physiologist had gained a new tool with the demonstration by Fritsch and Hitzig[6] in 1870 that parts of the cortex were electrically excitable and that localized muscle groups could be activated in this way. The cortex could now be explored with stimulating electrodes and the resultant movement observed. But the sensory systems still evaded the experimenter. One could stimulate a sensory receptor, but how could one know which region of the cortex received the message?

Caton, with his knowledge of Du Bois-Reymond's proof that activity in a nerve caused a change in its potential, thought of the idea that a similar phenomenon might occur in the brain on receiving a sensory input. He therefore applied to the British Medical Association for a grant to work on this problem. A sum was awarded, and at the 43rd meeting of the Association held in Edinburgh in August 1875, he made a preliminary report.[7]

This report, the first account in scientific literature that documents spontaneous electrical activity in the brain, appeared as a note in the Proceedings of the meeting of the British Medical Association at which Caton gave his demonstration. Burdon-Sanderson was in the Chair. The report, published in the *British Medical Journal*, reads, in its totality, as follows:

> The Electric Currents of the Brain. By Richard Caton, M.D., Liverpool. After a brief résumé of previous investigations, the author gave an account of his own experiments on the brains of the rabbit and the monkey. The following is a brief summary of the principal results. In every brain hitherto examined, the galvanometer has indicated the existence of

[4]R. Caton. On the electric relations of muscle and nerve. *Liverpool Med. Chir. J.* (1875).

[5]F. L. Goltz, Über die Verrichtungen des Grosshirns. *Pflüger's Arch. f. d. ges. Physiol.*, 13:1–44 (1876). (On the functions of the brain.)

[6]G. T. Fritsch and E. Hitzig. Über die elektrische Erregbarkeit des Grosshirns. *Arch. Anat. Physiol.*, 37:300–382 (1870). (On the electrical excitabilities of the brain.)

[7]R. Caton. The electric currents of the brain. *Br. Med. J.*, 2:278 (1875).

electric currents. The external surfaces of the grey matter is usually positive in relation to the surface of a section through it. Feeble currents of varying direction pass through the multiplier when the electrodes are placed on two points of the external surface of the skull. The electric currents of the grey matter appear to have a relation to its function. When any part of the grey matter is in a state of functional activity, its electric current usually exhibits negative variation. For example, on the areas shown by Dr. Ferrier to be related to rotation of the head and to the mastication, negative variation of the current was observed to occur whenever those two acts respectively were performed. Impressions through the senses were found to influence the currents of certain areas; e.g., the currents of that part of the rabbit's brain which Dr. Ferrier has shown to be related to movements of the eyelids, were found to be markedly influenced by stimulation of the opposite retina by light.

Caton's success in demonstrating that the technique of electrical events could be applied to the detection of sensory function in the brain was of such interest that it obscured the unexpected additional observations that he made, namely, that "feeble currents of varying direction pass through the multiplier when the electrodes are placed on two points of the external surface, or one electrode on the grey matter, and one on the surface of the skull." This was claimed by later scientists as the discovery of the electroencephalogram.

Two years later Caton published a much fuller account[8] of this work in a supplement to the *British Medical Journal* (1877), reporting experiments on "upwards of forty rabbits, cats, and monkeys, the rabbit having been principally employed." The paper is essentially devoted to localization of responses to the various sensory modalities; in exploring these he was successful in finding the visual receiving cortex, but all attempts to find the auditory area apparently failed.

Caton's observations in that paper are mostly on the effect of sensory stimulation on the standing potential difference between the external surface of the brain and a cut surface; but in some of his experiments, he applied both electrodes to the external surface of the brain, or one of them to the skull. As before, he used a Thomson reflecting galvanometer and described his electrodes as Du Bois-Reymond's nonpolarizable electrodes and said they were "small light electrodes . . . supported by small screw clamps, fixed firmly to the skull, in such a manner that no movement of the animal's body could affect the position of the electrodes on the brain." In the first section of this work, he summed up his results as follows:

> I. Facts observed relating to the Electric Currents themselves.
> (a) All the brains examined have shown evidence of the existence of electric currents.
> (b) If one electrode be applied to the external surface of the brain, and the other to the vertical section, a strong current passes through the multiplier, the external surface being usually positive to the vertical section.
> (c) If both electrodes be applied to the external surface, or one to the external surface of the skull, a feebler current passes through the multiplier.
> (d) The strength of the current varies at different points.
> (e) The current is usually in constant fluctuation; the oscillation of the index being generally small, about twenty to fifty degrees of the scale. At other times, great fluctuations are observed, which in some instances coincide with some muscular movements or change in the animal's mental condition.

In the section subtitled "Observations on the Relation to the Electrical Currents to the Function of the Brain," he reported that he:

> was led to suppose it probable that some such relation existed from the fact that fluctuations of the electric current often occurred coincidently with some movement of the animal's

[8]R. Caton. Interim report on investigation of the electric currents of the brain. *Br. Med. J.,* Supplement, L:62–65 (1877).

body or change in its mental condition; e.g., a variation of the current occurred when the rabbit awoke from sleep, or when anaesthesia was produced, or when death was occurring. The current usually fell to near zero after death.''

Ferrier's work[9] had brought localization of sensory function in the brain into prominence, and with the possibilities of his method as an aid, Caton proceeded to look for electrical changes in that part of the cortex, stimulation of which by Ferrier had evoked movements of the head, and also the area related to chewing movements. The results, as might be expected, were not too clear cut. He reported as follows:

> An examination was made of the currents of special areas. For example, the area pointed out by Professor Ferrier as related to rotation of the head was studied in six rabbits, with a view to discover if any change in the current occurred when the animal turned its head. In two rabbits out of six, variation of the current was observed whenever the head was turned towards the opposite side.
>
> The mastiscatory area was next experimented upon in eleven rabbits and two monkeys; and in four of the former, and one of the latter, marked variation was seen whenever the animal masticated. The remaining experiments were without result, either because the animals refused to eat, or from other causes to be spoken of below.

However, Caton did make the following interesting observation: ''In two rabbits, a point was discovered close to the masticatory area the current of which always showed variation when food was presented to the animal but before mastication commenced. This area was thought to be probably related to the perception of the odour of food.'' Some of today's experimenters might wonder whether or not he had perhaps stumbled on an electrical correlate of the salivary reflex.

Moving to the main goal of his experiments, namely, the localization of sensory functions in the brain, Caton proceeded to examine the effect of stimulation of the several afferent modalities—tactile, auditory, and visual. He found the following results:

> A number of experiments were made to ascertain if the electric currents of any part of the hemisphere were related to common sensation in the skin. The skin was stimulated in different parts of the body by means of a gentle interrupted current. Nothing results, excepting that some evidence was obtained indicating that the currents in and about the masticatory area were influenced by stimulation of the lips and cheeks. Gentle pinching of the lips and cheeks was also seen to be invariably followed by fall in the current of the same part of the brain. This was observed in six rabbits.
>
> Search was made to discover an area related to perceptions of sound. The electrodes were placed on various parts of the brain, and loud sounds were made close to the rabbit's ears by means of a bell, etc. No results were obtained.
>
> A similar search was made to discover an area related to impressions on the retina. A point was found on the posterior and lateral part of the hemisphere in which, in three rabbits out of seven experimented on, variation of the current was seen to occur whenever a bright light was thrown upon the retina.

In spite of the handicap imposed by his instrumentation with its lack of adequate amplification, there was enough hint of success in these experiments to encourage Caton to pursue this attack still further; as a later report witnesses, more striking results were eventually achieved. But at this stage of his work, he was cautious in his interpretations and very much aware that hazards of the operative procedure could introduce artifacts into his galvanometer readings. He stated:

---

[9]D. Ferrier. The localisation of the functions in the brain. *Proc. Soc.*, 22:229–232 (1874). Also in: *The Functions of the Brain*. Smith Elder, London, 1876.

The fact of so large a proportion of the experiments (more than half) being failures may be accounted for by the great difficulty encountered. Swelling and congestion of the exposed brain occurs sooner or later, and is accompanied by great disturbance of the electrical currents. If it occurs early, no observations of any value can be made. Haemorrhage, the development of currents in the electrodes, and other causes, render a large number of the experiments unproductive.

With praiseworthy caution Caton summarized his results at this time in a passage that shows he was not vehement:

The investigation thus far tends to indicate that the electrical currents of the grey matter have a relation to its function similar to that known to exist in peripheral nerves, and that the study of these currents may prove a means of throwing further light on the functions of the hemispheres.
Considering the comparatively small number of experiments yet made, and also the obscurity which involves the whole subject of the electrical currents of nerves, great caution is needful in drawing inferences from the facts above stated, and any such inferences must be considered merely provisional until many more observations have been made.

In 1875, when Caton was searching for a potential change evoked in the brain by sensory stimulation and of its intrinsic "feeble currents," there was no Physiological Society in England to whom he could report it and no specifically physiological journal in the English language. Consequently, it was to a medical group that he presented his first results and a medical journal that published his first two reports. These are probably the reasons why his discoveries were missed by physiologists, both in England and abroad, and why, years later when others rediscovered the same phenomena, they were unaware that he had anticipated their work.

The universities of Great Britain were slow to recognize physiology as an experimental science, other than as a part of medicine; as a result, its development lagged far behind that in France and Germany where the schools of Magendie and of Müller had gained renown by their achievements and by their rosters of famous pupils. Both Magendie and Müller had founded journals as outlets for research in physiology—Magendie, the *Journal de Physiologie Expérimentale* in 1821, and Müller, the *Archiv für Anatomie und Physiologie* in 1834 to supplement the *Duetsches Archiv für Physiologie* which had been started as early as 1795 by Reil in Halle.

In 1875, when Caton made his report, there were only two centers where physiology ranked a chair—at University College, London, and at Owens College, Manchester. Even Michael Foster at Cambridge held only the Praelectorship of Physiology at Trinity College. But the following year the Physiological Society was formed (Fig. 87). Interestingly enough, the motivation among physiologists in forming an official group was not primarily as an outlet for scientific reports but as a protection for their experimental procedures against the impending Antivivisection Act. Caton was one of the original members of the society, one of a group of famous scientists that included, among others, Burdon-Sanderson, Sharpey-Schäfer, Foster, Ferrier, and Langley. The names of all can be found in the minutes of the inaugural meeting where their signatures are preserved.

The need for a journal was soon felt. Anatomy had had its journal since 1866 and was still regarded as the leading basic science of medicine (the first Chair of Physiology, that at University College, London, was a Chair of General Anatomy and Physiology and its first Professor was an anatomist, Sharpey-Schäfer). But in 1878, one year after Caton's second publication on electrical currents in the brain, the *Journal of Physiology* was founded. Had this publication appeared in time to carry Caton's reports, it seems likely that

 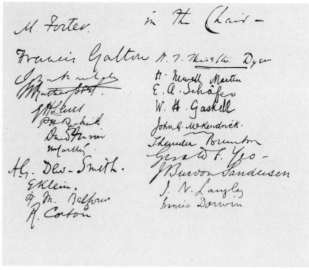

FIG. 87. Left: Richard Caton (1842–1926). A studio portrait taken when he was making his discovery of the electrical activity of the brain. (Photograph the gift to the author from his daughter, the late Miss Anne Caton). **Right:** The founding members of the English Physiological Society. In addition to Caton's signature, there are those of Foster, Burden-Sanderson, Ferrier, Schäfer, Gaskell, Langley, and others.

the controversey that broke out in the 1890s over priority of discovery would have been avoided.

It was about this time that Caton began to be drawn more closely into the activities that led eventually to the formation of Liverpool University in 1903. The first step was to get recognition of Liverpool as a college of Victoria University (of Manchester) and to open the Victoria University degree to Liverpool students. In 1878 the town's meeting appointed Caton as one of the members of the first Court of Governors. He was then 36 and in the middle of his electrophysiological research, at the same time holding a clinical appointment at the Northern Hospital. This marked an early step in his lifelong interest in education and in his devotion to university education in Liverpool, which was to lead eventually to his becoming Pro-Chancellor.

In 1882 he received recognition for his scientific work through an appointment to the Chair of Physiology in the medical school and for his services to the college by election to its Senate. Complete amalgamation of the Royal Infirmary School of Medicine with the college in Liverpool was not achieved until 1884, at which time Caton became Professor of Physiology in the University College of Liverpool, a position he held until his resignation in 1891.

During this period Caton's next report on the electrical activity of the brain appeared in the scientific literature.[10] In 1887 he visited the United States to attend the Ninth International Medical Congress in Washington, D.C., where he gave a paper entitled "Researches on Electrical Phenomena of Cerebral Grey Matter."[11] The communication was essentially devoted to the potentialities of the method for localizing motor and sensory functions. He

[10]R. Caton. Description of a new form of recording apparatus for the use of practical physiology classes. *J. Anat. Physiol.*, 22:103–106 (1887).

[11]R. Caton. Researches on electrical phenomena of grey matter. *Ninth International Medical Congress*, 3:246–249 (1887).

had looked not only for the receiving cortex for sensory impressions but also for signs of electrical change during a voluntary motor act.

He reported having experimented on the brains of 45 animals, namely, cats, rabbits, and monkeys, and he described his operating technique, electrodes, and instrumentation. Once again he reported:

> Under favorable circumstances I always obtained evidence of the existence of electrical currents of considerable energy—of much greater energy than those of nerve fibre. In some cases the currents were so powerful that shunts were needed. On applying one electrode to the external surface of the brain, and the other to the surface of a vertical section, vigourous currents passed through the galvanometer. The external surface was usually positive to the vertical section. If both electrodes were placed on the external surface a feebler movement of the galvanometer resulted.

In the interval between Caton's first and second publication, Ferrier's book *The Function of the Brain* had appeared and had rightly claimed the attention of the physiological world. Caton acknowledged the influence of Ferrier's findings on the design of his own experiments. He said:

> I obtained more definite results when experimenting on Ferrier's motor and sensory areas.
> 1. There is a region in the grey matter of the rabbit's brain, stimulation of which by the interrupted current causes rotation of the head to the opposite side. In the brain of the monkey there is also a corresponding centre. In several instances I found that by producing a sound, or by offering food on the one side of the animal experimented on, I would induce it to turn its head voluntarily to that side; when this movement was made, electrodes placed on the centre in question of the opposite hemisphere showed a fall in the current toward zero, in fact a negative variation, the movement of the needle exactly coincided with the movement of the animal's head to the opposite side. Probably the explanation is that the brain cells of the region were in a state of functional activity connected in some way with the head movement, and that a negative variation of the electric current occurred similar to that which is well known to occur in a nerve fibre when a reverse impulse traverses it.
> 2. It is difficult to induce a rabbit or a monkey to perform any definite voluntary act and to repeat the action frequently enough for the basing upon it of a physiological inference. The act of mastication is more easily induced than any other. A rabbit will frequently eat a piece of fresh lettuce, and a monkey will usually eat a raisin or a piece of raw potato as soon as it is offered him. I experimented, therefore, frequently on that centre of the brain which when stimulated causes masticatory movements. In half the animals used, I found that when the non-polarizable electrodes were placed on this centre, negative variations occurred invariably when the animal masticated, the variations lasting as long as mastication and ceasing when mastication ceased. In some instances it was evident that the thought or expectation of food caused the movement of the needle. If I showed the monkey the raisin but did not give it, a slight negative variation in the current occurred. When the electrodes were applied to this region, I found that sensory impressions made on the mouth or face caused a similar movement of the needle; for example, the introduction of the handle of a scalpel into the mouth, pinching lips or cheeks, or stimulation of skin of face by interrupted currents. It seemed from this experiment as though the centres for movement of jaw, for perception of sensory impressions from mouth and face, and for ideas of food derived through the eye, coincided or were closely adjacent to one another.
> The area associated with these functions appeared to be small. I frequently had to search for it for some time. If the electrodes were not upon it but merely near it, no relation was observed between mastication and the movement of the galvanometer.

Any impulse to assign his findings to movement artifacts is restrained by the final sentence of the passage quoted above. His account then moves on to his attempts to localize sensation. He was moderately successful with stimulation of a limb, though he failed to find a response to odors or sounds, but was markedly successful when he used light as a stimulus. Of particular interest are his results with what is now called intermittent photic stimulation:

> I tried the effect of alternate intervals of light and darkness on seven rabbits and four monkeys, placing the electrodes on the region stimulation of which causes movement of eyes. In three rabbits and two monkeys I found that light caused negative variation almost invariably. In those five experiments in which I was successful the relation between the intervals of light and darkness and the movements of the galvanometer needle was quite beyond question. If it partially shaded the animal's eye from the light, the effect on the electric current was diminished. The exact way in which the light produced its effect is not so easy to determine. It may have excited the visual centre especially, or it may have acted as a general excitant to the whole brain, or the result may possibly have been due to the heat radiated from the flame acting on the electrodes; I think one of the first two theories is more probable than the third.

The mention of a flame as the source of light used for the stimulus takes one immediately back to the gaslight era and is a reminder in this age of stroboscopes and electric flash bulbs of the primitive facilities that laboratories could offer a scientist in the 19th century.

An account of Caton's presentation in Washington[12] appeared in due time in the *Transactions of the Congress*, and an abstract was printed in the Russian language journal *Vrach* in November 1887.[13] A translation of the report of Caton's presentation which appeared in this Russian journal reads as follows:

> A paper was read by Dr. R. Caton, "On electrical phenomena in the grey matter of the brain." The author noted that the phenomena of cortical excitation could be elicited equally well by direct irritation and by drawing the animal's attention to food; the author's investigations can be summarized in general as follows—
> 1. Electrical currents in the ganglia of the grey matter are intensified when functional activity is reduced by anaesthesia or by approaching death, but after death they diminish and disappear.
> 2. In parts of the brain specific for certain functions, a negative oscillation is observed during such activity, a fact which serves as further evidence for localization of function.
> 3. The author's observations give considerable support to the concept that those parts of the brain that govern a given group of muscles also control the sensation of the corresponding area of the skin.

This abstract was no more successful in catching the eye of the Russian and Polish physiologists than those in the English language were in attracting the attention of Caton's countrymen.

Meanwhile, Caton had extended his exploration to electrical changes in nerves of the visceral organs on stimulation of the skin, for he felt this could have clinical significance. The experiments, which were made on frogs, were reported to the Liverpool Biological Society of which he was a member (and later the president). The abstract[14] of his report reads as follows:

> In studying the electric phenomena of nerve and muscle, it is well known that a change in the electrical state—the negative variation or current of action—occurs during frictional activity, and that in all nerves other than motor and secretory ones it forms the only objective phenomenon by which the transit of a nerve impulse can be recognized.

---

[12]Researches on electrical phenomena of cerebral grey matter. *Ninth International Medical Congress, Trans. Cong.,* 3:246–249 (1887). (The Congress at which Caton's paper was given was held at the then called Columbia University, which is now named George Washington University.)

[13]An abstract of Caton's paper given at the Ninth International Medical Congress (in Russian). *Vrach,* 8:881 (Nov. 5, 1887).

[14]On the variations produced in the electrical condition of viscera by stimulation of adjacent cutaneous nerves (abstract). *Proc. Liverpool Biol. Soc.,* 3:113–114 (1889).

Having a definite object in view relating to practical medicine, I was wishful to ascertain whether stimulation of the cutaneous nerves of the trunk had any influence on the nerves of the internal organs.

The only mode of ascertaining this was to find out if a negative variation in the electrical condition of these nerves or of the organ to which they are distributed occurs at the instant when the stimulation of the cutaneous surface takes place. The following experiment was therefore performed, and frequently repeated. A frog was rapidly and painlessly killed by destroying its brain; a small quantity of woorain was then injected into a lymph sac to prevent reflex spasm; a small coil of intestine was withdrawn through an opening in the left flank, and two non-polarizable electrodes were so arranged as to bring this coil of intestine into the circuit of a sensitive galvanometer. The needle at once showed a certain amount of deflection. This being noted, the skin of the abdomen was stimulated either by heat, by an induced electrical current, by a chemical irritant, or mechanically by pinching or otherwise, and coincidently a negative variation was shown by the galvanometer, indicating that a nerve impulse had travelled round by the intercostal nerves, and had produced an active condition of the nerves of the intestine. This experiment is one of some difficulty, and cannot reasonably be expected to succeed always; it has, however, succeeded so frequently that little doubt can be entertained of the fact it was intended to demonstrate.

During his years as professor, Caton contributed richly to the educational development of Liverpool College (he had also been Dean of the Medical Faculty), and this service continued and saw its fruition when Liverpool received its charter as a university in 1903. While on the teaching staff in physiology, he published several papers that revealed his goal of improving methods for instruction of students in his science. He wrote articles such as: "A new form of microscope for physiological purposes"; "Description of a new form of recording apparatus for the use of practical physiology classes"; and "On the teaching of hygiene in Government schools."

In the year 1891 Caton resigned the professorship, which at this date became an endowed chair, the Holt Chair of Physiology. Caton was followed by Gotch,[15] who held the appointment until 1895 when he was succeeded by Sherrington.[16] On resigning the Chair, Caton gave all his apparatus and material on his electrical research to the department, where the cause of their subsequent disappearance has remained a mystery.

After Caton's retirement from the Chair of Physiology in 1891, at the age of 49, his interests became more markedly directed to clinical problems, and he published on such subjects as enteric fever, typhoid fever, lead poisoning, and acromegaly (his patient was the first to be operated on for this condition). In 1896 he became President of the Liverpool Medical Institution.

In 1897 the International Medical Congress was held in Moscow. At this congress, Caton represented the University of Liverpool and was elected one of the eight British vice-presidents, but he did not read a paper.

At the time of the meeting in Moscow, Caton was, however, no longer working on the electrophysiology of the brain. The last note published by him on brain potentials[17] had appeared in the form of a letter to the German journal *Centralblatt für Physiologie* as the climax to a series of polemics that had appeared in 1890 over priority for success in recording the electrical activity of the brain. Claims came from the Polish scientist Adolf Beck,[18]

---

[15]Francis Gotch (1853–1913).

[16]Charles Scott Sherrington (1857–1952).

[17]R. Caton. Die Ströme des Centralnervensystems. *Cbl. Physiol.*, 4:785–786 (1891). Letter to the editor, written Feb. 1891. (The currents of the central nervous system.)

[18]A. Beck. Die Bestimmung der Localisation der Gehirn und Rückenmarkfunctionen vermittelst der elektrischen Erscheinungen. *Cbl. Physiol.*, 4:473–476 (1890). (The demonstration of localization of brain and spinal cord function by means of electrical phenomena in the central nervous system.)

from Ernst Fleischl von Marxow[19] (the Professor of Physiology in Vienna), and from Victor Horsley and Francis Gotch[20] in England. All claimed to have found the potential shift on sensory stimulation, but of them only Beck had found the "spontaneous" oscillations of the brain's potentials. None of them was aware that Caton had anticipated their work by 15 years. It was when the arguments among them had reached some acerbity that Caton wrote his letter to the editor drawing attention to the account of his work that had been published so long before.[21,22]

In 1904 he gave the Harveian Oration at the Royal College of Physicians, of which he had been a Fellow since 1888. He chose as his subject "Valvular Disease of the Heart," this going back to the interest of his first days as a young physician nearly 40 years before.

Richard Caton died in 1926, a figure much revered in the many fields to which his energy and versatility had contributed. His finding of ongoing "feeble currents" had not been followed up by the physiologists in England, possibly because of the belief of two very famous figures, John Burdon-Sanderson and Victor Horsley. Neither believed that the nerve cell itself could have electrical properties, and so attributed currents to underlying fibers. Both had worked at the West Riding Lunatic Asylum where Bevan Lewis had demonstrated fibers issuing from brain cells, and it was to these fibers that these skeptics assigned the electrical activity. They were in positions of great power, being members of the Royal Society and editors not only of its *Proceedings* but also of the journal *Brain*. But conditions were difficult in Europe, and reports began to come from Poland, from Russia, and from the Ukraine; and it was from these sources that at the early part of the 20th century the first photographs of the electroencephalogram were published.

Adolf Beck (1863–1942)

In 1886 a young student, who was to become one of Poland's outstanding scientists, joined the physiology department at the Jagiellonski University in Krakow and began his work for a doctorate in general medical science. This famous old university, founded in the 14th century, was named for Poland's ancient line of kings. There Adolf Beck came under the leadership of the distinguished physiologist Napoleon Cybulski.

Within a year Beck had published his first paper in collaboration with his professor,[23] but he was soon to branch out on research of his own in electrophysiology. His first independent work[24] was a demonstration of uniform excitability over the full length of peripheral nerve, thus disproving the "avalanche" theory of increasing activity from center to periphery proposed by Pflüger[25] to meet the criterion of decrementless conduction.

[19]E. Fleischl von Marxow. Mittheilung betreffend die Physiologie der Hirnrinde. *Cbl. Physiol.*, 4:538 (1890). (Report concerning the physiology of the cerebral cortex.)

[20]F. Gotch and V. Horsley. Über den Gebrauch der Elektricität für die Lokalisierung der Erregungserscheinungen im Centralnervensystem. *Cbl. Physiol.*, 4:649–651 (1891). (On the use of electricity for localizing arousal phenomena in the central nervous system.)

[21]N. Cybulski and Jelénska-Macieszyna. Prady cynnosciowe kory mózgowej. *Bull. Int. Acad. Cracovie*, Series B: 776–781 (1914) (in German). (Action currents of the cerebral cortex.)

[22]V. V. Pravdich-Nemmsky. Ein Versuch der Registrierung der elektrischen Gehirnerscheinungen. *Cbl. Physiol.*, 27:951–960 (1913). (Experiments on the registration of the electrical phenomena of the mammalian brain.)

[23]A. Beck and N. Cybulski. Badania poczucia smaku u osoby pozbawionej jezyka. *Rozpr. Wydz. mat.-przyr. Polsk. Akad. Um.* (Polish Academy of Sciences, Proceedings of the Faculty of Mathematics and Natural Sciences), 18:207–216 (1888). (Research on the sense of taste in an individual deprived of a tongue.)

[24]A. Beck. O pobudiwosci róznych miejsc tego samego nerwu. *Rozpr. Wydz. mat.-przyr. Polsk. Akad. Um.*, 15:165–195 (1888). (On the excitability of the various parts of the same nerve.)

[25]E. F. W. Pflüger. *Untersuchungen über die Physiologie des Elektrotonus.* Hirschwald, Berlin, 1859. (Studies on the Physiology of Electrotonus.)

In 1888 Beck was given the position of demonstrator and began the research that was to lead to his discovery of what was later to be named the electroencephalogram. He was clearly unaware of Caton's work, as was also his master, Cybulski, but it is of interest that he was led to it by the same desire to develop a method for the localization of function in the brain.

The interest of European scientists had been much stirred by the attempts of physiologists to localize function by the method of extripation, though, as Beck pointed out, there were many who, in reaction to Gall's theories, were averse to considering localization of the brain function as anything but an absurdity. At the time of Beck's work, Goltz[26] was one of those who held that it was the extent of cortex extirpated, rather than the site of the injury, that determined the character of functional loss, and at the International Congress of Physiology at Basel in 1889 he demonstrated a chronically hemidecerebrate dog in support of his view. The keen eye of Cybulski, who was present, detected contralateral impairments that had escaped Goltz's observation. On returning to Poland, Cybulski encouraged his young assistant to pursue the more promising electrical method.

The pioneer work of Fritsch and Hitzig[27] in the use of electrical stimulation of the cortex to evoke movements was for Beck, as it was for history, a turning point in the ideas relating to functions in the brain. In the previous century, convulsions had been evoked by stimulating the brains of frogs in some more primitive types of experiments by Fontana and by Caldani, Galvani's teacher at Bologna, but these experiments contributed nothing to localization of function. The experiment that came closest to demonstrating localization (in the same year as Fritsch and Hitzig's paper was published) was that stimulation of a frog's brain produced movement of the contralateral limbs. This was the work of a young student at the Medical Surgical Academy in St. Petersburg, I. A. Tyshletsky,[28] but as it was printed only as a thesis, it did not receive general recognition, and in addition, since it was work on frogs, it would be difficult to compare with the analogous experiment in the mammalian brain.

The problem of localization of sensory function was far more difficult to explore than that of the motor system. The outstanding attempts were those of Munk[29] whose ablation experiments in dogs convinced him that, as in man, the optic nerves from both eyes led to the occipital cortex and that bilateral homonymous hemianopsia was the consequence of the loss of one occipital lobe. He found he had to remove the temporal lobes bilaterally to produce deafness and that destruction of the hippocampal gyri abolished the sense of smell.

A more satisfactory method than extirpation suggested itself to Beck from his knowledge of peripheral nerve. By the same line of reasoning that Caton had used before, he argued that the "negative variation" found by Du Bois-Reymond to indicate activity in nerve (what is now called the action potential) might have its counterpart in the central nervous system. The task would be to demonstrate the activity of nervous centers while stimulating an afferent nerve. He wrote:[30]

[26]F. Goltz. *Über die Verrichtungen des Grosshirns*. Strauss, Bonn, 1881. (On the Functions of the Brain.)

[27]G. T. Fritsch and E. Hitzig. Über die elektrische Erregbarkeit des Grosshirns. *Arch. Anat. Physiol. Wiss. Med.*, 37:300 (1870). (On the electrical excitabilities of the brain.)

[28]K. S. Koshtoyants. The history of the problem of brain cortex excitability. Actes due 8 eme Congres Internat. d'Histoire des Sciences, pp. 862–864, 1956.

[29]H. Munk. *Über die Funktionen der Grosshirnrinde*. Hirschwald, Berlin, 1881. (On the Function of the Cerebral Cortex.)

[30]A. Beck. The Determination of Localization in the Brain and Spinal Cord by Means of Electrical Phenomena. Doctoral thesis published in *Polska Akademija Umiejetnosci*. Series 2:187–232 (1891). (In all future quotations from ths work the references will be abbreviated to Thesis.)

The question arises, are there any currents in the nervous centers of the brain and spinal cord? If so, are there changes in these currents during activity? And would the localising of such changes be of any help in demonstrating a state of activity of a focal nature in the central nervous system?

Beck thought this approach highly promising and based his optimism on the earlier work from Sechenov's laboratory on the electrical activity of the spinal cord. Sechenov, the great Russian physiologist who had trained with Du Bois-Reymond in Berlin and who had been responsible for introducing electrical techniques into the Russian schools of physiology, had recorded electrical potentials from the spinal cord and medulla. The work to which Beck referred was a report published by Sechenov as a preliminary communication in 1881.[31] In this paper Sechenov described changes in the demarcation potential between cut surface and longitudinal surface of the cord or medulla elicited by sciatic stimulation. He also found discharges in transected medulla (but not in the cord) which he believed to be spontaneous and indicative of an excitation related to the respiratory centers. This work was expanded by Sechenov in a later publication.

At the Third Congress of Russian Physicians and Biologists held in St. Petersburg in 1889, Verigo,[32] a pupil in Sechenov's laboratory, reported similar changes in demarcation current in the lumbar cord of frogs on stimulation of the leg and also of the frog's brain[33] when its hind legs were moved. This led Wedensky,[34] who was present, to suggest that this method might be used to delineate localization in the cortex. It is interesting that these suggestions were being made 14 years after Caton had first applied the method successfully.

These reports by Sechenov and by Verigo, together with the suggestion of Wedensky, encouraged Beck to pursue his experiments. His first series was on frogs, and when he wrote his thesis, he gave a detailed description of his techniques and apparatus together with the protocols. He used a modified form of Du Bois-Reymond's nonpolarizable electrodes. These were of clay soaked in 1 percent sodium chloride solution protruding from a glass tube filled with zinc sulfate, from which zinc wires made the connection to a Wiedermann galvanometer. Like Caton[35] before him, Beck was working without benefit of the invention of amplifiers. In order to compensate for the potential difference between the transected neuraxis and the longitudinal surfaces of the cord, he introduced a Daniel cell in parallel with a variable rheostat in one arm of the circuit to the galvanometer. Since he had no method available for photographing his galvanometer readings, the results throughout his experiments were expressed in terms of millimeters of deflexion. In fact, photography was still a rarity in the biological laboratory, although Marey[36] had begun to use it for recording of heart currents in 1876.

In his experiments on frogs, Beck carefully dissected out the brain and spinal cord down

---

[31]I. M. Sechenov. Galvanische Erscheinungen an der cerebrospinalen Axe des Frosches. *Pfüger's Arch. f. d. ges. Physiol.*, 25:281–284 (1881). (Galvanic phenomena in the medulla oblongata of the frog.)

[32]B. F. Verigo. Action currents of the frog's brain. Report of the Third Congress of Russian Physicians and Biologists, St. Petersburg (in Russian). *Vrach*, 10:45 (1889).

[33]B. F. Verigo. Action currents of the brain and medulla (in Russian). *Best. Klin. Sudev. Psychiatr. Nevropath.*, 7: (1889).

[34]N. E. Wedensky. Report of the Third Congress of Russian Physicians and Biologists, St. Petersburg (in Russian). *Vrach*, 10:45 (1889).

[35]R. Caton. The electrical currents of the brain. *Br. Med. J.*, 2:278 (1875).

[36]E. J. Marey. Inscription photographique des indications de l'électromètre de Lippmann. *C. R. Acad. Sci.*, 83:278–280 (1876).

as far as the lumbar region, laying them on a glass plate and leaving only the lower limbs fully intact, except for exposure of the sciatic nerve, which he stimulated with a Du Bois-Reymond induction coil. The preparation was covered with a Bell jar containing wet cotton to preserve a moist atmosphere and to delay deterioration.

Many experiments on 11 frogs in this series are recounted in detail, together with the measurements of galvanometer deflexion, these being changes in what modern electrophysiologists call steady potentials or direct current potentials to differentiate them from the oscillating potentials of the electroencephalogram. In all experiments in which he was successful in avoiding damage to the cord, Beck found the rostral portions of the neural axis negative to the more caudal ones even in the absence of a transection; in other words, he obtained what he interpreted as a resting current, as distinct from a demarcation current of injury. This was at the time when electrophysiology was being rocked by a controversy among its giants. Hermann,[37] a professor at Königsberg, maintained that there was no such thing as a resting current, that currents only came into being on injury to tissue. It was he who introduced the name "demarcation current." The opposition to Hermann's views was led by his old teacher, Du Bois-Reymond,[38] who championed the cause of the resting current. In the light of today's knowledge, one can recognize a confusion between resting potential difference and flow of current when a fall in resistance permits its passage.

From the mass of results present in his series of experiments on frogs, Beck interpreted an increase of negativity as indicating an activation of centers underlying the electrode on the brain, and a decrease as signaling activation of the spinal reflex center that lay below his more caudal electrode. With the hemispheres removed, his most constant finding on sciatic stimulation was a decrease in the negativity of the rostral electrode relative to the caudal one, irrespective of the latter's position on upper cord, medulla or optic lobes, either on cut or uncut surface. He concluded that stimulation of the nerve caused an increase of activity (and hence negativity) in the reflex centers of the lumbar cord that reduced the standing potential difference in the length of the neuraxis.

When the brain was intact and the upper electrode was on a hemisphere and the lower one on the cervical cord, Beck found an augmented negativity in some but not all experiments. However, on increasing his interelectrode distance by putting the lower electrode on the lumbar cord, the rise in negativity was multiplied about sevenfold, and this he took to indicate an increase of activity in the brain.

In addition, Beck concluded from these experiments that Hermann was wrong in maintaining that there is no such thing as a resting potential and that connection between two undamaged points of nerve would never produce a current. The demonstration of a standing potential difference between rostral and caudal parts of the uncut neuraxis was, Beck felt, evidence for a resting potential indicative of a different phenomenon from that evoked by afferent stimulation. Moreover, provided no cut was made, there were slowly developing "spontaneous" fluctuations of these resting potentials. Beck wrote:

> Such independent oscillations as we observed here never occur in the current between a cut surface and the longitudinal section of a nerve. It follows that the current we found in the central nervous system is an active current. To differentiate this from the current evoked by stimulation, I would propose the name: active independent current. The fact that the upper parts were always negative in respect to the lower ones suggests that the current was

[37]L. Hermann. *Handbuch der Physiologie*, Vol. 2.

[38]E. Du Bois-Reymond. Untersuchungen über thierische Elektricität, Reimer, Berlin. Vol. I, 1848; Vol. II, 1849. (Investigation into Animal Electricity.)

 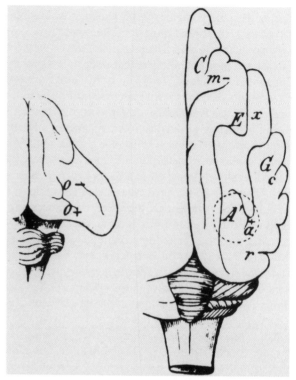

FIG. 88. Left: Adolf Beck (1863–1942). As a young man at the period of his doctoral thesis on the electrical activity of the brain. **Right:** Two of Beck's diagrams: On the *left,* the rabbit's brain on which he marked *(a⁻, a⁺)* the position of electrodes from which he recorded responses to light in his first experiment, and the more anterior position *((m⁻, m⁺)* where he found responses to stimulation of the hind leg. On the *right,* the dog's brain in which he marked the positions of electrodes that gave him a response to light *(m⁻, a⁺)* and a faint response to sound *(n, s).* (From: A. Beck. Thesis. 1890.)

the product or expression of an active state built up in the upper parts of the nervous system. In a word, we are dealing with a spontaneous excitation of the nervous centers.[39]

For examination of the brain's responses, Beck moved to warmblooded animals, employing both rabbits and dogs (Fig. 88). He removed the skull on one side and again used non-polarizable clay electrodes, except when the animal became too restless, in which case he substituted wick electrodes that moved with the brain. These were placed on two points of the cortex, and the current between them was observed both with and without afferent stimulation. In none of these experiments was there a cut in the brain, and hence the responses he obtained were not variations of demarcation current:

> The optic system ws stimulated with light, the auditory with sound, and the different sensory nerves of the skin with an induction current. For the light stimulation of the eye, I used a burning magnesium ribbon, which was moved by a special clock mechanism so that the flame reflected by a mirror into the eye remained constant. One could extinguish the flame by stopping the mechanism.[40]

[39]A. Beck. Thesis, p. 208.

[40]A. Beck. Thesis, p. 212. This was Beck's description of the brain's activity that, over 40 years later, would be called the electroencephalogram.

In his very first experiment (on a rabbit), Beck found an oscillating potential difference between two electrodes placed on the occipital cortex. The fluctuations ceased when he uncovered the animal's eyes and lit a magnesium flare, and they also ceased with stimulation of the hind leg. These oscillations rode on a direct current potential which increased with light stimulation, provided he had an electrode on the occipital regions. When the electrodes were placed frontally on the brain, stimulation of the hind leg also increased the standing potential and blocked the waves. A clap in the animal's ear also produced blocking, but he was unsuccessful in finding an electrode position where a sound evoked a response.

Thus, to Caton's discoveries, Beck had added yet another—that of desynchronization of cortical activity following afferent stimulation. This phenomenon, confirmed later by many, remained an empirical observation for almost 60 years, until the elucidation of the desynchronizing action on cortical potentials of the ascending reticular system. Beck's own description reads as follows:

> In addition to the increase or decrease in the original deviation during stimulation of the eye with light, rhythmic oscillations that have been previously described disappeared. However this phenomenon was not the consequence of light stimulation specifically for it appeared with every kind of stimulation of other afferent nerves.[41]

In a later part of his thesis in which he reviewed the inferences and interpretations to be drawn from his experiments, Beck returned to the discussion of the two effects of sensory stimulation he had found in the cortex, namely, the local shift in steady potential specific to the sensory system stimulated and the nonspecific blocking of the intrinsic oscillations. He wrote:

> An important event which occurred in nearly all the experiments on stimulation of the cerebral cortex by any of the afferent nerves (and especially on stimulation of the sensory nerves of the skin) was the arrest of the intrinsic oscillations of the functional current. Explanation of this phenomenon would be difficult were it not accompanied by the potential shift. I would simply say that there must have been an arrest at a certain point, a suppression of the intrinsic changes in the active state. In a word, it can be explained by suppression or blocking.[42]

He contrasted this type of "blocking" of the spontaneous oscillation with that produced by deep chloroform anesthesia, in which state both the intrinsic activity and the response are depressed.

Beck's attempt at an explanation of his findings has some similarities to the theory of dominata that was to be formulated in later years by the Russian physiologist Ukhtomsky,[43] namely, a state of inhibition developing in the central nervous system in an area surrounding a focus of excitation. In modern times some echo of this proposal can be found in the results derived by microelectrodes.

Moving from the rabbit to the dog, which he curarized for these experiments, Beck was again successful in evoking a potential swing as a response to light when at least one of his electrodes was on occipital cortex. In two of the experiments, he found a small deviation in response to a shout when one of the electrodes was on the temporal cortex, although in another experiment this stimulus caused only blocking.

[41]A. Beck. Thesis, p. 230.

[42]A. Beck. Thesis, p. 230.

[43]A. A. Ukhtomsky (1875–1942). *Collected Works* (in Russian). 5 vols. Leningrad, 1945–1954.

Beck felt the challenge of his results and of his proposal that the focal shift of potential on specific sensory stimulation signified increased activity in that cortical center. He argued that were this so, a locally produced increase of activity in that center should evoke the same potential shift. Consequently, he stimulated the cortex of a dog directly with a weak induction current and obtained a marked deviation:

> If my assumption concerning the relationship of the change in functional current to the origin of the active state in certain centers is correct, namely that the development of electro-negativity in an area of cortex really indicates the creation of an active state in centers located there, then on direct stimulation of that site an electro-negative swing should result . . . [This was theoretical reasoning; the experiments proved that it was right] . . . Since during stimulation of the eye by light there was a positive deviation of 21 mm, and on direct stimulation of the occipital cortex close to the negative electrode there was deviation of 80 mm in the same direction, does this not prove that this same cortical region went into an active state during stimulation by light? I obtained similar results on stimulating the leg and the corresponding part of the cerebral cortex, by which I mean that both gave a deviation in the direction of negativity.[44]

Beck warned his readers that his technique for localization of sensory functions in the cortex might give anomalous results if the second electrode lay near the motor cortex, for then the activity of the afferent response to the stimulus might mask the intitial sensory event that sparked it:

> After all the animal may simultaneously send a ''run-away'' impulse to its muscles and stimulation of the eye with light may actually evoke that impulse. Thus if one electrode lies on the occipital lobe, and the other on a part of the motor cortex corresponding to any of the four limbs, then the direction of potential change resulting from the light flash will depend on sensory centers or the impulse to move in the motor centers.[45]

It seems clear that Beck explored the electrical activity of the brain in greater detail than had Caton 15 years earlier. He thus made great contributions to the technique of localization of sensory functions in the brain as well as to knowledge of its electricity.

In reporting his results in the Polish language, Beck realized that his audience would be small, and he therefore published a short three-page summary of his results in the most widely read physiological journal of the day, the *Centralblatt für Physiologie*. In this brief report he summarized his findings as follows:

> Even in the very first experiment I noticed—and repeated experiments confirmed it—that the difference in potential between the electrodes when applied to two given points on the cortex of the hemispheres was not a stable level of potential; there was a continuous waxing and waning variation taking place which neither was related to the respiratory rhythm nor was it synchronous with the pulse, nor finally was it in any way dependent on movement of the animal, since it was present in curarized dogs.[46]

Beck reported in this paper having succeeded in localizing the visual cortex in both rabbits and dogs and noted that he had more trouble recording from the auditory cortex and mentioned that this was largely due to the difficulty in placing electrodes on the inner surface of the temporal lobe. He recorded reponses to stimulation of the skin in its fields of cortical representation.

Beck concluded by observing that the technique was not difficult and was capable of

---

[44]A. Beck. Thesis, p. 230.

[45]A. Beck. Thesis, p. 232.

[46]A. Beck. Die Ströme der Nervencentren (Letter to the editor), *Cbl. Physiol.*, 4:572–573 (1890). (The currents of the nervous system.)

improvement. He believed that it was a very valuable addition to the currently available methods for localization of brain function and (with considerable foresight) suggested that it could make many contributions to the neural sciences and to psychophysiology.

This brief German publication of Beck's unleashed a spate of claims for priority from scientists in several other countries. Cybulski, Beck's professor, realizing the new field opened up by this work, began to equip his laboratories more effectively for such studies. Balked by a limited budget from the university, Cybulski appealed to a prominant citizen of Krakow, who generously became the patron of this endeavor. Both Beck and his teacher were thereby able to pursue their electrophysiological studies. In the same year Beck wrote a scholarly appreciation of Hermann von Helmholtz and his contributions to physiology.[47]

Together with Cybulski, he extended the work on brain potentials to the study of monkeys, and this formed the subject of a report they made to the Third International Physiological Congress in Berne in 1895.[48] This was to be their last joint experimental work, for in the same year Beck was offered, and accepted, the Chair in the newly organized department of physiology at the University of Lwów[48a] in Galicia. This university was situated in a region occupied at different times by Austria, Russia, and Germany. At this period there was no free Poland; but, although these universities were under Austrian domination, Polish as the language of instruction had been reinstated in Krakow in 1870 and at Lwów in 1873, even though the cities were still generally known to the German-speaking world as Krakau and Lemberg. In 1872 the Emperor of Austria-Hungary signed the Charter of the Polish Academy of Science and Letters of Krakow, and this academy, together with the two universities, became the leaders of Polish culture and scholarship. It is in the journal of this distinguished academy, *Rozprawy Polska Akademii Umiejetnosci*, that many of Beck's published works are to be found, though many of them were published also in journals of the German language.

In the years just following his new appointment, Beck had to give a great part of his time to organizing and building up his new department; but, in spite of this, several of his scientific reports stem from this period. Beck's researches into the electrical potentials of the brain did not cease but were extended beyond the 19th century, including an attempt to locate the sensation of pain.

Also well into the 20th century, Beck continued research on the nervous system, including several with a colleague, G. Bikeles[48b], on the cerebellum.[49,50] His old professor, Cybulski, continued in Krakow to work on the electrical activity of the brain and with a colleague, Jelénska-Macieszyna, published some of the first photographs of the electroencephalographic potentials.[51,52] There is no photograph of theirs from the 19th century.

[47]A. Beck. Herman Helmholtz (in 2 parts). *Przegl. Lek.*, 33:548–550, 559–561 (1894).

[48]A. Beck and N. Cybulski. Elektrische Erscheinungen in der Grosshirnrinde beim Affen. 2nd International Physiology Congress, Berne, 1895. Abstract in *Cbl. Physiol.*, 9:474 (1895). (Electrical phenomena in the cerebral cortex of the ape.)

[48a]Since 1918 this institution has been named the Jan Kaziermetz University.

[48b]Gustav Bikeles was born in Lwów in 1861 and trained in Berlin and in Krafftlebing and Obersteiner.

[49]A. Beck and G. Bikeles. Eine Beobachtung über Reflexerscheinungen am Hintertier. *Pflüger's Arch. f. d. ges. Physiol.*,129:415–424 (1909). (An observation of reflex phenomena in the rear of the animal.)

[50]A. Beck and G. Bikeles. W wzajemnym stosunku cynnosciowym mózgu i mózkzku. *Rozpr. Wydz. mat.-przyr. Polsk. Akad. Um.*, Series III, 10B:457–471 (1911). (On the sensory activities of the central part of the cerebellum [vermis].)

[51]N. Cybulski and Jelénska-Macieszyna. Aktionsströme der Grosshirinde. *Cbl. Physiol.*, 11–12:407 (1918–1919). (Action currents of the cerebral cortex.)

[52]N. Cybulski and Jelénska-Macieszyna. Prady cynnosciowe kory nozgowej. *Bull. Acad. des Sci. Cracovie*, Series B: 776–781 (1914). (Action currents of the cerebral cortex.)

With Cybulski, Beck wrote the textbook *Physiology of Man*,[53] which became the standard teaching text in Polish universities. But these activities were not to last, for in 1914 the First World War broke out, and Lwów was taken by the Russians in their advance against Austria-Hungary. During this emergency Beck was once again asked to act as the rector of the university. The following year the Russians retreated in the face of the combined German and Austrian armies, and as a hostage for the university, they took Beck, the Rector, as well as nine other prominent citizens with them. He was taken to Kiev and held there.

While a prisoner at Kiev, Beck wrote a letter to Pavlov asking him to intervene on his behalf. Beck always in part attributed his release to Pavlov, for the following year (1916) an exchange of hostages was arranged and he was able to return to Lwów. Many years later he wrote a monograph,[54] putting on record the experiences and ordeals suffered by the university during the two years of the Russian invasion and the struggle to preserve a continuity of learning in the Polish tradition.

Even on Beck's return in 1916, it was not to a peaceful city, for Lwów was very much on the battlefront, almost surrounded by Ukrainian forces (the Ukraine was not yet a part of Russia). It is testimony to Beck's intense scientific drive that even from this stressful period there is one publication from him in the academy proceedings, a study in nerve physiology.[55]

The year 1919 is the great year in Polish history, for at the Treaty of Versailles a free and independent Poland came into being again, for the first time since 1795. This included the central part of Poland, which had been under Russian rule for 100 years. In the year of the liberation Beck had lost his great friend Cybulski. In a moving eulogy of his old teacher, Beck paid tribute to him as a scientist, a friend of the university, and a great human being.

Beck was an ardent Polish patriot and admirer of Pilsudski, the head of the new Polish state, but his allegiance was to his university in Lwów where he stayed to revive once again the program of physiological research to which he had already given over 20 years of his life. In 1930 at the age of 67, Beck retired with the title of *Professor honoris causa*, one of many distinctions: member of the Academy of Arts and Sciences in Krakow, of the Academy of Medical Science in Warsaw, as well as of other distinguished scientific and medical groups.

But the 20 years between the two World Wars was only a lull in Poland's war-torn history, and an even greater catastrophe was to engulf the country and with it Adolf Beck. In 1939 the Germans invaded Poland and with the occupation began their systematic extermination of its three and one-half million Jewish people. The Germans also began to suppress all educational institutions in the territories under their control. The German Governor (General Hans Frank) wrote at this time: "The Poles do not need universities or secondary schools. The Polish lands are to be changed into an intellectual desert." Only technical schools were encouraged, for the occupying forces needed skilled labor.

As the Germans closed in on Lwów, the danger to Beck increased, for he was Jewish. An old man now, rather than go into hiding, he chose to stay in the shadow of the univer-

[53]N. Cybulski and A. Beck. *Fizyologia Czlowieka*. 2 vols. Kraków, 1915. (The Physiology of Man.)

[54]A. Beck. Uniwersytet Jana Kazimierza we Lwowie podczas inwazji Rosyjskiej w roku, 1914, 1915. Lwów, Nakladem Sentau Akad. Uniw. Lwów, 1935. (The University of Jan Kasimir in Lwów during the Russian Invasion, 1914–1915.)

[55]A. Beck. O dwukierunkowem przewodzeniu nerwów. *Rozpr. Wydz. mat.-przyr. Polsk. Akad. Um.* Series III. 17B:1–13 (1917). (Two-directional conduction in nerves.)

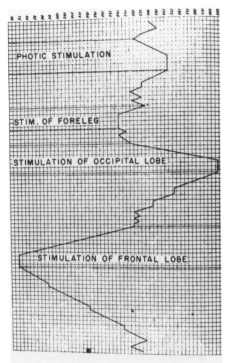

FIG. 89. **Left:** Beck's graph of shifts in steady potential caused by peripheral stimulation in an uncurarized dog. The galvanometer reading before stimulation oscillated in a range of 150 to 165 mm to the negative side of his scale. One electrode is on the visual cortex; the other, on the motor area for the foreleg. Stimulation with light (Draznienie okaswialtm) moves the galvanometer reading even more to negativity (i.e., 172). Stimulation of the foreleg (Draznienie konczyny przedniej) swings it in the opposite direction for the same electrode linkage. Electrical stimulation of the occipital cortex increases the galvanometer deflection to a reading over 200 (i.e., about 55 mm to the negative side of the baseline), whereas stimulation of the frontal motor cortex swings the reading of 75 mm (i.e., to the positive side of the original baseline). Beck made the note that he had not attempted to represent the time factor in these charts (in other words, the abscissa is abritrary). **Right:** Adolf Beck in rectoral robes wearing the ring given him in honor of 50 years of service to the University of Lwów and holding the textbook he wrote with Cybulski. Portrait by Stanislaw Batowski. (Photography by courtesy of Beck's daughter, Jadwiga Zakrzewska.)

sity to which he had given so many years of his life. Just before his 80th birthday, he became unwell, and while he was in the hospital for an ailment, the Germans came to take him to the extermination camp. Beck's son, a physician, had supplied all members of the family with capsules of potassium cyanide. Beck took his capsule and saved himself from the gas chamber.

The memory of this man, as a scientist and a humanist, is honored by his countrymen today, as it was in the years of his service to the university he loved so well. In 1934 he had been presented with a gold signet ring to mark 40 years of scientific work, and the following year his portrait had been painted for the university by Stanislaw Batowski (Fig. 89). This portrait, in which he is seen wearing the ring and holding a copy of the textbook he wrote with Cybulski, is one of the few material traces of Beck to survive the occupation. The ring, hidden by his daughter under the floor of her home in Warsaw, was found by her after the war in the ashes of the house, which like the rest of the city had been burned to the ground by Germans after the unsuccessful Warsaw uprising of 1944. The enamel had turned from red to black, but still legible on the ring are the words *Bene merenti facultas medica*, a fitting tribute to a fine scientist.

Ernst Fleischl von Marxow (1846–1892)

Unlike the four earlier reports of Caton's findings, Beck's short paper[56] in the *Centralblatt* describing his observations on the electrical activity of the brain immediately attracted the attention of physiologists and provoked letters to the editor claiming priority for having previously made these observations. The first of these came from Fleischl von Marxow, Professor of Physiology in Vienna.

In 1846 Ernst Fleischl, as he was usually called, was born in Vienna into a family unusual for their many intellectual interests. His uncle was the Bohemian physiologist Johann Czermak, a friend of Adolf Beck. On entering the University of Vienna, Fleischl came under the tutelage of Ernst von Brücke. Brücke had been one of the group with Du Bois-Reymond who, in their student days at the University of Berlin, had sworn to devote their endeavors to stamping out vitalism and to demonstrating that biological phenomena obeyed the laws of physics and chemistry. Another of these giants was also to be Fleischl's teacher, for in 1872 he went to Leipzig to study for a year at Carl Ludwig's famous institute.

On Fleischl's return to Vienna, Brücke appointed him as an instructor in physiology (one of his pupils was Sigmund Freud), and it was in this department that he spent the rest of his scientific life, succeeding his old teacher as a professor in 1876. In the same year he was appointed by the Austrian Government to be its delegate to the World Exhibition at Philadelphia with the special assignment of assisting the division of precision instruments. While in the United States, he crossed the country by the Pacific Railroad. Two years later he had a similar assignment to the Paris Exhibition, and for his services as an expert adviser to the government, he was awarded the honor as Knight of the Order of Franz Joseph. At the same time he was elected a corresponding member of the Imperial Academy of Sciences in Vienna.

In Vienna problems of physiological research filled his interest, and by 1883 he had begun experiments on the electrical activity of the brain, though none of these did he publish at the time. When Beck's article appeared in the German language in the *Centralblatt* in 1890, it caught Fleischl's eye, and he surprised the scientific world by revealing that he had written account of his findings in a sealed letter which he had deposited in the vault of the Imperial Academy of Sciences in Vienna seven years before (Fig. 90). The custom of depositing sealed envelopes containing scientific discoveries pending their confirmation was not unusual at some of the European universities (an example to turn up in more recent times was the sealed claim to the discovery of insulin before Banting and Best).

In his communication[57] to the *Centralblatt*, Fleischl quoted the official account of the formal opening of the letter by the secretary of the Academy and the letter was published in full in the *Centralblatt*. In translation it reads (in part):

[56]A. Beck. Die Ströme der Nervencentren. *Cbl. Physiol.*, 4:572–573 (1890). (Currents of the nervous system.)

[57]E. Fleischl von Marxow. Mittheilung betreffend die Physiologie der Hirnrinde. *Cbl. Physiol.* 4:538 (1890). (Report concerning the physiology of the cerebral cortex.)

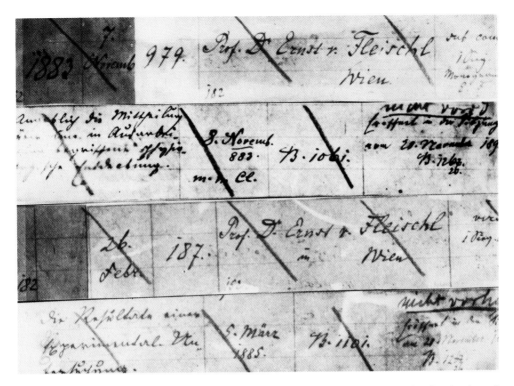

FIG. 90. Entries in the archives of the Imperial Academy of Sciences in Vienna registering the deposit in its vault in 1883 and the removal in 1890 of the letter sealed by Fleischl von Marxow in which experiments revealing the electrical activity of the brain were described. (Courtesy of Professor K. Pateisky.)

Vienna, Nov. 6, 1883

During the course of this year I have done a series of experiments on different animals. The results of these experiments seem to me to be of sufficient importance for me to secure priority for these findings by depositing this letter with the Imperial Academy.

If one connects two symmetrically localized points on the surface of the cerebral hemispheres to a sensitive galvanometer by means of nonpolarizable electrodes, one observes little or no movement of the galvanometer. Stimulation of a sense organ, however, whose central projections are located in one of the recording points, causes the needle to turn in a definite direction. Stimulation of this sense organ on the other side induces it to turn in the opposite direction.

For example, the experiments succeed very beautifully when one records bilaterally from the areas described by Munk[58] as centers for visual perception and then exposes alternately one or the other eye.

One notices that Fleischl confined his interest to the electrical responses evoked in the brain by sensory stimulation and makes no mention of ''spontaneous'' activity. In fact, in his sealed letter he stated that there is ''little or no movement on the galvanometer'' before stimulation is applied, and it is clear that his observations missed what is now known as the electroencephalogram. Fleischl apparently did not pursue this work and in just over a

[58]H. Munk. Über die Funktionen der Grosshirnrinde, Hirschwald, Berlin. 1881. (On the Functions of the Cerebral Cortex.)

year of making this claim, he died at the early age of 46, in the same year as Ernst von Brücke, his great teacher.

Fleischl's claim appeared on December 6, 1890, in the *Centralblatt* and drew a reply from Beck[59] in the following number, in which he expressed the view that surely the openly published report took precedence over the secret letter. Beck's letter was courteously phrased in rather flowery language, which nevertheless expressed the opinion that it was rather foolish to squabble about priority for the application of a known technique to a new question, for this scarcely could rate as a discovery. He was generous in giving credit for the basic discovery to Du Bois-Reymond, and for the idea of applying the method to the brain to Cybulski. It is noticeable that although he had made a discovery (of the ongoing electrical activity) he did not claim it as a separate finding. He wrote:

> Nature was held and still holds in her lap innumerable riddles under the seal of secrecy. It makes no difference for science whether these riddles are kept secret under the seal of Nature herself or under that of the Imperial Academy of Sciences in Vienna. Priority for discovery, in my opinion, belongs to the one who breaks the seal of Nature on a secret without them putting it under a new seal.
>
> But, in any case, the fight about priority, in this particular instance at least, seem superfluous. This is not a question of discovery but of the application to the solution of new problems of a method already known. I was drawn into these experiments by a prize contest on the following theme which had been set up by Professor Cybulski and some others of this Medical Faculty in October 1888. The subject of the theme for the prize was:
>
> "To determine whether it is possible to demonstrate a state of activity in nerve centers by using the so-called negative variation, and if positive results are obtained to find by its use: first, the localization of the reflex centres in the spinal cord, the visual centers in the cortex and the automatic centers in the medulla oblongata."
>
> In May, 1890, the Medical Faculty awarded the first prize to my work. As you can see from the above theme, the question of determining localization was conceived by Professor Cybulski and its solution based on facts already known. The priority for carrying out this idea doubtless belongs to the person who has worked on the Theme, who has got results and has published them.
>
> If one is going to discuss priority in this case it was certainly Professor Du Bois-Reymond who gave us the methods for experimenting on the electrical phenomena in the animal organism, Professor Hermann who demonstrated the existence of action currents in nerve, and Professor Sechenov who observed action currents in the medulla oblongata.
>
> In closing may I be permitted to express here my pleasure that my experiments so soon found confirmation from such an outstanding scientist as Professor Fleischl von Marxow.

One might have expected that this would be the end of the matter, but the following January, the editor of the *Centralblatt* published a letter he had received from the English physiologists Gotch and Horsley.[60] These two scientists had for some time been stimulating peripheral nerves and recording from the spinal cord and vice versa. They had also made some classical studies of the effects of stimulating the cortex and observing centrifugal effects. They wrote claiming priority over Beck and Fleischl, and in making this claim, Gotch and Horsley listed no less than 10 papers and demonstrations that, they declared, anteceded Beck's and Fleischl's work, yet, with one exception, in none of these is there

---

[59]A. Beck. Die Bestimmung der Localisation der Gehirn- und Rückenmark functionen vermittelst der elektrischen Erscheinungen. *Cbl. Physiol.*, 4:473–476 (1890). (The determination of localization of brain and spinal cord function by means of electrical phenomena.)

[60]F. Gotch and V. Horsley. Über den Gebrauch der Elektricität für die Lokalizierung der Erregungserscheinungen im Centralnervensystem. *Cbl. Physiol.*, 4:649–651 (1891). (On the use of electricity for localizing arousal phenomena in the central nervous system.)

any mention of recording from the brain. Even in the many experiments recounted in the lengthy Croonian lecture, published in the same year (but after their attention had been called to Caton's work), there is only one in which they recorded from the brain (the experiment is described as having been on a cat but is illustrated by a picture of a monkey's central nervous system).

In this single experiment, reported 16 years after Caton's work, and a year after Beck's, Gotch and Horsley were not very successful and stated that their results with retinal stimulation were capricious. They did not mention the ongoing oscillations and, from their publications, it would appear that they were recording demarcation currents and did not place both electrodes on the cortical surface. In acknowledging, in this later paper,[61] the work of Caton and of Beck, Gotch and Horsley expressed the opinion that the changes these scientists had observed in the electrical potentials of the brain were more likely to derive from white matter than from gray.

The controversy in the *Centralblatt* was cut short by a brief letter in German from Caton[62] drawing attention to his early work and quoting verbatim the first report of it that had appeared in 1875. In the course of this rather lengthy letter, Caton wrote:

> In the year 1875 I gave a presentation before the Physiological Section of the British Medical Association in which electrical currents of the brain in warm-blooded animals were demonstrated and, in addition, their undoubted relationship with regard to function was established. May I be permitted to draw your attention to the following publication (*Brit. Med. J.*, 1875, 2, 278).

At the end of this quotation he wrote:

> In the transactions of the 9th Medical Congress in Washington there is yet another published communication (Vol. 3, 246) under the title "Researches on electrical phenomena of cerebral grey matter."
>
> It is by no means my intention to detract from these learned physiologists, nevertheless I myself have made these observations, . . . as described above, I have published them, so I think it must be conceded that I am already an earlier discoverer.
>
> Respectfully,
>
> Richard Caton, M.D.
Professor of Physiology at Victoria University

Liverpool 1891

This letter was written in February 1891, and in the same year Francis Gotch succeeded to the Chair of Physiology in Liverpool from which Caton had resigned. That it was his own predecessor's work of which Gotch was ignorant adds another strange detail to the tangled story of the claims made with Horsley for priority in this field.

It is perhaps of interest to note the types of instruments being used by these early scientists whose results were drawn together by this controversy. Caton used a Thomson reflecting galvanometer, Beck a Wiedermann galvanometer, Gotch and Horsley a Lippmann capillary electrometer with optical magnification, though in their later work they followed Caton's lead and used a Thomson galvanometer.

What emerges from the melée of claims by the Austrian, English, and Polish workers is

---

[61]F. Gotch and V. Horsley. On the mammalian nervous system, its functions and their localisation determined by an electrical method. *Phil. Trans. R. Soc.*, 182B:267–326 (1891).

[62]R. Caton. Die Ströme des Centralnervensystems. *Cbl. Physiol.*, 4:758–786 (1891). (The currents of the central nervous system.)

FIG. 91. Two early workers in the electrical activity of the brain. **Left:** Ernst Fleischl von Markow (1846–1892). **Right:** Vasili Yakovich Danilevsky (1852–1939).

that almost certainly the first scientist to follow Caton in observing electrical potentials of the brain was not one of them, but a Russian, Danilevsky[63] (Fig. 91), who incorporated his results in a doctoral dissertation in 1876, printed as a thesis in 1877.

BIBLIOGRAPHY

Richard Caton (1842–1926)

Selected Writings

On the best methods of studying transparent vascular tissues in living animals. *J. Micr. Sci.* [New Ser.], 10:236–241 (1870).

Migration theory. *Q. J. Micr. Sci.* [New Ser.], 10:417 (1870).

Contributions to the cell-migration theory, *J. Anat. Physiol.*, 5:35–47 (1871).

On the electric relations of muscle and nerve, *Liverpool Med. Chir. J.,* (1875).

The electric currents of the brain, *Br. Med. J.*, 2:278 (1875).

Interim report on investigation of the electric currents of the brain, *Bri. Med. J.*, Supplement L:62–65 (1877).

---

[63]V. Y. Danilevsky. Investigations into the physiology of the brain (in Russian). Thesis. University of Kharkov, 1877. (A Lippmann electrometer indicates changing current by fluctuation of a mercury surface. This was probably too sluggish to react to ongoing brain potentials. The Thomson galvanometer could respond to the six-cycle activity of the rabbit's brain. [Note courtesy of Professor L. Geddes]).

Description of a new form of recording apparatus for the use of practical physiology classes, *J. Anat. Physiol.*, 22:103–106 (1887).

Researches on electrical phenomena of cerebral grey matter, *Ninth International Medical Congress, Trans Cong.* 3:246–249 (1887).

Vrach, 8:881 (1887). An abstract of Caton's paper given at Ninth International Medical Congress (in Russian).

On the variations produced in the electrical condition of the viscera by stimulation of adjacent cutaneous nerves (abstract). *Proc. Liverpool Biol. Soc.*, 3:113–114 (1889).

Die Ströme des Centralnervensystems (Letter to the editor, written Feb. 1891). *Cbl. Physiol.*, 4:785–786 (1891).

Typhoid and Salads (Letter). *Br. Med. J.*, 1171 (1891).

Remarks on the symptoms and treatment by intestinal antisepsis of enteric fever. *Br. Med. J.*, 2:165–167 (1892).

Two cases of lead poisoning. *Br. Med. J.*, 2:1371 (1893).

Notes on acromegaly. *Liverpool Med. Chir. J.*, 13:369–374 (1893). Notes on a case of acromegaly treated by operation. *Br. Med. J.*, 2:1421–1423 (1893) (with F. T. Paul).

Russia and the Twelfth International Medical Congress. *Liverpool Med. Chir. J.*, 18:119–130 (1898).

The Temples and Rituals of Asklepios. Lectures to the Royal Institution. Printed in *Otia Merseiana*, 1899.

Prevention of Valvular Disease of the Heart. Cambridge University Press, Cambridge, 1900.

Cardiac dilatation and hypertrophy. *Practitioner*, 15:35–42 (1902).

Acquired differences in structure and function between the right and left sides of the body (Inaugural address as president). *Proc. Liverpool Biol. Soc.*, 17:3–14 (1902).

Delphi, Delos, and Cos. Lectures published in *Aegean Civilizations*, edited by Sir Henry Lunn, pp. 143–161, 1925.

The temples, hospital and medical school of Cos. *Seventeenth International Medical Congress*, Section on History of Medicine, pp. 19–23, 1913.

Adolf Beck (1863–1942)

Selected Writings

Badania poczucia smaku u osoby pozbawionej jezyka. *Rozpr. Wydz. mat.-przyr. Polsk. Akad. Um.* (Polish Academy of Sciences, Proceedings of the Faculty of Mathematics and Natural Sciences), 18:207–216 (1888) (with N. Cybulski). (Research on the sense of taste in an individual deprived of a tongue.)

O pobudiwosci róznych miejsc tego samego nerwu. *Rozpr. Wydz. mat.-przyr. Polsk. Akad. Um.*,15:165–195 (1888). (On the Excitability of Various Parts of the Same Nerve.)

W sprawie lekarzy szkolnych. *Przegl. Lek.*, 28:29 (1890). (Concerning school doctors.)

Die Bestimmung der Localisation der Gehirn- und Rückenmarkfunctionen vermittelst der electrischen Erscheinungen. *Cbl. Physiol.*, 4:473–476 (1890). (The determination of localization of brain and spinal cord function by means of electrical phenomena.)

Die Ströme der Nervencentren (Letter to the editor). *Cbl. Physiol.*, 4:572–573 (1890). (The Currents of the Nerve Centers.)

Oznaczenie lokalizacyi z mózgu i rdzeniu za pomoca zjawisk elektrycznych. Presented October 20, 1890. *Rozpr. Wydz. mat.-przyr. Polsk. Akad. Um.*, Series II, I: 186–232 (1891). (Determination of localization in the brain and spinal cord by means of electrical phenomena.)

Dalsze badania nad zjawiskani elektrycznymi w korze mogowej u małpy i psa. *Rozpr. Wydz. mat-*

przyr. Polsk. Akad. Um., 32:369–375 (1891) (with N. Cybulski.) (Further Research on the Electrical Phenomena of the Cerebral Cortex in Monkeys and Dogs.)

Weitere Untersuchungen über die elektrischen Erscheinungen in der Hirnrinde der Affen und Hunde. *Cbl. Physiol.*, 6:1–6 (1892) (with N. Cybulski). (Further investigations on the electrical phenomena in the cerebral cortex of apes and dogs.)

Obecmy stan nauki o localizacyi czynnosci kory mózgowej. *Kosmos* (Journal of the Polish Biological Society in the name of Copernicus), 18:1–19 (1893). (The present scientific status of localization of functions in the cerebral cortex.)

O fizyologii odruchow wykład habilitacyjiny (in three parts). *Przegl. Lek.*, 33:189–191, 209–212, 223–225 (1894). (The physiology of reflexes.)

Hermann Helmholtz (in two parts). *Przegl. Lek.* 33:548–550, 559–561 (1894).

Zjawiska zyciowe i̇ sposoby ich badania (wykład wstepny). *Przegl. Lek.*, 34:633–637 (1895). (The vital phenomena and methods for their investigation—introductory lecture.)

O zmianach cisnienia krwi w zylach. *Rozpr. Wydz. mat.-przyr. Polsk. Akad. Um.*, 7:23–62 (1895). (On variations in venous pressure).

Elektrische Erscheinungen in der Grosshirnrinde beim Affen. 2nd Internat. Physiol. Cong., Berne, 1895. Abstract. *Cbl. Physiol.*, 9:474 (1895). (with N. Cybulski). (Electrical phenomena in the cerebral cortex of the ape.)

Dalsze badania zjawisk elektrycznuch w korze mózgowej. *Rozpr. Wydz. mat.-przyr. Polsk. Akad. Um.*, 12:174–257 (1896) (with N. Cybulski). (Further studies on the electrical phenomena of the dog's brain.)

Pomiary pobudliwosci róznych miejsc nerwu za pomoca rozbrojen kondenzatora. *Rozpr. Wydz. mat.-przyr. Polsk. Akad. Um.*, II:234–246 (1897). (Measurements of the excitability of various parts of a nerve by means of condenser shocks.)

Die Erregbarkeit verschiedener Stellen desselben Nerven. *Arch. Anat. Physiol.*, 415–425 (1897). (The sensitivity of various parts of the Same Nerves.)

Zur Innervation der Speicheldrüsen. *Cbl. Physiol.*, 12:33–37 (1898). (On the innervation of the salivary glands.)

Zur Untersuchung der Erregbarkeit der Nerven. *Arch. Physiol.*, 72:352–359, 1898. (On the investigation of the sensitivity of nerves.)

Badania nad unerwieniem gruczolow slinowych. *Rozpr. Wydz. mat.-przyr. Polsk. Akad. Um.*, 15:13–62 (1899). (Research into the Innervation of the Salivary Glands.)

Über die bei Belichtung der Netzhaut von Eledone moschata entstehenden Aktionsströme. *Arch. Physiol.*, 78:129–162 (1899). (On the action currents resulting in light falling on the retina of *Eledone moschata*.)

O zjawiskach elektrycznuch wywołanych przes oświetlenie siatkówki glowonoga Eledone moschata. *Kosmos*, (Lwów), 25:1–35 (1900). (On the electrical phenomena evoked by illuminating the retina of the cephalopod: *Eledone moschata*.)

O dzialaniu promieni radu na nerwy obwodowe. *Rozpr. Wydz. mat.-przyr. Polsk. Akad. Um.*, Ser. III. 5B:111–122 (1905). (The effect of radium irradiation of peripheral nerves.)

Zhawiska elektryczne kory mózgowej po czesciowem jej zniszczeniu. Przyzynek do lokalizacyi czucia bólu. *Rozpr. Wydz. mat.-przyr. Polsk. Akad. Um.*, Ser. III, 5B:319–355 (1905). (Electrical phenomena of the brain cortex after its partial destruction. A contribution to the localization of the sensation of pain.)

Über die Ermüdbarkeit des Nerven. *Arch. Physiol.*, 122:585–592 (1908). (On nerve fatigue.)

Eine Beobachtung über Reflexerscheinungen am Hintertier. *Arch. Physiol.*, 129:415–424 (1909) (with G. Bikeles). (An observation of reflex phenomena in the rear of the animal.)

Napoleon Cybulski (His Silver Jubilee at the University of Jagiellonski). *Lwów Medical Weekly* (1910).

O sensoryczmej cynnosci srodkowej czesci mózdzku (robaka). *Rozpr. Wydz. mat.-przyr. Polsk.*

Akad. Um., Ser. III, 10B:473–481 (1911) (with G. Bikeles). (On the sensory activities of the central part of the cerebellum [vermis].)

O.t.zw. odruchach dotykowych Monka i odruchu skórnym podeszwowym. *Rozpr. Wydz. mat.-przyr. Polsk. Akad. Um.*, Ser. III. 10B:686–698 (1911) (with G. Bikeles). (On the so-called Munk's tactile reflexes and the reflex of the skin on the sole of the foot.)

O ruchach odruchów rdzenowych i ruchach ogólnych (prychpalnych wedlug Monka). *Rozpr. Wydz. mat.-przyr. Polsk. Akad. Um.*, Ser. III, 10B:699–715 (1911) (with G. Bikeles). (On the spinal reflex movements and the phylogenetically older movements.)

Die sogenannten Berührungsreflexe Munks und die reflektorische Zehenbeugung bei Reizung der Fußsohle, *Arch. Physiol.*, 137:34–44 (1911) (with G. Bikeles). (The so-called touch reflexes of Munk, and reflex toe bending on stimulus of the sole of the foot.)

Zur Lehre Munks über Beginn und Reihenfolge in der Ausbreitung der Bewegungen bei Rückenmarksreflexen, wie bei Tatigkeit der sogenannten "Prinzipalzentren." *Arch. Physiol.*, 137:45–47 (1911) (with G. Bikeles). (On Munk's theory of the beginning and sequence in the spread of movements in spinal cord reflexes, as in the activity of the so-called "principal centers.")

Die Ausbreitung des Reflexbogens im Rückenmark festgestellt vermittels Untersuchung der Aktionsströme. *Arch. Physiol.*, 140:24–36 (1911) (with G. Bikeles). (The extent of the reflex arc in the spinal cord, determined by investigating the action currents.)

Über elektrische Erscheinungen im Zentralnervensystem des Frosches. *Arch. Physiol.*, 155:461–470 (1914). (On electrical phenomena in the central nervous system of the frog.) *Fizyologia Czlowieka.* 2 vols. Kraków, 1915 (with N. Cybulski). (Physiology of Man.)

O dwukierunkowem przewodzeniu nerwów. *Rozpr. Wydz. mat.-przyr. Polsk. Akad. Um.*, Ser. III, 17B:1–13 (1917). (Two-dimensional conduction in nerves.)

Napoleon Cybulski: Posthumous Reminiscences. Warsaw, 1919.

Uniwersytet Jana Kazimierza we Lwowie podczas inwazji Rosyjskiej w roku, 1914, 1915. *Nakładem Senatu Akad. Uniw. Lwów*, 1935. (The Univeristy of Jan Kasimir in Lwów during the Russian invasion 1914–1915.)

Suggested Readings

A classic in electroencephalography, Beck's doctoral thesis (English translation by W. Binetz and J. S. Barlow). *Acta Neuro. Exp.*, Suppl.3 (edited by M.A.B. Brazier) (1973).

Brazier, M. A. B. *A History of the Electrical Activity of the Brain.* Pitman, London (Macmillan, New York) 1961.

Zakrzewska, J. Beck. A daughter's memories of Adolf Beck. *Acta Neurol. Exper.*, 3:57–58 (1973).

Gambaroglii, K. Distinguished Polish Physiologist and Physician, Abram Adolf Beck (on the hundreth anniversary of his birth) (in Russian). *Klin. Med. Moscow*, 42:146–148 (1964).

Ernst Fleischl von Marxow (1846–1892)

Selected Writings

Untersuchungen über die Gesetze der Nervenerregung. IV. Abhandlung. Der interpolare Elektrotonus. *Wiener Akademische Sitzungsberichte* (1878). (Investigations on the laws of nerve arousal. IV: Interpolar electrotonus.)

Mittheilung, betreffend die Physiologie der Hirnrinde. *Cbl. Physiol.*, 4:537–540 (1890). (Report concerning the physiology of the cerebral cortex.)

Untersuchungen über Milzbrandinfection bei Fröschen und Kröten. *Fortschr. Med.*, 4, 2:45; *Cbl. Physiol.*, 5:245 (1891). (Investigations on anthrax in frogs and toads.)

Historisch-physiologische Notizen. *Cbl. Physiol.*, 19:543–582 (1891). (Historical note on physiology.)

CHAPTER XII

The Great Russian Schools Explore the Field of Neurophysiology

Ivan Mikhailovich Sechenov (1829–1905)

Sechenov came in the era when Russian scientists were going to Western Europe in their postgraduate days. In the previous period, when Peter the Great was planning the Academy of Sciences, he had encouraged scientists from Europe to come to Russia to help set science on its way. Among the men who came were the Swiss mathematicians, the two sons of Jean Bernoulli,[1] and Leonhard von Euler,[2] and later the distinguished German biologist, Karl Ernst von Baer.[3] By Sechenov's day, Russia was training its own scientists, but many were going abroad for the more advanced stages of their education.

Sechenov was born in 1829 in the middle of the Volga region of Russia. His mother was a peasant, but his father was an educated man of means, a retired army officer. Sechenov received a good education as a schoolboy at the Military Engineering School in St. Petersburg, and after a period of military service, including four years in Kiev in the Ukraine (where one of his fellow cadets was Dostoyevsky), he entered the University of Moscow in 1854. Moscow's university had been opened by the Academy of Sciences in 1755, largely through the influence of Lomonsov,[4] the talented scientist and poet. Sechenov trained there as a physician, qualifying in medicine in 1856. Then immediately, and on his own initiative, he went to Europe. As a medical student, he had been interested in physiology and had been dissatisfied with the sterile way it was being taught from books only, without any experimentation. In 1856 he went to the greatest center of physiology of the time, which was the laboratory of Johannes Müller in Berlin, though Müller himself was coming to the end of his life. Müller died in 1858, just two years after Sechenov had come

[1]Jean Bernoulli (1654–1705).

[2]Leohnard von Euler (1707–1783).

[3]Karl Ernst von Baer (1792–1876).

[4]Mikhail Vasileyvich Lomonsov (1711–1745).

to Berlin, but Sechenov was able to hear him lecture and to take his course. There he came under the tutelage of Müller's assistant, Du Bois-Reymond.

Du Bois-Reymond devoted his working life to the study of animal electricity, and during the period when Sechenov was in Berlin, the burning issue was whether or not nervous activity was an electrical event. Du Bois-Reymond had demonstrated unequivocally the action current of the nerve, the cornerstone on which electrophysiologists built. In 1858, however, Sechenov was not yet an electrophysiologist.

Sechenov spent four consecutive years in Europe, and from Berlin he went to Vienna. There he met and studied under Carl Ludwig. Ludwig was an older man than Sechenov, but they became intimate, lifelong friends. Their correspondence has largely been preserved, and some of it has been published. These letters show that Ludwig was a guiding influence on Sechenov throughout his life. Then, from studying with Ludwig in Vienna, where he pursued his work on alcohol and blood gases, Sechenov went to Helmholtz's laboratory in Heidelberg, where he was drawn into the master's work on the eye.

These men were the giants of their time. For one man to have had the experience of training with Müller, Du Bois-Reymond, Ludwig, Helmholtz, and with Bunsen, who was also in Heidelberg at this time, was an extraordinary opportunity for advanced learning. The experience Sechenov had during these four years undoubtedly shows how Russian neurophysiology stemmed from Western European roots.

In Europe Sechenov later went to Graz where he worked with Rollet,[5] and there his interest focused on experimental stimulation of the central nervous system using the experience he had gained in Berlin on the peripheral system. A publication resulted entitled (in translation) "On the electrical and chemical stimulation of the sensitive spinal cord of the frog."[6] Sechenov's experiments that proved so crucial to his future thinking were on the effect on reflex movements caused by salt crystals placed at various levels of the transected neuraxis. His preparation was the decapitated frog whose toe he dipped in acid (a procedure that had been developed by Türck).[7] He timed the interval between stimulus and onset of withdrawal of the frog's foot by counting the beats of a metronome. In this way he got some index of the degree to which application of the salt crystal to the brainstem slowed withdrawal. Sechenov interpreted lengthening of withdrawal time as inhibition of reflex activity. The selection of a salt crystal as a stimulus seems strange in the hands of a pupil of Du Bois-Reymond's and is reminiscent of Marshall Hall's use of it half a century earlier to study depression and augmentation of spinal reflexes.[8] Only later did Sechenov use electrical recording in his experiments on the "spontaneous" variations of spinal cord potentials that he regarded as signs of activity in the spinal centers.

While still in the West, Sechenov wrote another report on this experiment, a copy of which is in the library of the Josephenum in Vienna. It is entitled "New Studies in the Brain and Spinal Cord of the Frog." This was written jointly with a medical student named B. Paschutin and is dated 1865.[9]

---

[5]Joseph Pierre Rollet (1824–1894).

[6]I. M. Sechenov, Commentarri. *Acad. Sci. St. Petersburg,* 1863.

[7]Ludwig Türck (1810–1868). Über die Haut-Senisb der einzelnen Rückenmarksnervenpaare Denkschrift. *K. Akad. Wis. Wien. math-nat. Cl,* (29):299–326 (1868). (Investigation of the cutaneous distribution of the separate pairs of spinal nerves.)

[8]Marshall Hall. Ueber retrograde Reflextätigkeit im Frosche. *Arch. Anat. Physiol. Wiss. Med.,* 486–489 (1847). (On retrograde reflexivity in the frog.)

[9]I. M. Sechenov and B. Paschutin. Neue Versuche über die Hemmungsmechanismen im Gehirn des Frosches, als Erwiderung auf die im Laborat. des Herrn. M. Schiff ausgeführten Untersuchungen. *Bull. Acad. Sci.* (Series 3) 8:145–162 (1865). (New studies in the brain and spinal cord of the frog.)

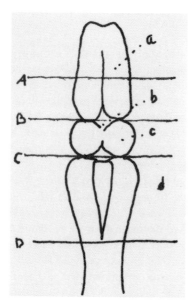 

FIG. 92. **Left:** Sechenov's sketch of his experiment showing levels of the frog's brain at which he made cuts. Cuts were made at *A*, *B*, *C*, and *D*. The sketch indicates *A*, the hemispheres; *B*, the optic thalamus; *C*, the quadrigeminae and *D*, the medulla oblongata. (From: I. M. Secheno. *Physiologische Studien über die Hemmungsmechanismen für die Reflextätigkeit des Rückenmarks in Gehirne des Frosches.* Hirschwald, Berlin 1863. [Physiological Studies on the Inhibiting Mechanisms for a Reflex Activity of the Spinal Cord in the Frog's Brain].) **Right:** Sechenov and his experiment. (From: *The Selected Works.* State Publishing House, Moscow, 1935.)

In 1860 Sechenov went home and introduced neurophysiology into Russia. He took back with him a Du Bois-Reymond induction coil and a galvanometer. These were the first instruments of this kind in his country. He began to teach electrophysiology, and his lectures at the Medico-Chirurgical Academy, where he was appointed Assistant Professor of Physiology, proved immensely popular not only with his students but also with his superiors. The Academy of Sciences gave him a prize for his lectures, in recognition of the quality of instruction that he was giving.

Sechenov concluded his report[10] of his finding of a special inhibitory center by saying that although this was discovered in frogs, there must be a generalized phenomenon in the physiology of the central nervous system of all animals, but he did not himself seek this. A universal significance of his discovery by experiments on frogs (Fig. 92) is evident by the fact that these very experiments supply the foundation of his later remarkable monograph *Reflexes of the Brain*.[11] L. N. Simonov,[12] one of his followers, provided the first experimental proof of the broader significance of the phenomena discovered by Sechenov, for in 1866 he published an article in which he demonstrated the presence of mechanisms inhibiting the reflex processes also in mammals. The article appeared in the *Military-Medical Journal* under the title (in translation) ''Experimental proof of the existence of central

[10]I. M. Sechenov. *Physiologische Studien über die Hemmungsmechanismen für die Reflextätigkeit des Rückenmarks in Gehirne des Frosches.* Hirschwald, Berlin, 1863. (Physiological Studies on the Inhibiting Mechanisms for a Reflex Activity of the Spinal Cord in the Frog's Brain.)

[11]I. M. Sechenov. *Reflexes of the Brain* (in Russian). St. Petersburg, 1866.

[12]I. N. Simonov. *Military-Medical Journal* (Part 97), 2:1–31, 67–92 (1866). (Experimental Proof of the Existence of Central Inhibition of Reflexes in Mammals.)

inhibition of reflexes in mammals.'' Neither man had any concept of the anatomical path this would demand.

Although at this stage his own experimental evidence seemed slender, Sechenov must have pondered its meaning in much wider terms, for on his return to Russia a year later, he published as a series of articles the essay that proved to be so influential in Russian physiology. This essay, *Reflexes of the Brain,* was later published as a book after a stormy period during which efforts were made to suppress its publication and censure its author. This opposition was stirred by Sechenov's assertion that all higher brain function was a material reflex consisting of three sectors: an afferent initiation by sensory inflow, a central process entirely subject to physical laws, and an efferent component resulting in a muscular contraction. All reactions, however, might be described in common parlance as pleasure, fear, distress, or other descriptive terms, and they were, according to Sechenov, in essence muscular in expression. During the passage of the inflow through the central portion of the arc, there could either be excitation, which would augment the reflex motor response (as in so-called emotional states), or inhibition, which would decrease the reflex movement, the resultant being "rational" controlled behavior. It is interesting that Sechenov conceived that inhibition could be learned and that with maturity an increase in the degree of inhibition exerted was achieved.

Thus, Sechenov concluded that all human behavior was a balance between inhibition and excitation operating mechanically at the central link of the reflex arc. A so-called willed movement, according to him, only apparently lacked the first component of the arc, its afferent inflow being material memory traces left by external stimuli in the past. In elaborating this part of his theory, Sechenov approached the concept of the conditional reflex, for he postulated that the memory trace of a past sensory experience could be evoked by the recurrence of any fraction of it, even if this fraction were quite insignificant and unrelated in its apparent meaning. This is essentially the principle underlying the formulation of conditional reflex theory, namely, the potency of an indifferent external stimulus, provided it is repeatedly time locked to the original experience. One further point should be noted in this early attempt to relate mental processes to brain physiology: Sechenov believed that man had the special faculty of increasing the degree of inhibition exerted at the central link until a level of total inhibition of the efferent discharge was reached, and he held that thought was an example of this condition.

The Council of the Censorial Department in Moscow declared that the book must be destroyed, but Sechenov had a powerful friend in the Minister of Justice who succeeded in getting the proceedings dropped. There seemed to have been a publications censor, and it was his responsibility to see to it that nothing was published that would tend to oppose the prevailing orthodoxy. Sechenov came into trouble because of the obvious materialistic character of his book, which had been prohibited by the Censorial Committee of St. Petersburg.

The move to censure came during the period when he was most concerned with the physiology of the central nervous system that he was then teaching. He was evidently a very independent and courageous person who went his way if he thought it was right, irrespective of the storms around him. He was disturbed by the weakness of Russian textbooks in physiology, so he wrote a book called *A Course in Physiology of the Nervous System.* At the same time he undertook the translation into Russian of some of the textbooks that he had used in Germany. He translated, for example, Hermann's *Handbook of Physiology.* When he had translated this into Russian, he went on to translate Willy Kühne's book on physiological chemistry. He was anything but an ivory tower scientist. On the contrary, he was a great believer in broadening education and was always trying to get more people into the fold, more people to be interested in science. He even supported and believed in the education of women. One of his students became the first woman doctor in Russia, and she also became Madame Sechenov.

About this time (1870), Sechenov took a very strong stand because of the treatment that was given to E. Metchnikov,[13] who had been named for a Chair in Zoology at the Medico-Chirurgical Academy but had been turned down. Sechenov suspected that there was some anti-Semitic feeling and, in protest, he resigned from his own Chair. Metchnikov moved to Odessa and later to the Pasteur Institute in Paris. In 1908 he received the Nobel Prize jointly with Paul Ehrlich.

Sechenov left Russia and went to the Ukraine hoping to work in Odessa, but on leaving Moscow he was not immediately appointed as a professor at the University of Novorossiisk in Odessa. This finally came to him in 1871.[14] The delay was because of political reasons, for Sechenov's views were thought to be too materialistic. The kind of work Sechenov pursued in Odessa was very close to the interests of Claude Bernard, being on the role of carbon dioxide in the blood. In his autobiography written years later he said:

> For almost five years I was engaged here with the problem concerning the state of CO_2 in the blood, and this apparently simple problem required for its solution not only experiments with eight primary component parts of the blood but also with various combinations of one with the other, and to an even larger extent, experiments with a great many saline solutions.

During his years at Odessa Sechenov made a visit (in 1872) to Claude Bernard in Paris who reinforced his interest in blood gases.

Sechenov, in spite of his reputation for using a salt crystal as a stimulus to the nervous system, had used electrical stimulation in his research project pursued in 1870 (though not published until two years later).[15] He had presumably done the work in Moscow using the electrical equipment there that he had brought from Du Bois-Reymond's laboratory, though the paper was published in Odessa and was read to the Novorossiisk Society of Natural Sciences. In the paper the decrease of effect in response to excessively frequent stimulations of the nerve was noted, and the inference is made that the process of inhibition can develop not only in the central nervous system but under certain conditions—also in a peripheral nerve. In very flowery language Sechenov wrote:

> All the infinite variety of efferent cerebral activity necessarily converges to a single form: muscular contraction. When a child chuckles on seeing a plaything, when Garibaldi smiled while being expatriated for his great love for his country, when a young girl trembles in thinking of her first love, when Newton creates his laws of the universe and writes them down on paper—in all cases the final act is a muscle movement.

While in Odessa Sechenov wrote about more of the experimental work he had done in Europe. This he sent to Pflüger's *Archiv* in 1872. In this paper his familiarity with the work of Heidenhain and that of Bernstein is evident. The problem examined was the development of tetanus and the comparative influences of frequency and strength of the applied

---

[13]Elie Metchnikov (1845–1916). (There is now a Metchnikov Institute of Animal and Human Physiology at Odessa.)

[14]D. S. Voronstov, V. M. Nikitin, and P. M. Serkov. *Sketches from the History of Physiology in the Ukraine.* Akad. Sci. Ukrainian SSR, Kiev, 1959. (In Ukrainian).

[15]I. M. Sechenov. Einige Bemerkungen über das Verhalten der Nerven gegen sehr schnell folgende Reize. *Pflüger's Arch. f.d. ges. Physiol.*, 5:114–119 (1872). (Some remarks on the behavior of nerves following rapidly repeated stimuli.)

FIG. 93. Left: Ivan Michailovich Sechenov (1829–1905). Working in the laboratory on the analysis of blood gases. **Right:** While training with Ludwig, Sechenov taught himself the art of glass blowing, using this table, which has been preserved in Leipzig. (From: Cesnakova and Lindemann. *Wiss. Zwit.,* 2:315– 323, 1970.)

current. Also, while he and Metchnikov were in Odessa, they reported some work they had done together in Moscow in 1868. This research was on the action of stimulating the vagus nerve and was later published in the *Centralblatt* in 1873.[16]

Sechenov left the Ukraine in 1876, having received an offer of a professorship in St. Petersburg. There he taught a course in experimental physiology that was really quite revolutionary in its content, for he included chemistry and physics and integral calculus, differing greatly from the courses of his student days. He contributed greatly toward the teaching of this subject in Russia. He himself at this period went back to his old interest in blood gases (Fig. 93) and between 1877 and 1888 he published 24 papers in the field. It was this subject in which he published most, but there are other works of his that are of special interest to neurophysiologists. He wrote an essay called *The Elements of Thought,* which has been translated into English. There is also another essay he wrote entitled *Who Must Investigate the Problems of Psychology, and How?* At the University of Novorossiisk Sechenov had several students who later distinguished themselves in the field of neurophysiology; one of these was Bronislav Fortunatovich Verigo,[17] who was later to be a pioneer in electroencephalography.

Sechenov's activities were manifold and not restricted to the laboratory. He took an active interest in social reform and in the education of the working man. The part of his scientific life that was concerned with the nervous system marks him as Russia's first great neu-

---

[16]E. I. Metchnikov and I. M. Sechenov. *Centralblatt,* January 1873. (Concerning electrical and chemical stimulation of sensory cerebrospinal nerves in the frog.)

[17]Bronislav Fortunatovich Verigo (1860–1925).

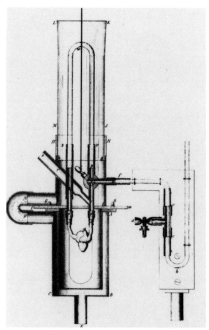

FIG. 94. **Left:** Elie Cyon (1842–1912). Professor at the University of St. Petersburg. **Right:** Apparatus designed and used when Cyon was working with Ludwig in Leipzig. The purpose was to pass current through the surviving frog's heart. (From: E. Cyon and Carl Ludwig. *Arch. Physiol. Acad. Leipzig,* 1:128, 1867.)

rophysiologist. In 1888 he stepped down, because of an objection of the Ministry of Education to his election to the Academy of Sciences. He became Instructor in Physiology in Moscow and then advanced to Professor there from 1891 to 1901. From 1903 to 1904 he was an instructor in the school of teaching by electrical methods in Moscow. He died there in 1905. The 30-year younger Pavlov succeeded him in the Chair of Physiology and wrote an obituary.

Elie Fadeevich Cyon (1842–1912)

One of the earliest Russian physiologists to go to Western Europe was Elie Cyon who worked in Paris and Berlin, though the major tie he made was with Carl Ludwig. Cyon, schooled in Chernigov, went at the early age of 17 to the University of Kiev (named for Prince Vladimir). In 1864 he moved to the University of Berlin to take courses in medicine, and he received a degree as a doctor of medicine from St. Petersburg in 1865.

Cyon was only 23 when he first went to Ludwig's laboratory but returned many times because they shared an intense interest in the nervous control of the heart (Fig. 94). Cyon had studied this subject in Du Bois-Reymond's laboratory where he worked with his brother M. E. Cyon, who later became a doctor.

In 1866 a paper authored by Ludwig and Cyon on this subject won a prize from the Académie Française. It was published in 1867.[18] They were also concerned with the result

[18]E. Cyon and C. Ludwig. Die Reflexe eines sensiblen Nerven des Herzens auf die motorischen der Blutgefässe. Arbeiten aus der Physiol. *Anstalt Leipzig,* 128 (1866) and *Arch. Physiol. Acad. Leipzig,* 1:128 (1867). (The Reflex Action of the Sensory Nerve in the Heart on the Motor Nerves of the Blood Vessels.)

of central excitation of the vagus. They cut this nerve in a rabbit and demonstrated a fall in blood pressure when they stimulated the central end. They called this afferent unit the "depressor nerve." Other experiments of Cyon's that he carried out with Ludwig were on the effect of curare on the results they were getting by stimulating the depressor nerve. Their final conclusions were reported by Ludwig in their paper as follows:

> We now know of a new process by which the separate parts of the circulatory system adjust to each other. By this process we see that the main motor of the blood stream is able to adjust the resistances which it itself has to overcome. We may thus say that the heart, if lacking in propulsion power, or if overfilled, can not only change its rate of beat, but also reduce the resistance to its emptying.

The importance of these experiments that Cyon carried out in Ludwig's laboratory was that they were able to establish a regulatory mechanism for the blood pressure. The achievement of Cyon's work with Ludwig was the recognition of what is now called "feedback" regulation.

Cyon returned to St. Petersburg in 1868 and worked there in Sechenov's department. Sechenov thought very highly of him, and in a letter to the Medical Surgical Academy in 1870, he wrote, "Doctor Cyon, Professor of the Saint Petersburg University, is the one most deserving to take my seat." And indeed, when Sechenov left for Odessa, Cyon succeeded to his position.

Cyon brought with him from Berlin a determination to evict vitalism from physiology, and he attacked Pflüger[19] for his concept of a "soul" in the cerebrospinal fluid. He taught his students that "all neural activities can be traced to physical and chemical laws." We hear an echo here of Du Bois-Reymond.

Close as Cyon was to Sechenov, he had a major disagreement with him. When Sechenov developed his idea of reflexes in the brain, Cyon maintained that they did not occur above the spinal level. In all matters, teaching or psychology, Cyon distanced himself from that approach, deeming it unphysiological. His strong views did not, however, change the developing ideas of one of his pupils, Ivan Petrovich Pavlov.

In 1875 Cyon resigned from the university owing to severe disagreements with both staff and students. Unable to continue active research, he moved in 1880 back to Paris where he opened his own laboratory and continued to work on the nervous system. In the 1890s Cyon went to Switzerland to work with Adolf Oswald and added the endocrine system to his interests.[20]

Cyon continued to write in his later years, expanding his interests to such subjects as perception of space and the role the vestibular system plays.[21] On the latter he had corresponded with Helmholtz. Cyon's views on the role of the labyrinth raised both dissent (from Bechterev) and an agreement (from Ukhtomsky and later in 1959 from Beritoff).

Cyon died in Paris in 1912, alone and unpopular. Even the death notices reflected antagonism, including one by Metchnikov that described him as having "an angry personality and inability to elevate, even slightly, his moral point of view." Ukhtomsky was more

---

[19]E. Pflüger. Die sensorischen Functionen des Rückenmarks der Wirbelthiere nebst einer neuen Lehre über die Leitungsgesetze der Reflexionen. Hirschwald, Berlin, 1853. (The Sensory Functions of Spinal Cord Vertebrates, together with a New Theory of the Laws of Transmission of Reflexes.)

[20]E. F. Cyon and Adolf Oswald. Über die physiologischen Wirkungen einiger aus der Schilddrüse gewonnener Producte. *Pflüger's Arch. f. d. ges. Physiol.*, 83: (1901). (On the physiological effects of some products extracted from the thyroid gland.)

[21]E. Cyon. *The Ear—Orientation Organ in Time and Space* (written in French).

generous; "Cyon," he said, "should be remembered as a man with a great talent for teaching and research."

Sherrington[22] was kinder. In his famous Yale lectures, published as *The Integration of the Nervous System,* he gave Cyon[23] (whose adoption of the German honorific of "von" he repeated) the sole credit for discovering the depressor nerve.

Vasili Yakovich Danilevsky (1852–1939)

One of the most active biologists in Czarist Russia was Danilevsky. He was a Ukrainian born in Kharkov to the family of a watchmaker. He was educated first at the University of Kazan, the university where a young man named Vladimir Ilyich Ulyanov was to spend a brief period studying law before being expelled for his part in a student protest. This young law student was to become known to the world as Lenin.

Danilevsky finished his training at the university, which was founded in 1804, in Kharkov. He graduated there in 1874 and received a medical degree in 1876. Danilevsky's dissertation was entitled "Investigations into the physiology of the brain," and because of its strongly mechanistic approach, it caused considerable strong debate, but he was finally successful in defending it.

Following the pattern of young Russian scientists, Danilevsky went abroad first to study with Adolf Fick[24-26] in Würzberg, who was applying physics to the problems of muscle physiology. With this experience in biophysics, Danilevsky moved on to work with Fick's most famous pupil Ludwig, who was then in Vienna. Unfortunately, there his goal to do research on the brain was not encouraged by his chief. On his return to Kharkov Danilevsky entered a long, productive life devoted to teaching and research in physiology. In the latter he concentrated on neurophysiology (particularly on the brain), muscle physiology, and blood pressure. A great number of publications emerged from this work.

Danilevsky's experiments expanded on those of Hitzig in that he used electrical stimulation of the cerebral cortex as presaged by his early dissertation, for he explored centers on the cortex that on stimulation caused changes in heartbeat, respiration, pupil changes, and eye movements. Some of these had been reported earlier by Hitzig.[27]

Danilevsky was naturally attracted to the ideas of his contemporary Sechenov, who held that the brain was the substratum of all experience. Danilevsky had displayed the brain to have its own "spontaneous" activity and its response to metabolic change. In the 14th section of this thesis he stated that:

> Study of the electromotive properties of the brain caused by change in its metabolic condition (respiratory, circulatory changes) and by stimulation should serve as one of the chief goals in the physiology of the nervous system, for the definite changes in these properties provide an indicator of activity in a given part of the brain.

[22]C. S. Sherrington. *The Integration of the Nervous System.* Yale University Press, 1906.

[23]E. F. Cyon. *Berisachs Akad. Wiss. Leipzig,* 1865.

[24]A. E. Fick. *Untersuchungen über electrische Nervenreizung.* Braunschweig, Wiewig, 1864. (Investigations into the Electrical Stimulation of Nerves.)

[25]A. E. Fick. Über die Aenderung der Elasticität des Muskels während der Zuckung. *Pflüger's Arch. f. d. ges. Physiol.,* 4:301–315 (1871). (On the Change in Elasticity of the Muscle during Spasm.)

[26]A. E. Fick. Über die Wärmeentwicklung bei der Muskelzuckung. *Pflüger's Arch. f. d. ges. Physiol.,* 16:59–90 (1878). (On the generation of heat during muscle contraction.)

[27]E. Hitzig. Über die galvanischen Schwindelempfindungen und eine neue Methode galvanischer Reizung der Augenmuskeln. Verhandl. der Berl. med. Gesellsch vom. 19 Jan. 1870 in Berlin. *Klin. Woch.,* No. 11, 1870. (On galvanic sensations of faintness, and a new method of galvanic stimulus of the eye muscles. Meeting of the Berlin Medical Society, Jan. 19, 1870.)

Fifteen years later, when so much controversy broke out in the *Centralblatt für Physiologie*[28] as to who could claim priority for first observing these phenomena, Danilevsky wrote a fuller account to the editor of that journal. He reported in the German language the experiments he had done in 1876 in the course of his work for his thesis, stating clearly that he had at that time been unaware of Caton's previous work.

Danilevsky had used curarized dogs and had recorded their brain potentials "both superficially and deeply" by using nonpolarizable electrodes and a sensitive Du Bois-Reymond galvanometer. He too found a negative variation with various kinds of sensory stimulation and observed that these effects were obtained from the posterior lobes of the brain and in most cases from the side opposite to the one in which the excitation had been applied.

Because of the increasing interest in the subject of localization of sensory function in the brain, Danilevsky published a longer and more detailed paper in the journal *Fiziologicheskiy Sbornik*[29] that he had founded in 1888 in Kharkov. He described in some detail the experiments he had made on five dogs in 1876, and in reference to what is now called the electroencephalogram, he noted that the galvanometer needle oscillated as the circuit to the brain was closed, although he had not yet applied any stimulus. He wrote:

> . . . immediately the circuit was connected, independent or spontaneous currents appeared and the magnetic needle deviated to a greater or lesser degree, even though the animal had not yet been subjected to any external stimulation.

In this long paper of 1891 Danilevsky gave full credit to Caton, about whose original statement he made the comment that it was "distinguished by its unwarranted brevity." He also included long accounts of Beck's and Fleischl von Marxow's findings. In describing once again his own original experiments, he noted that he did not extend them further since they appeared merely to corroborate Caton's findings.[30,31]

It is interesting that on looking back over the intervening 15 years, Danilevsky expressed considerable uncertainty as to how the results should be interpreted. He stated that his original high hope had been to find an electrical reaction of the brain that correlated with emotional, psychic processes evoked in the animal by certain external stimuli. He expressed disappointment at having achieved so few of these goals; although he definitely observed marked changes on stimulation of his curarized animals, he noted that they showed great variability.

Danilevsky's original experiments had included acoustic stimuli—a shout, a whistle, and a gunshot—all of which evoked changes in the standing potential, though not always of the same polarity. He interpreted this variability as being in part because of electrode placement and of the condition of the animal. He obtained responses to cutaneous stimulation that were very complex, and his results and conclusions were clearly hampered by his having at that time no idea of the topography of the receiving areas, so that his electrode placement was often far from optimal for picking up responses. However, he was able to conclude:

---

[28]A. Beck. Die Bestimmung der Localisation der Gehirn- und Rückenmarkfunctionen vermittelst der elektrischen Erscheinungen. *Cbl. Physiol.*, 4:473–476 (1890). (The determination of localization of brain and spinal cord function by means of electrical phenomena.)

[29]Essays and works from the laboratories of A. Y. and V. Y. Danilevsky (in Russian). *Fiziologicheskiy Sbornik* (Physiological Collection). Vol. 1, Kharkov, 1888; Vol. II., Kharkov, 1891.

[30]V. Y Danilevsky. Zur Frage über die elektromotorischen Vorgänge im Gehirn als Ausdruck seines Tätigkeitszustandes. *Cbl. Physiol.*, 5:1–4 (1891). (On the question of electromotor processes in the brain as the expression of its state of activity.)

[31]V. Y. Danilevsky. Electrical phenomena of the brain (in Russian). *Fiziologischeskiy Sbornik* (Physiological Collection), 2:77–88 (1891).

And so, the above examples from my experiments conducted fifteen years ago, undoubt-edly show that the processes of excitation evoked in the brain by the direct influence of external sensory stimulation are accompanied by characteristic electrical phenomena. Hence we are justified in recognizing it as a highly probable hypothesis that physiological func-tional activity in nerve, brain (and other cells) is intimately related to the electrical reaction, as is already established in the case of the nerve fibre. Hence, as already stated above, the study of the electrical phenomena of the brain provides a possible means for the investiga-tion of the objective material processes that are the substrate of subjective, psychic part of the cortex (either in the psychomotor region or in the case of light in the occipital region, or with sound in the temporal region), this electrical reaction is elicited demonstrating that a sensory stimulus has evoked a process of excitation in a certain part only of the brain; other regions of the cortex remain inactive. Under the influence of chloroform narcosis these electrical oscillations weaken and even vanish entirely.[32]

During his many years of scientific work at the University of Kharkov, in which he rose to the position of Professor of Physiology, Danilevsky not only continued his interest in the nervous system, as witnessed by his research into such diverse aspects of it as hypno-tism and the effects on the peripheral nerve of various pharmacological substances, but he extended his investigations to the energetics of muscular contraction, including the mea-surement of the mechanical equivalent of heat.

At the same university, and occupying an equivalently important position in biochemis-try, was his brother, Alexander Yakovlevich Danilevsky. In 1888 the brothers brought out a volume of the collected works from their two laboratories, and in 1891 this was followed by a second volume.[33] Through his brother's speciality Danilevsky became interested in the significance for the organism of the so-called chemical regulators, and a large number of his investigations touched on problems concerning the role of the products of metabolism in the regulation of vital activities. In the case of the central nervous system, he studied the roles of lecithin, cholesterol, and some of the hormones.

Contrasting with these interests was his research on protozoology and parasitology with special emphasis on the malaria parasite. He brought this work together in two volumes entitled *Research in Comparative Parasitology of the Blood*. He held an interest in lower animals all his life and in 1892 wrote an article[34] on the physiology of the central nervous system in the amphioxus, an article that drew a comment from Sigmund Freud[35] in Vienna, who was himself working on invertebrates.

Many years later, in 1915, when Danilevsky came to write his textbook *Human Physiol-ogy,* knowledge of the electrical activity of the brain had increased sufficiently for him to make more specific statements about topography and to call attention to spontaneous activ-ity "of peculiar rhythm." In summarizing this work for his readers, he acknowledged the contributions of Caton, Beck, Cybulski, Fleischl von Marxow, Kaufman, Larionov, and Neminsky, as well as Sechenov's work on the spinal cord.

The progressive views of Danilevsky were unacceptable to the Czarist government. Re-pressions poured in upon him continuously, and as a consequence, Danilevsky lost the Chair of Physiology in 1901. Only after the February Revolution in 1917 was Danilevsky again restored to the Chair, and after October he had the possibility of once again carrying on scientific work.

[32]V. Y. Danilevsky. *Human Physiology* (in Russian). Vol. II, Part 2, 1915, pp. 1347–1348.

[33]A. Y. Danilevsky and V. Y. Danilevsky. Essays and works from the laboratories of A. Y. and V. Y. Dani-levsky (in Russian). *Fiziologischeskiy Sbornik.* Vol. 1, Kharkov, 1888; Vol. 2, Kharkov, 1891.

[34]V. Y. Danilevsky. Kymorheonomische Untersuchungen. *Pflüger's Arch. f. d. ges. Physiol.,* 51:240–252 (1892). (Kymorheonomic investigations.)

[35]S. Freud. Über den Bau der Nervenfasern und Nervenzellen beim Flusskrebs. *Sitz. Akad. Wiss.,* 85:9–46 (1882). (On the structure of nerve fibers and nerve cells in the river crayfish.)

FIG. 95. Two scientists concerned with the electrical activity of the brain: a supporter and a disbeliever. **Left:** Napoleon Cybulski (1854–1910). The great Polish scientist, teacher of Beck and producer of some of the first photographs of the electroencephalogram. **Right:** A disbeliever, S.Tchiriev of Kiev who held that all the electrical charges observed were artifacts.

Extremely prolific in both his work and his writings (which number over 200), Danilev-sky received many honors in his lifetime, among which was election in 1926 to the Academy of Sciences of the Ukrainian S.S.R. He died in 1939, the year that brought devastation to his native Ukraine as part of the overwhelming tragedy of Europe's Second World War, which included destruction of his beloved university building at Kharkov.

Ivan Romanovich Tarkanov (1848–1909)

Tarkanov was a Georgian, having been born in Tblisi on the Kura River. He was a pupil of Sechenov in St. Petersburg where he developed his great interest in electrophysiology. It was during his work on cutaneous sensation that he discovered the psychogalvanic reflex (which for a long time was known as the "Tarkanov effect").

Tarkanov had studied at Krakow and there had made a lifelong friend in Napoleon Cybulski.[36] Cybulski was Professor of Physiology in Krakow in Poland. It was he who encouraged his pupil Adolf Beck[37] to search for the electrical activity of the brain, and it was he who in the 20th century gave the scientific community almost the first photographs of the electroencephalogram,[38] for he was the first to have a camera in his laboratory (Fig. 95). Just a year previously, Pravdich-Neminsky[39] had published some very clear photographs. However, this was not Cybulski's major interest, for like many who trained in

[36]Napoleon Nicodemus Cybulski (1854–1919).

[37]Adolf Beck (1863–1942).

[38]Cybulski and Jelenska-Macieszyna. *Bull. Acad. Sci. Cracovie,* Series B, 776–781, 1914.

[39]V. V. Pravdich-Neminsky. *Zbl. Physiol.,* 27:951–960 (1912).

FIG. 96.   Left: Ivan Romanovich Tarkanov (1848–1909). The discoverer of the psychogalvanic reflex. **Right:** Alexander Filippovich Samoilov (1867–1930). Prominent electrophysiologist and friend of Einthoven. (From: N. A. Grigorjan. *A. F. Samoilov (1867–1930).* Moskow, 1963.)

Sechenov's laboratory, he followed the master's preoccupation with the blood.[40] In 1883 Cybulski visited Tarkanov, and together with Anrep, they published an article on the phrenic nerve.[41]

Tarkanov, with his interest in the psychogalvanic reflex, followed closely the work of other electrophysiologists and especially that of his student Verigo. When he heard his pupil give his paper on the action currents of the brain and spinal cord of frogs at the Third Congress of Russian Physicians and Biologists in 1889,[42] Tarkanov alerted the audience to the fact that Verigo had "opened up a whole series of new experiments—galvanometric investigation of the hemispheres during the action of voluntary movements would apparently be evidence for the motor functions of their anterior parts as postulated by Goltz."

Among others trained in Sechenov's laboratory was Alexander Filippovich Samoilov (Fig. 96) of Odessa, later himself a prominent electrophysiologist. Samoilov was born in 1867, and after training for a period in St. Petersburg with Pavlov, he moved in 1894 to Moscow, where Sechenov had moved in 1888, and there he became an assistant to him. Writing at Kazan in the last years of his life, Samoilov[43] gave this spirited description of his famous teacher:

[40]N. Cybulski. Die Bestimmung der Stromgeschwindigkeit des Blutes in den Gefässen mit dem neuen Apparat—Photohämotachometer. *Pflüger's Arch. f. d. ges. Physiol.,* 37:552–561 (1885). (The determination of the speed of blood flow in the vessels using the new machine–the photohaemotachometer.)

[41]R. von Anrep and N. Cybulski. Ein Beitrag zur Physiologie der Nervi phrenici. *Pflüger's Arch. f. d. ges. Physiol.,* 31:243–247 (1883). (Contribution on the physiology of the nervi phrenici.)

[42]Bronislav Fortunatovich Verigo (1860–1925). Action currents of the frog's brain. Report of the Third Congress of Russian Physicians and Biologists, St. Petersburg (in Russian). *Vrach,* 10:45 (1889).

[43]A. F. Samoilov. *Selected Writings and Lectures* (in Russian). *Akad. Nauk.,* Moscow, 1946.

His external appearance, good manner, gentleness and pleasantness of speech were as fine as I had been led to expect of this charming and fascinating man, I. M. Sechenov. If I say that he was an elderly gentleman of average height, well built, with a lined face and a strange, almost greenish pallor, this is quite an inadequate description of him as seen in person. His eyes and his penetrating look cannot be described in words. His face was very mobile and extremely expressive. His moods were changeable; when well and in a good humor his face was fine and gentle, for he loved to be kind and always praised kindness in other people. He was fundamentally a kind person, but his hot temper, inflammable nature, over anxiousness and sometimes even his mistrustfulness, made him unable to keep his natural kindness. When, as judged by his strict conscience, the government praised him unduly, he lost his temper in his anger and indignation, his eyes became inflamed and darted around.

Samoilov long outlived his teacher, dying in 1930. Tarkanov was also gone, having died in 1909.

Bronislav Fortunatovich Verigo (1860–1925)

One of the most industrious electrophysiologists of his time was Bronislav Fortunatovich Verigo who graduated from the Vibersy Academy of his hometown and later qualified in natural sciences in 1877 from the University of St. Petersburg, where he worked in Sechenov's laboratory. He also came under the influence of philosophers and developed an interest in the materialistic philosophy of the Russian school (Chernischevsky, Pisarev) that lasted all his life. In St. Petersburg Verigo worked in the laboratories of Sechenov and Tarkanov, and he attained his medical degree from the Military Medical Academy in 1886. His thesis was "On the action on nerve of galvanic currents, interrupted and constant." His purpose was, as he said, "to explain the physiological phenomena of electrotonus."

Verigo's work began to receive recognition when he gave a presentation to the Third Congress of Russian Physiologists and Biologists in 1889. His paper was "On the Action Currents of the Frog's Brain."[44] He had previously noted a "spontaneous" activity in the frog's brain causing "a continuous oscillation of the needle that did not lend itself to explanation." In the report he gave at the congress as reported in *Vrach*,[45] he was quoted as follows:

> B. F. Verigo gave a paper "On the action currents of the frog's brain." Professor I. N. Sechenov, investigating for the first time the nerve currents of the medulla oblongata, noticed it in periodic oscillations. These have been interpreted as produced by periodic excitation in the medulla oblongata.
>
> The speaker studied the nerve currents of the brain as well as of the spinal cord in frogs. For his experiments he employed a circuit connecting the cord and the hind limbs so that he could observe the occurrence of motor impulses in the central nervous system. When the galvanometer was connected to two points on the spinal cord, one of which lay in the thoracic region and the other on the lumbar swelling, the speaker obtained an oscillation of the needle on each reflex movement of the legs, indicating the presence of a current. The direction of the current was always a negativity at the lumbar swelling. Because during a reflex movement of the hind legs it was the lumbar region that was in the state of greatest excitation, the speaker assumed that a negative electrical potential is related generally to the foci of greatest excitation. Proceeding from this idea one sees that the galvanometric method can be employed to study localizations in the brain, as was suggested by Professor

---

[44]B. F. Verigo. Action currents of the brain and medulla (in Russian). *Vest. Klin. Sudev. Psychiatr. Nervopath.*, 7 (1889).

[45]B. F. Verigo. On the action currents of the frog's brain (in Russian). *Vrach*, 10:45 (1889). (Verigo's report to the Third Congress was published more fully the same year.)

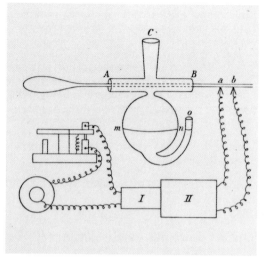

FIG. 97.   Left: Bronislav Fortunatovich Verigo (1860–1925). Right: One of Verigo's experiments to test the action of chemicals on conduction in nerve. The liquids being tested are put into the vessel to a level *m–n*. The nerve receives the fumes between *A* and *B* and is stimulated between *a* and *b* until it fails to respond. (From: B. F. Verigo. *Pflüger's Arch. f. d. ges. Physiol.*, 76:552–607, 1899.)

> Tarkanov. On connecting the hemispheres of the frog with the circuit of the galvanometer, Dr. Verigo obtained a continuous oscillation of the needle that did not lend itself to explanation; however, it was nevertheless possible to be certain that on each movement of the legs, the anterior part of the hemisphere became relatively negative, electrically, to the posterior part.[46]

Danilevsky spoke in the discussion that followed Verigo's paper[47] and referred to some of the difficulties and lack of success in attempts being made in similar experiments at Kharkov. His remarks are not reported in full, but they presumably reflected the views he published later.

In 1894 Verigo was appointed as Professor of Anatomy and Physiology at the Ukrainian University of Novorossiisk in Odessa. His field was essentially electrophysiology, but he also inherited from Sechenov an interest in blood gases. He stayed in Odessa for 20 years until he ran into trouble up north. Verigo was instrumental in building up an important center of physiological research and teaching, equipping it with apparatus for chemical and electrical research. His own work was in the relation between conductivity and excitation in the nerve (Fig. 97). His experiments led him to state that "between excitation and conductivity there is an undoubted connection, so that both these most important properties of nerve can in no case be considered as independent." He was responsible for defining cathodic depression. He concluded from his experiments that the impulse could leap across an inactive section of nerve. In other words, he foresaw saltatory transmission in nerves (it was in the 1940s that this was firmly established by Tasaki[48]).

---

[46]This discovery of the continuous "oscillation" of the currents of the brain is certainly of interest to future electroencephalographers.

[47]B. F. Verigo. Action currents of the brain and medulla (in Russian). *Vest. Klin. Sudev. Psychiatr. Nevropath.*, 7 (1889).

[48]I. Tasaki. *Nervous Transmission.* Thomas, Springfield, Ill., 1953.

After a few years at Odessa, Verigo was able to make a visit to the laboratory at the Pasteur Institute in Paris of Metchnikov (who himself had moved there from Odessa when he lost his Chair for political reasons). A publication by Verigo appeared from this visit and reflected Metchnikov's major interests, interests which brought him the Nobel Prize in 1908. Verigo's interest in phagocytosis continued in the entire Odessa period of his activity. He obtained data concerning the significance of phagocytosis in protecting an organism against infection and concerning the role of the liver and spleen in the struggle against such infections as Siberian plague and cholera. His works on the problem of immunity were awarded a prize by the Russian Academy of Sciences.

Working in the field of general biology, Verigo took the position of consistent materialism and Darwinism. He rose in strong opposition to various manifestations of vitalism and in particular against the theory of A. Veisman, fashionable at that time, concerning embryonic plasma. These works of Verigo had great significance in the struggle of materialism in native biological sciences. His views got him into great trouble with the authorities, but he survived several threats of expulsion. He was very progressive in his views and tried to spread knowledge of physiology outside the university (as had Sechenov). These views got him into trouble with the Czarist government, but in spite of this he was elected Dean of the Medical Faculty in 1906, though confirmation was denied by the Minister of Education.

While at the University of Odessa, Verigo brought out two volumes of his extensive textbook on physiology, *Foundations of the Physiology of Man and Higher Animals*. The first volume, published in 1905, included an introduction to the study of the physiology of vegetative processes and a chapter on blood and circulation. The second included the physiology of lymph, respiration, digestion, and secretion, and was published in 1910.

Verigo's position at the University of Odessa ended in 1914 by dismissal for his progressive views. He returned then to his old institute in St. Petersburg, the Military Medical Academy, where he was allowed, once again, to lecture. With the fall of the Czarist government in 1917, he was offered resumption of his professorship in Odessa, but he preferred to stay with his recent appointment at the newly created University of Perm, of which he had been one of the founders.

Towards the end of his life, he was again approached for returning to Odessa, but his health failed; and he died in 1925 away from the university he loved so much. Great neurophysiologists who followed him there included Beritoff and Vorontsov.

Vladimir Mikhailovich Bechterev (1857–1927)

Vladimir Mikhailovich Bechterev was born in Sarali in Viatka in 1857. After training in St. Petersburg at the Military Medical Academy, he went to Berlin and Paris, as did so many of his generation. He had started out with an interest in neurophysiology, and this led him in 1884 to the laboratory of Flechsig[49] in Leipzig, but he also visited Charcot[50] and Wundt;[51] their influence may have led to his interests branching away from neurophysiology to the field of psychology and eventually to psychiatry, in which he became famous. On return to his homeland, he opened a psychiatric clinic in Kazan where a Chair in Psychiatry was made for him. This resulted in his appointment in 1893 at his old school, the

---

[49]Paul Flechsig (1847–1929).

[50]Jean Martin Charcot (1823–1893).

[51]Wilhelm Wundt (1832–1920). *Gründer der Psychologie*. Leipzig, 1896.

Military Medical Academy in St. Petersburg. He held this appointment until his death in 1927.

Bechterev's interest in the brain started from a neurophysiological point of view, and this viewpoint was an encouragement to the younger men who came to work with him. His particular interest was tracing how sensory information reached the cortex (Fig. 98). Two of his early papers[52,53] are on this theme. In this work he ran into arguments with the Swedish neurophysiologist Hensen[54] and a controversy broke out in the *Archiv*.[55]

Bechterev was also interested in the problem of balance and the role of the acoustic nerve on the semicircular canals.[56] He tested his theories through experiments[57,58] on animals in his early days in the laboratory. In all these experiments his motivation was to clarify how the nervous system interacted with its environment. Bechterev's interest in the mechanism of balance led to research involving experiments on the spinal cord in which he cut the dorsal columns in an attempt to identify the cells in the cord to which they brought excitation and their role in the maintenance of balance. His report as given in the *Centralblatt*[59] reads in part:

> He sketches experiments of severing sections of the spinal cord inasfar as they illustrate the physiological significance of the dorsal columns, and then proceeds to the results of experiments of severing the whole dorsal column in doves, rabbits and dogs, compared with severing only Goltz's column in dogs. Severing both sides produced the expected loss of balance, falling to one side, and inability to keep a constant direction while walking. It was found in all three animals that sense of touch and pain were not affected. B. also claims that the dog's muscle sense was also unaffected, but does not state how he tested this. He only found hyperaesthesia if there was an allergic reaction following the cut with "visible redness of the grey matter, spread of vessels with plasmatic exudate into the tissue, and lack of clarity of the cell elements."
>
> After severing only the inner bundle in dogs he found less serious balance disturbance, and no loss of touch or pain sensation. After cutting the dorsal columns on one side only in dogs he found the animals were unable to walk in a straight line. The dorsal columns

[52]V. M. Bechterev. Ueber den Verlauf der die Pupille verengenden Nervenfasern im Gehirn und über die Localisation eines Centrums für die Iris und Contraction der Augenmuskeln. *Pflüger's Arch. f. d. ges. Physiol.*, 31:60–87 (1883). (On the path in the brain of those nerve fibers which cause the pupil to narrow, and on the localization of a center for the iris and for contraction of the eye muscles.)

[53]V. M. Bechterev. Zur Physiologie des Körpergleichgewichts. Die Function der centralen grauen Substanz des dritten Hirnventrikels. *Pflüger's Arch. f. d. ges. Physiol.*, 31:479–590 (1883). (On the physiology of body balance. The function of central gray matter of the third ventricle.)

[54]V. Hensen. Bemerkungen zu dem Aufsatz: Ueber den Verlauf der die Pupille verengenden Nervenfasern im Gehirn. *Pflüger's Arch. f. d. ges. Physiol.*, 31:309–340 (1883). (Remarks on the article "On the path of nerve fibers in the brain which narrow the pupil.")

[55]V. M. Bechterev. Ueber die Bemerkungen von V. Hensen zu meinem Aufsatz "Ueber den Verlauf der die Pupille Verengenden Nervenfasern im Gehirn." *Pflüger's Arch. f. d. ges. Physiol.*, 33:240–242 (1884). (On remarks by V. Hensen on my article "On the path of nerve fibers in the brain which narrow the pupil.")

[56]V. M. Bechterev. Ergebnisse der Durchschneidung des N. acusticus, nebst Erörterung der Bedeutung der semicirculären Canäle für das Körpergleichgewicht. *Pflüger's Arch. f. d. ges. Physiol.*, 30:248–312 (1883). (Results on severing the N. acusticus, together with a discussion of the significance of the semicircular canals for body balance.)

[57]V. M. Bechterev. Zur Physiologie des Körpergleichgewichts. Die Function der centralen grauen Substanz des dritten Hirnventrikels. *Pflüger's Arch. f. d. ges. Physiol.*, 31:479–590 (1883). (On the physiology of body balance. The function of central gray matter of the third ventricle.)

[58]V. M. Bechterev. Über die Verbindung der sogenannten peripheren Gleichgewichtsorgane mit dem Kleinhirn. *Pflüger's Arch. f. d. ges. Physiol.*, 34:362–388 (1884). (On the Connection of the So-called Peripheral Organs of Balance with the Cerebellum.)

[59]V. M. Bechterev. On the phenomena produced by cutting the spinal cord in animals and the relation of these to balance (in German). *Cbl. Physiol.*, 2:58–59 (1890).

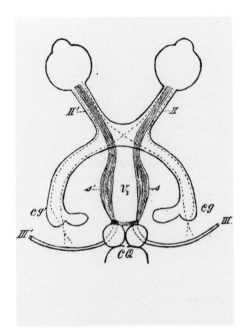

FIG. 98. Left: From Bechterev's studies of the visual system showing the path of fibers that control the pupils. V_3 = 3rd ventricle; *cg* = corpora geniculata; *II* = optic nerves; *III* = oculomotor nerves. (From V. M. Bechterev. *Pflüger's Arch f. d. ges. Physiol.*, 31:60–87, 1883.) **Right:** Median section through the brain stem, showing the optic thalamus, third ventricle, the medulla oblongata, and 4th ventricle. (From: V. M. Bechterev. *Pflüger's Arch. f. d. ges. Physiol.*, 31:479–590, 1883.)

are thus centripetal transmission pathways, which send impulses to the cerebellum and the hemispheres.

The report of Bechterev's original paper published in *Pflüger's Archiv f. d. ges Physiologie* in 1890 drew adverse criticism in the *Centralblatt*. The report read (in translation):

> We were not able to determine on what evidence Bechterev based his assertion that following cutting through the dorsal columns that muscle sensitivity remained intact. The inability of brainless animals to stand when the skin of the legs has been removed, cannot simply be explained by lack of local signals necessary for coordination. More probably, it is due to pain. The article reviewed seems to offer little of value on the undoubtedly worthwhile subject of the conditions of coordination.

Bechterev's interests were essentially in the brain itself (Fig. 99), and in 1885 he published a paper[60] on the results of destruction of the motor cortex in animals. In this paper Bechterev took issue with Schiff.[61] It reads (in translation):

> The opinion which I expressed in a recent article (*Neurologia. Centrabl.* 1883, no.16), that in the so-called motor cortex of the cerebral hemispheres there are contained genuine centers of movement, had already been earlier expressed by many authors, especially Ferrier and Duret.[62] However, Professor Schiff has raised some objections to this (*Pflüger's*

[60]V. M. Bechterev. Wie sind die Erscheinungen zu verstehen, die nach Zerstörung des motorischen Rindenfeldes an Thieren auftreten. *Arch. Anat. Physiol.*, 35:137–145 (1885). (How are those phenomena to be understood, which occur after destruction of the motor cortex in animals.)

[61]Moritz Schiff (1823–1896).

[62]C. Duret. Observation sur un enfant né sans anus, et auquel il a été fait une ouverture pour y suppléer. *Rec. Med. Paris*, 4:45–50 (1798).

 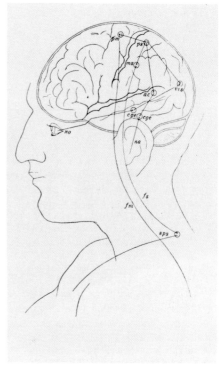

FIG. 99. **Left:** Vladimir Mikhailovich Bechterev as a young man when he was working on the sensory tracts to the brain. (Photograph by gift of Professor N. P. Bechtereva). **Right:** Beckterev's schema of pathways by which reflexes could be travelling in the brain.

Arch., 33, 264–270, 1884). These objections have led me to write the following observations. Schiff has long maintained that in the area of the hemisphere surface which is capable of being stimulated there are, in reality, no motor centers, but rather that the motor disturbances observed after extirpation of this area are caused by inhibition of the sense of touch, while the muscle twitchings which occur on electrical stimulus of this area are of a reflex nature.

Accordingly Schiff maintains that there is no reason to assume paralysis in animals with destroyed motor areas of the hemispheres, since movement is impossible "only when certain stimuli to movement lose their normal effect." He here states that many animals can, following removal of the excitable area "move their front paws easily and quickly forward in climbing a steep incline" but "are unable to do this when it comes to grasping and holding onto a piece of food." He continued by noting that these animals not only use their affected paw in swimming, but also make strong swimming motions if they are lifted into the air. "Where the movement mechanism is strongly and perfectly maintained, as here, and continues to obey a certain series of stimuli; but where all stimuli which presuppose a subjective or objective sense of touch do not bring forth a response, and where after anomalies cannot be observed, then we should not seek the deficiency in movement" (*Pflüger's Arch.*, 33, sec. 5 & 6, p. 266, 1884).

Another objection consists in Schiff's assertion that the animals operated on suffered a loss of sense of touch at the contralateral extremities which was continuous up to the death of the animal, and was even evident after the animal had regained the ability to stretch out the other extremity and use it as a paw.

Schiff supports this objection with an experiment carried out on a dog, of which he had destroyed the motor area of one hemisphere. He was convinced of a loss of sense of touch at the contralateral extremities by observing crossed touch reflexes on pulling the dog's hair.

If in fact everything were as Schiff maintains, then his opinions would long have become accepted scientific fact. Everyone recognizes that disturbances of movement would have to be seen as the result of changes in the sphere of sensitivity if it were proven that:

(1) These disturbances can only be explained by the sole inhibition of sensitivity.
(2) Such an inhibition of sensitivity had actually been observed in the animals operated on.

Neither (1) nor (2) can be proven by using animals in which an isolated destruction of the motor or sensitive area of the hemispheres is carried out. Neither the observations of Schiff, other authors, or myself, allow the motor disturbances caused to be explained by inhibition of sensitivity or sense of touch.

Goltz has pointed out that excluding the usual clumsiness in the mastery of the contralateral extremities, animals are deprived of the ability to use these extremities as a paw. This phenomenon can also be observed following destruction of the motor area alone, while making up only a partial element of the total disturbance of movements in the animal. These disturbances which I observed in my own experiments, consist in a more or less total loss of all those voluntary or involuntary movements which do not belong to the category of associated movements such as the movements of the extremities in walking, running, climbing and so-called swimming movements.

For example, a dog who has had one hemisphere extirpated exhibits only unimportant disturbances in walking (at least on even ground) even immediately after the chloroform anaesthetic has worn off. However, if the animal remains at rest, then it is incapable of improving an unusual position of its extremities, or of lifting its paw if the paw is hanging off the table. It is not able to give its master its paw when commanded, and cannot use it to scratch. Similar phenomena can be observed in cats and other higher animals.

If a certain movement becomes impossible, or is affected to a great degree, we are accustomed to refer to it as a paralysis. I can therefore not agree with Schiff's assertion that following destruction of the sensitive cortex area no paralysis occurs.

The question as to the cause of such disturbances of movement is however somewhat more difficult to answer. Goltz, who observed the animals he operated on in the most careful manner, does not attribute the loss of ability to use the extremity as a paw to sensitivity disturbance, and offers an, in my opinion, more plausible explanation. "If I touch the animal's right paw, and encourage him to give it to me, I can see by the look on his face that he has understood what I want, and is trying to comply. Finally, in frustration, he offers his left paw instead. Between the organ of will and the nerves that carry out this will, there has developed an insurmountable resistance."

On the other hand, Schiff holds such phenomena to be not the expression of motor paralysis, on the grounds that the animal operated on was supposedly able to carry out the same movements under different circumstances.

However, in my opinion, the use of the paw in climbing and swimming cannot be compared to the stretching of the paw for food, or on command of an observer. The first acts mentioned form part of a coordinated act, while the others are directly dependent on the animal's will.

If Schiff believes (as I infer from his own words) that the animal loses the ability to perform the latter acts due to a lack of sensory impressions, then we must ask why the animal does not use muscle sense and sight instead? And are there facts available to support the assertion that loss of feeling following brain infections or lack of feeling, impressions alone cause a more or less complete inability to perform certain movements?

As to the objection contained in the evidence that in animals with extirpation of the sensitive brain area, disturbances of feeling occur, I would propose that neither Schiff's earlier experiments, nor the new experiment described in his latest article can effectively disprove my opinions, for the following reasons:

In my article cited above, I say the following: "I could never convince myself of the existence of any sensitivity disturbances, if the lesion did not exceed the limits of the sensitive area of the hemisphere surface." Anyone who wants to check this assertion, must first be clear as to my definition of the limits of the sensitive area of the hemisphere surface, since various authors give different definitions. On p. 411 of my article I define the motor centers as being restricted almost exclusively to the gyrus signoides, while I place the centers of feeling for skin and muscles in the areas lying directly over the fossa sylvii.

In order to arrive at such conclusions,[63] extremely accurate brain lesions were obviously necessary. In this regard I therefore extirpated only the area of the hemisphere surface whose stimulus with weak electrical charge produced muscle twitching in the members. Other authors have also followed this procedure. How, however, does Schiff proceed to disprove my conclusions?

In his early experiments Schiff confined himself to an isolated destruction of the sensitive hemisphere region alone, or so I conclude from his own words: "The objection of Luciani and Lemorgne (see centri encefalici, Sperimentale, 1877) is justified, in relation to many of my experiments. My results with extirpation are very often obtained from another place as exactly the one whose stimulation produces twitching, as I have previously mentioned in agreement with Goltz" (Schiff I p. 235).

In his latest experiment Schiff also does not restrict himself to the extirpation of the area in which I surmise the motor centers are located. Now the right sulcus cruciates of the animal is exposed, and the whole of the gyrus sigmoides lying in front of it. The brain material lying underneath the latter is removed to a width of 5 mm!

If one considers that in destroying the brain material, one also creates an inflammation which affects the neighboring parts of the brain (in my opinion up to at least a distance of 3 to 4 mm), then it is clear that Schiff has removed far more than just the motor region. I am therefore not surprised that Schiff noted numbness in the contralateral extremities.

To prove my own point, I shall now describe an experiment I carried out on some cats.[64] Since cats are very sensitive animals, a light touching on any part of a cat's head (provided the touching is unexpected), will produce a closing of the eyes and a pulling of the ears. If the cat is particularly nervous, then this reaction will occur on touching the fur on one of its parts. In addition, cats are exceptionally sensitive to water. I placed my experimental cat under anaesthetic, and exposed its left gyrus sigmoides; then I determined the motor centers for the front and rear extremities using weak current. I then separated these motor areas off with an incision, and removed the grey matter with a spoon. The extirpation did not exceed the limits of the gyrus sigmoides, and in backward and outward direction the limit of the destroyed area was 2–4 mm. within the convolution.

After the operation the animal exhibited noticeable movement disturbances in the contralateral extremities, especially immediately following the operation. Both right extremities were used very clumsily in walking. In standing the front extremity was placed awkwardly, and often slid under the body of the animal. The animal tries to correct the position of the limb, but usually fails, as long as no change of location is involved. Climbing remains possible. Single voluntary movements of the right front paw seems impossible. The animal continued to close its eyes and pull in its ears when touched around the head, or on the fur of the left or right extremities. The animal remains very sensitive to water.

The motor disturbances in the cat quickly improved, so that after two days only careful observation could detect any clumsiness in the animal's use of its right extremities.

I find these results totally convincing. It is obviously quite impossible to explain the motor disturbances in the animal by inhibition of sensitivity, if in reality such an inhibition of sensitivity cannot be experimentally proven.

Bechterev's views on the motor cortex are reported here in detail to illustrate that this man, who became Russia's most prominent psychiatrist, had given considerable time to

[63]V. M. Bechterev. Zur Physiologie des Körpergleichgewichts. Die Function der centralen grauen Substanz des dritten Hirnventrikels. *Pflüger's Arch. f. d. ges. Physiol.*, 31:479–590 (1883). (On the physiology of body balance. The function of central gray matter on the third ventricle.)

[64]V. M. Bechterev. *Die Energie der lebenden Organismen und thiere, psychobiologishe Bedeutung.* Bergmann, Wiesbaden, 1902. (The Energy of the Living Organism and Animals and its Psychobiological Significance.)

exploring the neurophysiology of the brain when it came to be revealed that the brain itself emitted electricity. Bechterev was a strong admirer of Caton, to whom he gave priority of an important discovery. Bechterev gave a detailed account of the followers of Caton and their experiments (Fleischl, Danilevsky, Verigo). He discussed the report of Gotch and Horsley and Wedensky's use of the telephone to detect current flow. Bechterev said that Wedensky was not the inventor of his method, for it had been used by Hermann and by Brücke.

At this time (1889) when Bechterev moved to the Chair of Psychiatry in St. Petersburg, there were only seven established universities under Russian influence: St. Petersburg and Moscow in Russia; Kharkov, Kazan, and Kiev in the Ukraine; Dorpat in Estonia; and Krakow in occupied Poland. One at Odessa, known then as the University of Novorossiisk, should be mentioned, though it had no medical school at that time. New universities were just opening up at Tomsk in Siberia and at the city of Saratov. Competition for Chairs was therefore acute and had been rendered more so by the recent reorganization of the universities by the Ministry of Education. Previously, the retiring professor had had the privilege of naming his successor, approval by the Ministry being little more than a formality. But in the late 1880s, a new regime was created by which a list of candidates for a Chair had to be submitted by the faculty, and in the case of the medical schools, a diploma or license to practice medicine was not enough; the candidate had to have the higher degree of Doctor of Medical Sciences. A similar ruling would apply today, for the practicing physician's degree does not qualify him for an academic appointment in the U.S.S.R.

The professorship in St. Petersburg, to which Bechterev came, was the oldest Chair of Psychiatry in Russia, having been founded in 1867. The first holder, Valinsky, was succeeded in 1877 by Mierjievsky, a benefactor to his colleagues by virtue of the journal he founded as an outlet for work in the field of nervous and mental diseases called (in translation) the *Journal of Clinical and Legal Psychiatry and Neuropathology*.

In common with the practice of the Russian universities, a clinic was run in close conjunction with the academic department, and it was in this clinic of Bechterev's that his pupils pursued the study of the electrical activity of the brain. Since the medical school in St. Petersburg was the Military Medical Academy, the Ministry of War had been responsible for building and financing the clinic, which functioned solely as a research institution. Only cases being studied as part of a research program were admitted; when the investigation came to an end (and even during vacations), the patients were moved to another hospital.

As was the pattern for these clinics attached to the medical schools in Russia at the time, there were ample laboratories closely associated with the wards. A contemporary account, written by a visiting French professor, describes each clinic as possessing "a vast laboratory, plentifully supplied with instruments and everything that is not only quite essential, but even everything that could possibly be useful for the study and research work in the field of investigation offered by the patients in the clinic. These laboratories comprise a certain number of rooms, usually spacious, well lit and well ventilated, and assigned one to microscopy, another to chemical research, yet others to photography, to electrotherapy and to electro-diagnosis."

Bechterev's eminence as a clinician and experimentalist made him the outstanding figure in Russian neurology. His clinic with its laboratories became a magnet for students who were drawn to the search for a physiological basis for psychic events. He survived the First World War, dying in 1927.

Vladimir Elimovich Larionov (1857–1919)

The surge of work in England on cortical localization had centered on the motor system for finding zones that, on stimulation, produced muscular movements and for detecting at the cortex the arrival of impulses from the periphery. The site of the ability shared only by man—speech—had of necessity passed from the neurophysiological laboratory to the clinic and the autopsy room. But those more psychologically minded had an urge to find the centers for the sensory systems. The ambitious (and successful) attempt to locate with exactitude the center for hearing was made by a student, Vladimir Elimovich Larionov in Bechterev's laboratory in Moscow, where Bechterev himself was interested in tracking the central paths of the sensory systems and especially the visual system.[65] Larionov made some of the earliest observations on the localization of tones in the auditory cortex.

Larionov, the son of a physician, graduated from the University of Kazan in the Ukraine in 1881, and then he went to St. Petersburg to the Military Medical Academy where he earned his doctor's degree in 1898. He worked as an assistant to Bechterev, and wrote his doctoral thesis, "On Cortical Centers of Hearing."[66]

Larionov's first attempts to localize tones were by extirpation experiments, and the map that he published of the points he located for the various tones of the octave in the dog's brain is of great interest for comparison with today's work in this field, accomplished by modern techniques. Impressed by demonstrations of the applicability of electrical techniques to localization of sensory functions in the brain, for which he gave full priority to Caton, Larionov decided to adopt the galvanometric method. He used tuning forks for his stimuli and obtained permission to use a Wiedermann d'Arsonval galvanometer in the Department of Physics. His preliminary experiments were so promising that Bechterev decided to obtain one of these instruments from Paris and thus equip his own laboratories for electrophysiological research. Bechterev was very receptive to this approach, for he himself envisaged the nerve cells of the brain as so many Leyden jars whose discharge supplied the currents recordable by these techniques.

Larionov was intrigued by some experiments that Sechenov had done in 1882, experiments which had drawn the derision of Tchiriev. In his experiments on the electrical activity of the medulla and spinal cord, Sechenov had shown this to vary when a loud sound was made. He used copper tubes of three different sizes, which on being struck, sounded three different octaves. Sechenov had concluded that the changes he observed were in some way related to vibrations of the air, but Larionov wondered whether they might not have entered the central nervous system through the organs of hearing.

Larionov's technique was to lead off in unipolar fashion, the reference lead being on muscle fascia. With his stigmatic electrode (made of a hair soaked in 0.6 percent saline), he explored the temporal gyri for responses to tuning forks of various tones. He reported finding "a center of hearing for a low tone (tuning fork A) in the posterior lower part of the second temporal gyrus. For tuning fork a^i he found responses in the third temporal gyrus, and for C^3 in the posterior half of the fourth temporal gyrus."[67]

[65]V. M. Bechterev. Ueber den Verlauf der die Pupille verengenden Nervenfasern im Gehirn und über die Localisation eines Centrums für die Iris und Contraction der Augenmuskeln. *Pflüger's Arch. f. d. ges. Physiol.,* 31:60–87, (1883). (On the path in the brain of those nerve fibers which cause the pupil to narrow, and on the localization of a center for the iris and for contraction of the eye muscles.)

[66]V. E. Larionov (1857–1919). On the cortical centers of hearing in dogs (in Russian). *Obozr. Psychiatr. Nevol. St. Petersburg,* 2:419–424 (1897).

[67]V. E. Larionov. On the cortical centers of hearing in dogs (in Russian). *Oboxr. Psychiatr. Nevrol. St. Petersburg,* 2:419–424 (1897).

In a report written in the last year of the 19th century, Larionov wrote of these researches as follows (in translation):

> Last year, I carried out several experiments in this direction with a Wiedermann-d'Arsonval galvanometer at the Military Medical Academy in the physics laboratory of the highly respected Professor N. G. Egorov, with the kind cooperation of his assistant, N. N. Georgievsky, and also in the physiology laboratory of Professor Pavlov under the direction of V. I. Vartanov. From the experiments and the literature collected at the time, it appeared that these experiments were worth continuing, since, besides the simple fact of the existence of currents in the cortical centers of hearing during their active excitation by peripheral auditory activity they could provide valuable data regarding the location of tone centers, i.e., in other words, for confirmation of the tonic scale lying in three temporal gyri, which I explained. Therefore, this year I resumed my experiments on a Wiedermann-d'Arsonval galvanometer ordered from Paris especially for this purpose by V. M. Bechterev.

It is of interest to find Pavlov supporting electrophysiological research in his laboratory. Larionov[68] went on to give the views of Bechterev as to the intimate mechanisms and followed this with his own:

> The highly respected Professor V. M. Bechterev[69] considers the multiple series of nerve cells in the brain similar to Leyden jars, whose discharges lay down the conditions for nerve currents.
>
> It seems to me that from this point there is already an easy transition to a recognition of a line of batteries in nerve cells of the brain, made up of small elements connected to each other in series, which is necessary, according to the laws of physics, for surmounting the enormous resistance of our bodies. With the development of knowledge about neurons, it becomes clear that there is a possibility of a locking of the current during the neurons' approach to energy perhaps with hyperthermia of the brain during functioning of its central organs and, on the other hand, the possibility of disconnection under the reverse condition.
>
> Of course, all of this is in the realm of hypothesis and supposition, but there is some justification for these suppositions on the basis of the existence of currents in the organism, such as: in muscles, in the skin, in nerves, in the spinal cord, the medulla oblongata, and the brain, concerning which galvanometric investigations give us a clear understanding.

In 1899 Larionov published a very long review of the development of electrophysiology, especially of the brain.[70] Du Bois-Reymond had just died, but Larionov tackled the dispute between him and Hermann and gave a detailed account of what is now called electroencephalography, quoting Caton, Fleischl von Marxow, Beck, and the Russian workers, Trivius, Danilevsky, Tarkanov, and Verigo. He followed this with a summary of his own work, closing with the following summation:

> In this way, galvanometry gives very important results for explaining the functions of the brain. It is clear that the posterior sensory half of the brain normally has electronegative potential which, under stimulations of peripheral sensory organs, changes to positive, i.e. the descending currents become ascending. In the anterior motor area, in all probability, the relation of potentials is reversed; for final clarification of this, however, still further experiments are needed.

[68]V. E. Larionov. On galvanometric measurements of currents in the cortex of the temporal gyrus during stimulation of the peripheral organs of hearing (in Russian). *Nerolog. Vestnik. Kazan,* 7:44–64 (1899).

[69]V. M. Bechterev. *Les Voies de Conduction du Cerveau et de Moelle.* 2 vols. Storck, Lyon, 1900. (Conducting Tracks of the Brain.)

[70]Über die musikalischen Centren des Gehirns. *Pflüger's Arch. f. d. ges. Physiol.,* 76:608–625 (1899). (On the musical centers of the brain.)

Larionov added another technique to his electrical exploration; he ran a large series of experiments—again on dogs—in which he tested their hearing after specific cortical excisions. He had been encouraged by Bechterev to test the results published by Munk.[71] Larionov wrote as follows:

> In testing Munk's results I also wanted to determine whether in the temporal lobes one might determine a gradation of the position of the fine tone centers.
>
> For the experiments well-bred dogs with good hearing reactions for tones, noises and commands were selected. These reactions consisted of contraction of the stimulated ear, turning of the head, eyes and rump to the origin of the sound. They have already been used by the following authors: Autenrieth and Kerner, Esser, Flourens, Preyer, Ferrier, Munk, Ewald, Luciani and Sepilli, Corardi and Stepanow.
>
> The auditory reactions were tested many times, before and after the operation, and noted by the symbols $+$, $++$ and $+++$ (good, better, best) on both sides. The dogs were also tested for their sight, taste, smell and feeling of every sort, including muscle feeling. After the operation the dogs lived on for months under observation, after being tested for hearing and other feelings.
>
> By partially removing small sections of the cortex of the temporal lobes I became convinced that the tone centers are strictly graduated, that is, that there is the same tone scale as in the cochlea. However, in the latter rows of strings of differing lengths and turning exist which precipitate the tones, while in the temporal lobes there are situated groups of sensitive cells.
>
> It was further found that following the destruction of the second convolution the dogs were unable to hear deep sounds from A^1 to e. After destruction of the third convolution they could no longer hear e to c^2. After destruction of the rear half of the fourth convolution (gyrus angularis) they could not hear tones higher than c^2. After destruction of the cortex of a whole temporal lobe hearing on the opposite side diminishes greatly, on the same side only slightly. This points to an incomplete crossing of the auditory fibers in the brain. Therefore most run from the temporal lobe to the opposite ear, but some run to the ear on the same side. This was also demonstrated by the use of the microscopic apparatus of Marchi.
>
> I also determined that the tone scale of sensitive elements runs from deep to high tones in the rear quarter of the 2nd convolution from top to bottom, continues in the rear third of the third convolution from bottom to top and finally runs in the rear half of the fourth convolution from top to bottom, as can be seen in figure 1 (p. 619).
>
> In control experiments with destruction of frontal and parietal lobes hearing did not suffer. It must also be added that following partial removal of a hearing center there is complete deafness to all sounds and noises in the opposite ear for one to two days, partial deafness in the ear on the same side. Hearing then returns to normal on both sides, apart from a perception of certain tones, which is slightly diminished on the same side, severely impaired or non-existent on the opposite side. After several months dogs with these injuries become almost totally deaf in both ears:
>
> 1. Due to the spread of the softening process around the injured point.
> 2. Due to the degeneration of the associating and commissural auditory fibers which join the temporal lobes of both hemispheres through the corpus callosum.
> 3. Due to the degeneration of the auditory projection fibers of both hemispheres.
>
> This was shown both by microscopic investigation of the brains of experimental animals, and by the experiences of Muratow[72] who demonstrated the degeneration of commissural fibers from the affected areas of one hemisphere to the corresponding areas of the other hemispheres.

---

[71]H. Munk. Zur Physiologie der Grosshirnrinde. *Klin. Wiss., Berlin,* 14:595 (1877). (The physiology of the cerebral cortex.)

[72]Muratow. *Arch. Anat. Physiol.* p. 108, 1893.

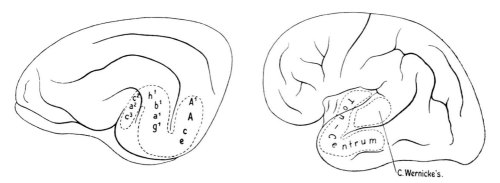

FIG. 100. **Left:** Larionov's mapping of the musical tones in the dog's cortex. **Right:** His transposing of these findings to the human cortex. (From: *Pflüger's Arch. f. d. ges. Physiol.*, 76:608–625, 1899).

Larionov went on to consider what these results might mean for the brain of man (Fig. 100). He wrote (in translation) as follows:

> If one now applies these centers to the brain of man, then the rear quarter of the 2nd convolution of the dogs, according to Turner and Ferrier,[73,74] will correspond to the 2nd convolution of man, the rear third of the third convolution of the dog will correspond to the first temporal convolution of man, and the rear half of the fourth convolution of the dog will correspond to the rear cross convolutions of the island.
>
> The fourth convolution of the dog has become hidden in man, and forms the so-called island—the island of Reil—because of the great development of the frontal and temporoparietal convolutions (the front and rear associations centers of Prof. Flechsig[75,76]).
>
> On surveying the literature concerning the auditory pathways we became convinced of the veracity of the determination of the hearing centers shown above, since these pathways are both in dogs and in humans in the convolutions shown above. Prof. Flechsig has determined that in embryo human brains the ending of the hearing pathways lies either in the cross convolutions or in the rear half of the island, and in the temporal convolution. My microscopic investigations, following Marchi's methodology, show that also in dogs' brains the endings of the hearing pathways are situated in the tone centers.[77]

Another student in Bechterev's laboratory who pursued the study of electrical activity of the brain was Solomon Abramovich Trivius, a less energetic figure, whose contributions in this field did not extend beyond his work for a doctoral thesis. Trivius was born in Berdyansk in 1874 and graduated from the University of Kiev in 1897. He came to the Military Medical Academy to take his doctor's degree, which he gained in 1900. Under the guidance of Bechterev, he investigated the electrical potentials of the cortex evoked by photic stimulation,[78,79] work in which he received considerable help from Larionov. He consis-

[73]D. Ferrier. *The Croonian Lectures on Cerebral Localisation.* Smith Elder, London, 1890.

[74]D. Ferrier and W. A. Turner. An experimental research upon cerebro-cortical afferents and efferents. *Phil. Trans. roy. Soc.,* 190:1–44 (1898).

[75]Paul M. Flechsig. *Grosshirn und Organ der Seele.* Engelmann, 1896. (Brain and Soul.)

[76]P. Flechsig. *Neurol. Cbl.,* 7:295 (1897).

[77]V. E. Larionov. Two cases of transcortical sensory and motor aphasis with retention of the musical capability (in Russian). Report in meeting of doctors of Bechterev's neuro-psychiatric clinic, January 28, 1895.

[78]S. A. Trivius. Action currents in the cortex of the cerebral hemispheres of the dog under the influence of peripheral stimulation (in Russian). *Obozr. Psikh. Nervol. St. Petersburg,* 4:791–795 (1899).

[79]S. A. Trivius. Action currents in the cerebral cortex under the influence of peripheral stimuli (in Russian). Dissertation from the Laboratory of Physiological Pathology in the Clinic on Nervous and Mental Diseases of Academician V. M. Bechterev. St. Petersburg, 1900.

tently found a response which correlated with the intensity of the light stimulus. He found white-yellow light to give the largest responses. He attempted to find responses in the brain to olfactory stimulation but failed.

Trivius used unanaesthetized, uncurarized dogs because he believed that "the velocity of nervous conduction is modified by curarization, also it is not necessary to use artificial respiration which causes a disturbance of the animal as well as a noise." In relating some further experiments he described how "the brain of the unanaesthetized dog, complete above the level of the spinal cord, is removed and placed on a plate." He noted movements of the galvanometer needle as he made this manipulation, but then he remarked "that the action currents disappear. Spraying the brain with normal saline however caused variations in potential but they did not seem to follow any law." To the modern worker these techniques seem alarmingly crude, but in an age when the source of currents in animal tissue was still something of a mystery, quite primitive methods were used to determine their biological origin.

In his long doctoral thesis for Bechterev's department, Trivius was extremely critical of the stimulating equipment and the components of the recording instruments. He spared none of the previous experimenters; his doubts included Du Bois-Reymond whose electrodes he claimed introduced errors. None of the earlier workers from Matteucci to Gotch and Horsley escaped criticism. One experimenter whose results he confirmed was Larionov.

Just after the turn of the century, some expressions of doubt appeared as to the reliability of the results obtained by the early workers on the electrical potentials of the brain. The dissenting voice was that of Tchiriev at the Imperial University of St. Vladimir at Kiev who, in reviewing the work of Caton and of Beck, cast doubt on their interpretation of the potential changes they had found as being due to brain activity. He thought they could all be explained as currents of injury or artifacts of movement. He made the same criticism in regard to Sechenov's work on the potentials of the spinal cord and would not accept them as neural in origin.

In challenging Sechenov's findings of oscillating currents in the frog's spinal cord, Tchiriev noted that these had been recorded between electrodes placed on the longitudinal surface and a transection, respectively, and asserted with some emphasis that: "These discharges are incontestably the currents of injured cells at the surface of the transverse cut in the spinal cord; to regard them as the manifestation of physiological centers in the cord is absolutely impossible."

Tchiriev stated that he very carefully repeated the experiments of those workers who found brain potentials that varied with stimuli, such as light, sounds, and electrical stimulation of the skin and paws. He used curarized rabbits and dogs and failed to find any variation potential whatsoever. His results were equally negative whether he used a Wiedermann galvanometer or a Lippmann electrometer. He obtained similarly negative results with uncurarized animals when anaesthetized by chloral hydrate or chloroform. Only when the animals emerged from anaesthesia and either struggled or moaned did he obtain the potential changes on sensory stimulation that others had claimed.

Tchiriev concluded that the fluctuations observed when electrodes were applied to the surface of the brain were due to the action of the air on the exposed surface, and that the reactions to sensory stimulation claimed by others were present only during movements of the animal and were therefore probably due to movement of blood in the cerebral vessels and of lymph in the subarachnoid lymphatic spaces.

Tchiriev was equally discouraging about the electrocardiogram and reached the iconoclastic conclusion that:

the process of excitation of various tissues and organs, either muscular or nervous, and even the propagation of excitation within these structures, does not depend on any electrical change that can be detected outside them. They are not electrical or physical processes but phenomena sui generis, dependent on the vital activity of the given tissue and belonging to the same category of physiological phenomena as metabolism, growth, preservation of the species, psychic processes, etc., in providing an explanation for which, physiology, today as in other times, falls short. It is here that we reach the limits of our knowledge—the famous Ignorabimus by Emil Du Bois-Reymond.[80]

Professor Tchiriev seems to have been a controversial skeptic who did not hesitate to impugn the work of such eminent physiologists as Sechenov and Wedensky. Wedensky came under fire for his use of the telephone to listen for activity in the nerve and brain. Tchiriev tried the technique[81] for himself by listening to a live nerve and to a dead one. Sometimes the latter gave the louder noise. He concluded that the sounds Wedensky heard were not biological in origin. A young "Privatdozent" in St. Petersburg came to the defense of Wedensky's interpretations and analyzed the errors in Tchiriev's experiments. Wedensky himself also entered the fray and the controversy occupied several pages of the current journals for some time.

Fortunately, the discouraging pronouncements from Tchiriev in Kiev did not hold others back, and research on the electrical activity of the brain was continued by workers in laboratories in Poland and in Russia, though noticeably not in Kiev.

Nikolas Yevgenevich Wedensky (1852–1922)

One of Sechenov's most distinguished students was Nikolas Yevgenevich Wedensky. Born in 1852 he, like most of the young Russian scientists, went to Europe for training with Du Bois-Reymond, Hoppe-Seyler,[82] and Krönecker.[83] It is not surprising that he came back with an interest in neurophysiology. He began to work at first on peripheral nerve and muscle, but he extended his work to the brain. He developed a technique by which he listened to the action current in nerve by telephone (Fig. 101). He used this first on muscle and then moved to nerve[84] and to the heart.[85] He was present at the Congress in Berlin at which discussion arose about electrical effects in the brain. Wedensky reported[86] that he had listened to action currents in the brain of dogs and cats and that he heard "faint roars and noises which apparently arose spontaneously"[87] and that he could "provoke noises by stimulating the sciatic nerve when listening to the cortex." Wedensky remarked that he connected "the telephone to a hemisphere of the brain. The oscillations of its nerve cur-

[80]S. Tchiriev. Propriétés électromotrices du cerveau et du coeur. *J. Physiol. Pathol. Gen.*, 6:671–682 (1904).

[81]S. Tchiriev. Le téléphone comme indicateur d'une excitation nerveuse. *J. Physiol. Pathol. Gen.*, 5 (1902).

[82]Ernst Felix Immanuel Hoppe-Seyler (1825–1895). *Physiologische Chemie*. Hirschwald, Berlin, 1881. (Physiological Chemistry.)

[83]Edmond Krönecker (1870–1926).

[84]N. Y. Wedensky. Zur Methodik der telephonischen Beobachtungen über die galvanischen Muskelwirkungen während willkürlichen Tetanus. *Akad. Nauk., St. Petersburg*, 28:290–292 (1883). (On the Methodology of Telephonic Observations of Galvanic Muscle Effects during Voluntary Tetanus.)

[85]N. Y. Wedensky. Die telephonischen Erscheinungen am Herzen bei Vagusreizung. *Akad. Nauk. St. Petersburg*, 29:289–291 (1884). (The Telephonic Phenomena at the Heart during Stimulus of the Vagus.)

[86]N. Y. Wedensky. Report on the Third Congress of Russian Physicians and Biologists, St. Petersburg (in Russian). *Vrach*, 10:45 (1889).

[87]This detection of "faint roars and noises" has been interpreted as the sounds of the electroencephalogram. (Vorontsov, D.S. *The Famous Russian Physiologist N. E. Wedensky* [in Ukrainian]. Ukrainian Academy of Sciences.)

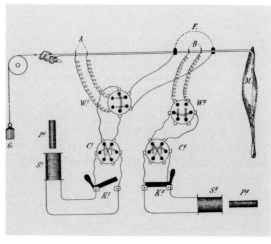

FIG. 101. **Left:** Nikolai Yevgenovich Wedensky (1852–1922). **Right:** One of the circuits used by Wedensky with which he listened to the action currents of nerve and muscle by telephone. (From: *Pflüger's Arch. f. d. ges. Physiol.*, 100:1–4, 1903.)

rents actually activated the telephone, but the sounds received were very faint, almost imperceptible.''

Bechterev, in his overview[88] of this subject, would not give credit to the use of the telephone to detect current flow. Bechterev said that Wedensky was not the inventor of this method, for, he said, it had been used by Hermann and Brücke, and Helmholtz[89] had listened to muscle sounds.

Wedensky, the leader of one of the great schools of Russian physiology, was appointed to the Chair of Physiology at St. Petersburg, a position he held to the end of his life. His major work was on the electrical properties of peripheral nerve, and his name is immortalized in the concept of Wedensky inhibition[90] (Wedensky's word for it was ''pressimum''). He brought all this work together in a book published in 1884, but although he had a large school of students, the use of the telephone in the study of excitation did not spread, for it gave detection of activity but not measurement. However, Wedensky's own summary[91] of his technique is of interest:

> The results of our attempts to use the telephone in this direction come to the following: action currents of nerve centers excite the telephone by perceptible means; but its indic-

---

[88]V. M. Bechterev.

[89]H. L. F. Helmholtz. Über das Muskelgeräusch. *Arch. Anat. Physiol. Wiss. Med.*, 25–43 (1864). (On Muscle Sounds.)

[90]Note: When a nerve fiber is stimulated to give an action potential, this is followed by a period of subnormality. If stimulated again in this period there is a lack of response. This is the phenomenon discovered by Wedensky and named for him, ''Wedensky inhibition.'' This is a presynaptic inhibition following sustained depolarization.

[91]N. Y. Wedensky. Investigations on nerve centers. Chapter 3. *Telephone Investigations on Electrical Phenomena in Muscle and Nerve Apparatuses* (in Russian), 1884. (In the first chapter investigations on muscle are presented; in the second chapter, investigations on nerve.)

ations here, during its present sensitivity, are not so evident as to enable them to be considered right now as subjects of detailed investigations.

This result explains why not as many experiments were conducted as the importance of the subject of investigation would have demanded: the experiments conducted convinced us that first of all it is necessary to fulfill a technical task—to obtain a telephone of great sensitivity and only then can one successfully proceed to further investigations. This technical problem will be solved in the future by one means or another, but at any rate it demands a great deal of time and experimental design.

However, one fact concerning the possibility of using the telephone to investigate phenomena of this field already in itself deserves to be reported, although in general terms. When the experiment is staged for the first time using the telephone to investigate action currents of the central system, then of course the most natural idea is to transfer to it those phenomena which are observed in galvanometric investigations. Now these phenomena may be observed on the galvanometer with such certitude like the well-known form of "violent movements" upon removal of certain parts of the brain and under such exact systematic conditions which exist in respect to the electrical phenomena of muscles and nerves. Two years ago Professor Sechenov[92] found that if a frog's medulla oblongata is removed together with the spinal cord and led off to the galvanometer by the usual principles of leads for muscles and nerves, then on the galvanometer from time to time, periodically, without any artifical stimulation, negative waves of the current of the lead-off preparation are observed. This periodic effect on one and the same preparation varies both according to the distribution in time and intensity, but it has the general characteristics of all negative waves (as conclusively shown when in the cited investigations the existence of similar phenomena of muscles and nerves) and may be cited in connection with periodic excitation developing in the medulla oblongata. It is produced by reflex means (by stimulation of nn. ischiadici, prepared in connection with the brain) and elicited even more continually than the reflex contraction of the central apparatus itself. When it develops without artificial stimulation, the author calls it "spontaneous discharges" in view of the explosive nature of the whole process. It is important for us to add to this that before the described form of effects is established, the given preparation soon usually produces a steady and rather quick decrease of the initial deflection with noticeable jerks or accelerations (in the sense of a compensating current) upon the application of a new transverse section. The general picture and comparison of these last observations with the first bring to mind that they also can be considered as an expression of a kind of uninterrupted excitation on the fresh preparation. We will refer to this stage of the effect later on as *first* to differentiate it from the *second* when the effects produce typical characteristics of negative waves with big pauses between them.

Wedensky ended this long chapter of his book as follows:

> Thus, all these experiments completely convinced us that action currents of the nerve centers are able to act on the telephone. In regard to the periodicity of nerve centers, these experiments are not able to give a clear answer. The observed phenomena are either so weak or appear in such an uncertain way by the nature of their form, the sound symptom of which is a rumble, that to consider them from this angle would be premature. But the very possibility of observing the phenomena already by our present sensitive telephone allows us to expect a lot from it for this little studied and little developed field, and therefore it has seemed to us worthy to report on it in its preliminary form.

Wedensky's influence was very great, and when he succeeded Sechenov in the Chair, one of his goals was to inaugurate a school of physiology in Russia analogous to that which Du Bois-Reymond had created in Germany. Among his many students, the account of whose work belongs to the 20th century was Alexei Alexeivich Ukhtomsky.

Ukhtomsky (1875–1942) was a student of Wedensky at the University of St. Petersburg,

---

[92]I. M. Sechenov. Galvanische Erscheinungen an dem verlängerten Marke des Frosches. *Pflüger's Arch. f. d. ges. Physiol.*, 27:524–566 (1882). (Galvanic phenomena in the medulla oblongata of the frog.)

and he succeeded his teacher in the Chair of Physiology there. Ukhtomsky's efforts were similarly directed to the investigation of excitatory and inhibitory states, as these could be analyzed in peripheral nerve and applied by analogy within the central nervous system, to which he contributed also the theory of dominant action. Born the son of a prince, Ukhtomsky became, in the latter part of his career, a prominent Soviet physiologist. His name is commemorated today in the Ukhtomsky Institute of Physiology in Leningrad.

Wedensky died in 1922 leaving behind him a strong school of neurophysiology.

BIBLIOGRAPHY

Ivan Mikhailovich Sechenov (1829–1905)

Selected Writings

Einiges über die Vergiftung mit Schwefelcyankalium. *Arch. Path. Anat. Physiol. Klin. Med.,* 14:356 (1858). (On Poisoning with Sulfur Cyanide.)

A note on moderators of reflex movements in the frog's brain, and investigations on the centers which inhibit reflex movement in the frog's brain (in Russian). *Med. Herald,* nos. 1, 2, and 3, 1863.

A supplement to the theory of nerve centers which inhibit reflex movements (in Russian). *Med. Herald,* nos. 34, 35, 1863.

Physiologische Studien über die Hemmungsmechanismen für die Reflextätigkeit des Rückenmarks im Gehirne des Frosches. Hirschwald, Berlin, 1863. (Physiological Studies on the Inhibiting Mechanisms for a Reflex Activity of the Spinal Cord in the Frog's Brain.)

Note sur les modérateurs des mouvements reflexes dans le cerveau de la grenouille, introduction by Claude Bernard. *Acad. Sci. Paris.,* 1863. (Notes on the Inhibiting Mechanisms of Reflex Movements in the Brain of the Frog.)

Neue Versuche am Hirn und Rückenmarksnerven des Frosches. Hirschwald, Berlin, 1865. (with B. Paschin.) (New Experiments on the Brain and Spinal Cord of the Frog.)

Notiz über die erregende Wirkung des Blutes auf die cerebrospinalen Nervencentra des Frosches. *Bull. Acad. Sci.* Series 3. 8:380–384 (1865). (Notes on the Stimulating Effect of the Blood on the Cerebrospinal Nerve Centers of the Frog.)

Über die electrische und chemische Reizung der sensiblen Rückenmarksnerven des Frosches. Graz, 1868. (Electrical and Chemical Stimulation of Spinal Nerves in the Frog.)

Einige Bemerkungen über das Verhalten der Nerven gegen sehr schnell folgende Reize. *Pflüger's Arch. f. d. ges. Physiol.,* 5:114–119 (1872). (Some remarks on the reaction of nerves to an extremely rapid series of stimuli.)

Über die Absorptiometrie in ihrer Anwendung auf die Zustände der Kohlensäure im Blute. *Pflüger's Arch. f. d. ges. Physiol.,* 8:1–39 (1874). (On absorptiometric in its use on the states of carbonic acid in the blood.)

Notiz die reflexhemmenden Mechanismen betreffend. *Pflüger's Arch. f. d. ges. Physiol.,* 10:163–164 (1875). (Notes concerning the reflex inhibiting mechanism.)

Galvanische Erscheinungen an der cerebrospinalen Axe des Frosches. *Pflüger's Arch. f. d. ges. Physiol.,* 25:281–284 (1881). (Galvanic phenomena in the cerebro-spinal axis of the frog.)

Hemmung spontaner Stromesschwankungen an dem verlängerten Marke des Frosches. *Zbl. Wiss.,* 20:177 (1882). (Inhibition of spontaneous current variations in the medulla oblongata of the frog.)

Galvanische Erscheinungen an dem verlängerten Marke des Frosches. *Pflüger's Arch. f. d. ges. Physiol.,* 27:524–566 (1882). (Galvanic phenomena in the medulla oblongata of the frog.)

Notiz über den Nierenblutkreislauf. *Pflüger's Arch. f. d. ges. Physiol.,* 31:411–414 (1883). (Notes on blood circulation in the kidneys.)

Notiz über Ausgleichung der Schliessungs—und öffnungsinductionsschläge. *Pflüger's Arch. f. d. ges. Physiol.*, 31:415–516 (1883). (Note on the equalisation of opening and closing induction surges.)

Phénomènes galvaniques dans l'axe cérébrospinal de la grenouille. *Bull. Acad. Imp. Sci. St. Pétersburg*, 28:43–45 (1883). (Galvanic phenomena in the cerebrospinal cord of the frog.)

Selected Works. Translated by A. A. Subkov. State Publishing House for Biological and Medical Literature, Moscow, 1935.

Über die Tätigkeit von Galvani und Du Bois-Reymond auf dem Gebiet der Tierelektrizität (in Russian). *Arb. Physiol. Inst. Moscow Univ.*, 5:5 (1899). (On the activity of Galvani and Du Bois-Reymond in the field of animal electricity.)

Suggested Readings

Arkhangelski, G. V. *First Native Investigations on Electroencephalography* (in Russian). Moscow, 1949. (Includes a chapter on Sechenov.)

Autobiographical Notes. English translation by Kristan Hanes. American Institute of Biological Sciences, 1965.

Kostoyantz, K. S. *Sechenov.* U.S.S.R. Academy of Science, Moscow, 1945.

Kostoyantz, K. S. The history of the problem of brain cortex excitability. Acts du International Congress of the History of Science, pp. 862–864, 1956.

Razran, Gregory. Russian Physiologists' Phychology and American Experimental Psychology: A Historical and Systematic Collation and a Look Into the Future. *Psychol. Bull.* 63, 1:42–64 (1965).

Selected Physiological and Psychological Works. Foreign Language Publishing House, Moscow, 1962.

Selected Works. Translated by A. A. Subkov. State Publishing House, Moscow, 1935.

A Source Book in the History of Psychology, edited by R. I. Herstein and E. J. Boring. Harvard University Press, Cambridge, Mass., 1965.

Veronstov, D. S., Miktin, V. M., and Serkov, P. M. *Sketches from the History of Physiology in the Ukraine* (in Ukrainian). Academy of Science, Ukrainian SSR, Kiev, 1959.

Elie Fadeevich Cyon (1842–1912)

Selected Writings

Die Reflexe eines sensiblen Nerven des Herzens auf die motorischen der Blutgefässe. *Arb. Physiol., Anstalt Leipzig,* 1:128 (1867) (with C. Ludwig). (The reflexes of a sensitive heart nerve on the motor nerves of the blood vessels.)

Über die Wurzeln, durch welche das Rückenmark die Gefässnerven für die Vorderpfote aussendet. *Sitzungsbert. Math. Phys. K. Ges. Wiss. Leipzig,* 73:(1868). (On the roots by means of which the spinal cord sends out functioning nerves for the front paws.)

Über die Function der halbzirkelförmigen Canäle. *Pflüger's Arch. f. d. ges. Physiol.*, 8:306–327 (1874). (On the function of the semi-circular canals.)

Zur Lehre von der reflectorischen Erregung der Gefässnerven. *Pflüger's Arch. f. d. ges. Physiol.*, 8:327–340 (1874). (On the theory of the reflex stimulus of the blood vessel nerves.)

Über den Einfluss der Temperaturänderungen auf die centralen Enden der Herznerven. *Pflüger's Arch. f. d. ges. Physiol.*, 8:340–346 (1874). (On the influence of temperature changes on the central ends of the cardiac nerves.)

Über den Einfluss der hinteren Wurzeln auf die Erregbarkeit der vorderen. *Pflüger's Arch. f. d. ges. Physiol.*, 8:347–348 (1874). (On the influence of the dorsal roots on the sensitivity of the anterior roots.)

Ueber die Innervation der Gebärmutter. *Pflüger's Arch. f. d. ges. Physiol.*, 8:349–351 (1874). (On the innervation of the womb.)

Methodik der physiologischen Experimente und Vivisektionen. Gissen, 1876. (Methodology of physiological experiments and vivisection.)

Beiträge zur Physiologie des Raumsinns. Dritter Theil. Täuschungen in der Wahrnehmung der Richtungen durch das Ohrlabyrinth. *Pflüger's Arch. f. d. ges. Physiol.*, 33:139–250 (1884). (Contributions to the physiology of the sense of space. Third part. Delusions in the perception of directions through the ear labyrinth.)

Zur Frage über die Wirkung rascher Veränderungen des Luftdruckes auf den Organismus. *Pflüger's Arch. f. d. ges. Physiol.*, 69:92–97 (1898). (On the question of the effect of rapid changes in air pressure on the organism.)

Beiträge zur Physiologie der Schilddrüse des Herzens. *Pflüger's Arch. f. d. ges. Physiol.*, 70:126–280 (1898). (Contributions to the physiology of the thyroid gland of the heart.)

Ueber die physiologische Bestimmung der wirksamen Substanz der Nebennieren. *Pflüger's Arch. f. d. ges. Physiol.*, 72:370–371 (1898). (On the physiological determination of the active substance of the suprarenal bodies.)

Über die physiologischen Wirkungen einiger aus der Schilddrüse gewonnener Producte. *Pflüger's Arch. f. d. ges. Physiol.*, 83:1901 (with Adolf Oswald). (On the physiological effects of some products extracted from the thyroid gland.)

Suggested Reading

Kvasov, D. G., *In Memory of Elie Fadeevich Cyon* (in Russian). (For the fiftieth anniversary of his death.) *Sechenov J. Physiol. Leningrad*, 1962.

Vasili Yakovich Danilevsky (1852–1939)

Selected Writings

Ein Beitrag zur Physiologie des Muskelstimming. *Cbl. Med. Wiss.*, 12:721–725 (1874). (A contribution to the physiology of muscle stimulation.)

Experimentelle Beiträge zur Physiologie des Gehirns. *Pflüger's Arch. f. d. ges. Physiol.*, 11:128–138 (1875). (Experimental contributions to the physiology of the brain.)

Investigations into the Physiology of the Brain. Doctoral thesis. University of Kharkov (in Russian), 1877.

On the summation of electrical stimulations of the vagus nerves (in Russian). *Akad. Nauk. Mem. St. Petersburg*, Ser. 7, 35:69–77 (1879).

Thermophysiological investigation of muscles (in Russian). *Akad. Nauk. Mem. St. Petersburg*, Ser. 7, 35:57–68 (1879).

Ueber die Wärmeproduction und Arbeitsleistung des Menschen. *Pflüger's Arch. f. d. ges. Physiol.*, 30:175–199 (1883). (On heat production and efficiency in man.)

Zur Physiologie des tierischen Hypnotismus. *Cbl. Med. Wiss.*, 23:337–344 (1885). (On the physiology of animal hypnotism.)

Substitution, physiologique réciproqué de l'activité cérébrale et des impulsions extérieures. *Arch. Slav. Biol.*, 2:199–216 (1886).

Sur les microbes de l'infection malarique aiguë et chronique chez les oiseaux et chez l'homme. *Ann. Inst. Pasteur*, 12:754 (1890).

Physiological Collection. Essays and Works from the Laboratories of A. Y. and V. Y. Danilevsky (in Russian). Vol. 1, Kharkov, 1888; Vol. 2, Kharkov, 1891.

Researches in Comparative Parasitology of the Blood (in Russian). 2 vols. Vol. 1, Kharkov, 1888; Vol. 2, Kharkov, Darre, 1891.

Electrical phenomena of the brain (in Russian). *Fiziologischeskiy Sbornik* (Physiological Review) 2:77–88 (1891).

Zur Frage über die elektromotorische Vorgänge im Gehirn als Ausdruck seines Tätigkeitszustandes. *Cbl. Physiol.*, 5:1–4 (1891). (On the question of electromotor processes in the brain as the expression of its state of activity.)

Über die physiologische Wirkung des Cocains auf wirbellose Thiere. *Pflüger's Arch. f. d. ges. Physiol.*, 51:446–454 (1892). (On the physiological effect of cocaine on invertebrate animals.)

Kymorheonmische Untersuchungen. *Pflüger's Arch. f. d. ges. Physiol.*, 61:235–264 (1895). (Kymorheonomic investigations.)

Ueber die blutbildende Eigenschaft der Milz und des Knochenmarks. *Pflüger's Arch. f. d. ges. Physiol.*, 61:264–275 (1895). (On the blood-forming peculiarity of the spleen and the bone marrow.)

Sensation and Life (in Russian). Breitigam, Kharkov, 1895.

Human Physiology (in Russian). 2 vols., Moscow, 1915.

Ivan Romanovich Tarkanov (1848–1909)

Selected Writings

Ueber die Bildung von Gallenpigment aus Blutfarbstoff im Thierkörper. *Cbl. Physiol.*, 12:761–762 (1874). (On the formation of gall pigment from blood pigment in the animal body.)

Bermerkung zu Dr. William Stirling's Arbeit: Ueber die Summation electrischer Hautreize. *Pflüger's Arch. f. d. ges. Physiol.*, 12:307–308 (1876). (Comment on Dr. William Stirling's work: On the summation of electrical skin stimulation.)

Ueber die Verschiedenheiten des Eiereiweisses bei befiedert geborenen (Nestflüchter) und bei nackt geborenen (Nesthocker) Vögeln, und Ueber die Verhältnisse zwischen dem Dotter und dem Eiereiweiss. *Pflüger's Arch. f. d. ges. Physiol.*, 31:368–376 (1883). (On the differences in the albumin of birds born feathered [nest-fleers], and those born naked [nest-dwellers], and on the relationships between the yolk and the egg white.)

Ueber die Verschiedenheiten des Eiereiweisses bei befiedert geborenen (Nestflüchter) und bei nackt geborenen (Nesthocker) Vögeln, und Ueber die Verhältnisse zwischen dem Dotter und dem Eiereiweiss (Biologisch-chemische Untersuchung). *Pflüger's Arch. f. d. ges. Physiol.*, 33:303–378 (1884). (On the differences in the albumin of birds born feathered [nest-fleers], and those born naked [nest-dwellers], and on the relationships between the yolk and the egg white [biological-chemical investigation].)

Ueber automatische Bewegungen bei enthaupteten Enten. *Pflüger's Arch. f. d. ges. Physiol.*, 33:619–622 (1884). (On automatic movements in beheaded ducks, Preliminary Report.)

Über die willkürliche Acceleration der Herzschläge beim Menschen. *Pflüger's Arch. f. d. ges. Physiol.*, 35:109–137 (1885). (On the voluntary acceleration of the heart rate in man.)

Aus einem Schreiben an den Herausgeber, betreffend die willkürliche Acceleration der Herzschläge. *Pflüger's Arch. f. d. ges. Physiol.*, 35:198–199 (1885). (From a letter to the editor, concerning the voluntary acceleration of the heart rate.)

On the galvanometric observations on the human skin (in Russian). *Vest. Klin. Sudev. Psychiatr. Nevropath.*, 1:73 (1889).

Bronislav Fortunatovich Verigo (1860–1925)

Selected Writings

Ueber die secundären Erregbarkeitsänderungen an der Kathode eines andauernd polarisirten Froschnerven. *Cbl. Med. Wiss.*, 52:945 (1882). (On the secondary action change at the cathode of a permanently polarized frog nerve.)

Die secundären Erregbarkeitsänderungen an der Kathode eines andauernd polarisirten Froschnerven. *Pflüger's Arch. f. d. ges. Physiol.*, 31:417–479 (1983). (The secondary sensitivity changes at the cathode of a constantly polarized frog nerve.)

Ueber die gleichzeitige Reizung des Nerven an zwei Orten mit Inductionsschlägen. *Pflüger's Arch. f. d. ges. Physiol.*, 37:519–548 (1885). (On the simultaneous stimulation of the nerve in two places with induction shock.)

Action currents of the frog's brain. Report on the Third Congress of Russian Physicians and Biologists, St. Petersburg (in Russian). *Vrach*, 10:45 (1889).

Action currents of the brain and medulla (in Russian). *Vest. Klin. Sudev. Psychiatr. Nevropath.*, 7: (1889).

Ueber das Harnack'sche aschenfreie Albumin. *Arch. Physiol.*, 48:3, 4, S 127 (1891); *Cbl. Physiol.*, 5:6 (1891). (On Harnack's ash-free albubin.)

Zur Frage über die Wirking des Sauerstoffs auf die Kohlensäureausscheidung in den Lungen. *Pflüger's Arch. f. d. ges. Physiol.*, 51:321–361 (1892). (On the question of the effect of oxygen on carbon dioxide secretion in the lungs.)

Ueber das Vorkommen des Pentamethylendiamins in Pankreasinfusen. *Pflüger's Arch. f. d. ges. Physiol.*, 51:362–366 (1892).

Zur Frage über die Beziehung zwischen Erregbarkeit und Leitungsfähigkeit des Nerven. *Pflüger's Arch. f. d. ges. Physiol.*, 51:551–552 (1892). (On the question of the relationship between the sensitivity and conductivity of the nerve.)

Ueber die Reizung des Nerven mit dreiarmigen Elektroden. *Pflüger's Arch. f. d. ges. Physiol.*, 76:517–530 (1899). (On the stimulus of nerves with three-armed electrodes.)

Fundamentals of the Physiology of Man and Higher Animals. 2 vols. (in Russian). Odessa, 1905, 1910.

Fundamentals of General Biology. 2 vols. (in Russian). Odessa, 1912.

Vladimir Mikhailovich Bechterev (1857–1927)

Selected Writings

Ueber die functionelle Beziehung der unteren Oliven zum Kleinhirn und die Bedeutung derselben für die Erhaltung des Körpergleichgewichts. *Pflüger's Arch. f. d. ges. Physiol.*, 29:257–265 (1882). (On the functional relationship of the lower olivary bodies to the cerebellum and its significance for the maintenance of balance.)

Ergebnisse der Durchschneidung des N. acusticus, nebst Erörterung der Bedeutung der semicirculären Canäle für das Körpergleichgewicht. *Pflüger's Arch. f. d. ges. Physiol.*, 30:248–312 (1883). (Results on severing the N. acusticus, together with a discussion on the significance of the semicircular canals for body balance.)

Ueber den Verlauf der die Pupille verengenden Nervenfasern im Gehirn und über die Localisation eines Centrums für die Iris und Contraction der Augenmuskeln. *Pflüger's Arch. f. d. ges. Physiol.*, 31:60–87 (1883). (On the path in the brain of those nerve fibers which cause the pupils to narrow, and on the localization of a center for the iris and for contraction of the eye muscles.)

Bemerkungen zu dem Aufsatz: Ueber den Verlaud der die Pupille verengenden Nervenfasern im Gehirn. *Pflüger's Arch. f. d. ges. Physiol.*, 31:309–340 (1883). (Remarks on the Article "On the path in the brain of those nerve fibers in the brain which narrow the pupil.")

Zur Physiologie des Körpergleichgewichts. Die Function der centralen grauen Substanz des dritten Hirnventrikels. *Pflüger's Arch. f. d. ges. Physiol.*, 31:479–590 (1883). (On the physiology of body balance. The function of central gray matter of the third ventricle.)

Über die Bemerkungen von V. Hensen zu meinem Aufsatz "Ueber den Verlauf der die Pupille verengenden Nervenfasern im Gehirn." *Pflüger's Arch. f. d. ges. Physiol.*, 33:240–242 (1884).

(On remarks by V. Hensen on my article "On the path of nerve fibers in the brain which narrow the pupil.")

Über die Verbindung der sogenannten peripheren Gleichgewichtsorgane mit dem Kleinhirn. *Pflüger's arch. f. d. ges. Physiol.*, 34:362–388 (1884). (On the connection of the so-called peripheral organs of balance with the cerebellum.)

Wie sind die Erscheinungen zu verstehen, die nach Zerstörung des motorischen Rindenfeldes an Thieren auftreten. *Arch. Physiol.*, 35:137–145 (1885). (How are those phenomena to be understood which occur after removal of the motor regions from animals.)

Über eine bisher unbekannte Verbindung der grossen Oliven mit dem Grosshirn. *Neurol. Zbl.*, 4:194–196 (1885). (On a hitherto unknown connection of the olives with the brain.)

Über die Längsfaserzüge der Formatio reticularis medulla oblongatae. *Neurol. Zbl.*, 4:327–346 (1885). (On the longitudinal fiber sections of the formatio reticularis medulla oblongata.)

Die Leitungsbahnen im Gehirn und Rückenmark. Fisher, Leipzig, 1894. (The Transmission Pathways in the Brain and Spinal Cord.)

Über corticale Centren beim Affen. Neurol. *Central. Physiol.*, 140 (1898). (On the cortical centers in apes.)

Les Voies de Conduction du Cerveau et de la Moelle. Storck, Lyon, 1900. (The Conduction Paths of the Brain and the Medulla.)

Osnovy Ucheniya o Funktisiyaxh Mozga. St. Petersburg, 1906. (Elementary Teaching of Brain Functions.)

Die Funktionen der Nervencentra. 3 vols. Fisher, Jena, 1908–1911. (The Functions of the Nerve Centers.)

Objective Psychologie oder Psychoreflexologie. Die Lehre von Associationreflexen. Teubner, Leipzig, 1913. (Objective Psychology or Psychoreflexology. The Theory of Association Reflexes.)

Reflexologie des Menschen. Allgemeine Grundlagen der Reflexologie des Menschen. Deuticke, Leipzig, 1926. (Reflexology of Man. Bases of the Reflexology of Man.)

Mozg I Ego Deyatel 'Nost.' First edition. Moscow and Leningrad, 1928. (The Brain and its Functions.)

Central Principles of Human Reflexology: The Objective Study of Personality. Translated by E. and W. Murphy. International Publishers, New York, 1932.

Vladimir Elimovich Larionov (1857–1919)

Selected Writings

Über galvanometrische Messungen der Ströme in der Rinde der Schläfenwindungen bei Reizung der peripheren Gehörorgane. *Montsschr. Psychiatr.*, 464 (1889). (On galvanometric measurements on the currents in the cerebral cortex of the temporal convolution during stimulus of the peripheral auditory organs.)

On the cortical centers of hearing in dogs (in Russian). *Obzr. Psychiatr. Nevrol., St. Petersburg,* 2:419–424 (1897).

The auditory tracts (in Russian). *Nevrolog. Bestnik,* 3:162–203 (1898).

Über die musikalischen Centren des Gehirns. *Pflüger's Arch. f. d. ges. Physiol.*, 76:608–625 (1899). (On the musical centers of the brain.)

On galvanometric measurements of currents in the cortex of the temporal gyrus during stimulation of the peripheral organs of hearing (in Russian). *Nevrolog. Bestmik Kazan,* 7:44–64 (1899).

Anatomical and other principles of science concerning the association centers of the brain (in Russian). *Vopr. Nerv. Psikh. Med. Kiev.,* 7:76 (1903).

Nikolas Yevgenevich Wedensky (1852–1922)

Die telephonischen Wirkungen der erregten Nerven. *Akad. Nauk. St. Petersburg,* 28:290 (1883). (The Telephonic Influences of Aroused Nerves.)

Zur Methodik der telephonischen Beobachtungen über die galvanischen Muskelwirkungen während willkürlichen Tetanus. *Bull. Acad. Imp. Sci. St. Petersburg,* 28:290–292 (1883). (On the methodology of telephonic observations of galvanic muscle effects during voluntary tetanus.)

Die telephonishen Erscheinungen am Herzen bei Vagusreizung, *Bull. Acad. Imp. Sci. St. Petersburg,* 29:289–291 (1884). (The telephonic phenomena at the heart during stimulus of the vagus.)

Wie rasch ermüdet der Nerv. *Zbl. Med. Wiss.,* 22:65–68 (1885). (How quickly does the nerve tire.)

Report of the Third Congress of Russian Physicians and Biologists, St. Petersburg. *Vrach,* 10:45 (1889).

Suggested Reading for the 19th Century Russian Neurophysiologists

Archangelevski, G. V. *First Native Investigations of Electroencephalography (Danilevsky, Sechenov and Verigo)* (in Russian). Moscow, 1949.

Erste Vaterländische Untersuchungen über Elektroenzephalographie (in Russian). State Publisher of Medical Literature, Moscow, 1949.

Brazier, M. A. B. *A History of the Electrical Activity of the Brain.* Pitman, London, 1961.

Brazier, M. A. B. Russian contributions to the understanding of the central nervous system. In: *Central Nervous System and Behavior,* pp. 23–136, Josiah Macy Foundation, New York, 1959.

Brazier, M. A. B. The historical development of neurophysiology. Handbook of physiology. *Am. Physiol. Soc.,* 1:1–58 (1959).

Faytel'berg-Blank, V. R. Notes on the development of physiology in the Ukraine. New data on the life and scientific activity of B. F. Verigo, an outstanding scientist (in Russian). *Fiziologichnyi Zhurnal,* 11(2), 1965.

Merkulov, V. L. The work of German and Austro-Hungarian scientists in the physiology department at the Institute of Experimental Medicine under the direction of I. P. Pavlov in 1902–1908 (In Ezhegodnik for 1960) (in Russian). Leningrad, 1961.

Sammelband: Danilevskij, V. Ja.; Secenov, I. N.; Verigo, B. F.

Ukhtomskii, A. A. *Collection of Works.* Typescript of translation of selected passages from volume VL (in Russian), edited by V. L. Merkulov. Leningrad, 1962.

Vorontsov, D. S. *Sketches from the History of Physiology in the Ukraine* (in Ukrainian). Academy of Science, Ukrainian SSR, Kiev.

Envoi

The 19th century was, for Europe, a stormy century: France had seen Napoleon come and go; Bismarck had brought power to Prussia; Italy was no longer divided; and Poland was occupied by Prussia, Austria-Hungary and Russia. Against this disturbed background, a fine flush of scientific expansion developed and, for this, Europe takes the credit; for no stream of neurophysiology had yet come from the New World.

In the years under review in this century, there were young scientists still getting their training and beginning to publish their first papers. Among them were men who were to dominate their country's physiology in the quarter-century to follow—in England, Sherrington who had trained with Langley and Goltz; in Russia, Pavlov who had trained in Ludwig's laboratory in Leipzig; and in Poland, Marceli Nancki, for whom an institute is now named. All had begun to publish in the 19th century. A great century of progress lay ahead, though torn in its first half by two great wars that channeled much research into the arts of war and tended to dim the opportunities for research in Europe.

In a century which had opened with the first glimpse of electricity in the peripheral neuromuscular systems, knowledge had grown to the realization that even the brain itself was a producer of electricity, and no longer could there be any doubt that this subject was a science—the vitalism with which the century had opened having faded away.

Indexes

Author Index

Pages that contain illustrations or portraits are indicated by italicized page numbers.

Subject Index

Pages that contain illustrations and portraits are indicated by italicized page numbers.

Steady potential, shifts in, caused by peripheral
    stimulation, *203*
Stereoscopic vision, 69
Stereotaxic instrument, 170,*171*
Stereotaxic methods, for localization of nervous
    structures, 163
Stimulation
    afferent, desynchronization of cortical activity
        following, 199
    of central nervous system, 213
    of cortex, 200
    electrical, *see* Electrical stimulation
    faradic, 73
    of hind leg, *198,*199
    of motor cortex, 99–100
    of muscle, circuits in, *97*
    of nerve, muscle currents evoked by, 186
    peripheral
        cortical responses to, *172*
        shifts in steady potential caused by, *203*
    photic, intermittent, 192
    of skin, 192–193
    see also Galvanism
String galvanometer, 165
Sturm and Drang, 6
Sugar, in blood, 53
Surgery
    brain, 172–173
    of spinal cord, 173
Sydenham Society, 35,69
Sylvian fissure, 153
Synapse, 39,103

T

Teleology, 98
Telephone, and action current in nerve,
    232,239–241,*240*
Temperature, effect of, on response of muscle,
    90
Tetanization, in nerve, 97–98
Therapy, electrical, 1,2,*3,*4,*4,*5

Thomson reflecting galvanometer, 186,187,207
Tone(s)
    localization of, in auditory cortex, 233–234
    musical, in cortex, mapping of, *230,*234
    produced by contraction of muscle, 70–71
Tonus of skeletal muscle, 91
Tuning forks, 234

U

Ukhtomsky Institute of Physiology, 242

V

Vagus, central excitation of, 219
Vagus inhibition, of heart, 103
Ventral root, muscle spindles and, 119–120
Visceral organs, electrical changes in nerves of,
    192–193
Vision, comparative physiology of sense of,
    58–59,*58*
Visual center, of brain, *175*
Visual cortex, localization of, 200
Visual system, 175–176,*229*
Vivisection, 44,46,47,189
Voltaic pile, 2,16,17,*19,*22,*30,*129,
    130–131,*156*

W

Wedensky inhibition, 240
White line of Gennari, 129–130,*130,*131
West Riding Lunatic Asylum, 138,139,*139,*
    159,166,168,194
Wiedermann d'Arsonval galvanometer, *53,*
    234,238
World Exhibition (Philadelphia; 1876), 204

Y

Young-Helmholtz Theory, 8–9